Sustainable Civil Infrastructures

Editor-in-chief

Hany Farouk Shehata, Cairo, Egypt

AF774430

Advisory Board

Khalid M. ElZahaby, Giza, Egypt
Dar Hao Chen, Austin, USA

Steering Editorial Committee

Dar Hao Chen, Texas A&M University, USA
Jia-Ruey Chang, National Ilan University, Taiwan
Hadi Khabbaz, University of Technology Sydney, Australia
Shih-Huang Chen, National Central University, Taiwan
Jinfeng Wang, Zhejiang University, China

About this Series

Sustainable Infrastructure impacts our well-being and day-to-day lives. The infrastructures we are building today will shape our lives tomorrow. The complex and diverse nature of the impacts due to weather extremes on transportation and civil infrastructures can be seen in our roadways, bridges, and buildings. Extreme summer temperatures, droughts, flash floods, and rising numbers of freeze-thaw cycles pose challenges for civil infrastructure and can endanger public safety. We constantly hear how civil infrastructures need constant attention, preservation, and upgrading. Such improvements and developments would obviously benefit from our desired book series that provide sustainable engineering materials and designs. The economic impact is huge and much research has been conducted worldwide. The future holds many opportunities, not only for researchers in a given country, but also for the worldwide field engineers who apply and implement these technologies. We believe that no approach can succeed if it does not unite the efforts of various engineering disciplines from all over the world under one umbrella to offer a beacon of modern solutions to the global infrastructure. Experts from the various engineering disciplines around the globe will participate in this series, including: Geotechnical, Geological, Geoscience, Petroleum, Structural, Transportation, Bridge, Infrastructure, Energy, Architectural, Chemical and Materials, and other related Engineering disciplines.

More information about this series at http://www.springer.com/series/15140

Shanzhi Shu · Liangcai He
Yao Kai
Editors

New Developments in Materials for Infrastructure Sustainability and the Contemporary Issues in Geo-environmental Engineering

Proceedings of the 5th GeoChina International Conference 2018 – Civil Infrastructures Confronting Severe Weathers and Climate Changes: From Failure to Sustainability, held on July 23 to 25, 2018 in HangZhou, China

Springer

Editors
Shanzhi Shu
Kiewit Corporation
Englewood, CO, USA

Liangcai He
Amec Foster Wheeler
Los Angeles, CA, USA

Yao Kai
Department of Civil and Environmental Engineering
National University of Singapore
Singapore, Singapore

ISSN 2366-3405 ISSN 2366-3413 (electronic)
Sustainable Civil Infrastructures
ISBN 978-3-319-95773-9 ISBN 978-3-319-95774-6 (eBook)
https://doi.org/10.1007/978-3-319-95774-6

Library of Congress Control Number: 2018948641

© Springer International Publishing AG, part of Springer Nature 2019
This work is subject to copyright. All rights are reserved by the Publisher, whether the whole or part of the material is concerned, specifically the rights of translation, reprinting, reuse of illustrations, recitation, broadcasting, reproduction on microfilms or in any other physical way, and transmission or information storage and retrieval, electronic adaptation, computer software, or by similar or dissimilar methodology now known or hereafter developed.
The use of general descriptive names, registered names, trademarks, service marks, etc. in this publication does not imply, even in the absence of a specific statement, that such names are exempt from the relevant protective laws and regulations and therefore free for general use.
The publisher, the authors and the editors are safe to assume that the advice and information in this book are believed to be true and accurate at the date of publication. Neither the publisher nor the authors or the editors give a warranty, express or implied, with respect to the material contained herein or for any errors or omissions that may have been made. The publisher remains neutral with regard to jurisdictional claims in published maps and institutional affiliations.

Printed on acid-free paper

This Springer imprint is published by the registered company Springer Nature Switzerland AG
The registered company address is: Gewerbestrasse 11, 6330 Cham, Switzerland

Contents

Introduction

Infrastructure is essential to facilitate economic development and to reduce poverty worldwide. The diverse infrastructures we are planning and building today will shape our communities tomorrow; however, all infrastructures should be designed and delivered with environmental sustainability. The challenges to achieve sustainable infrastructure development can be seen in our civil engineering projects of transportation, communication, sewage, water and electric systems. The challenges in our civil infrastructures include our design for extreme temperatures, floods, droughts, earthquakes, and other geohazards. Delivering environmentally sustainable infrastructures efficiently has been at the core of many problems for civil engineering practitioners, and extensive researches have been performed. The volume of the proceedings includes the instigations presenting on new advances in materials for infrastructure sustainability and addressing the contemporary issues in geo-environmental engineering, which will provide benefits to both civil engineering practitioners and researchers. Experts from the various engineering disciplines around the globe will participate in this event, including geotechnical, geological, earthquake, geoscience, petroleum, structural, infrastructure, energy, chemical and materials, and other related engineering disciplines. This volume is part of the proceedings of the 5th GeoChina International Conference on Civil Infrastructures Confronting Severe Weathers and Climate Changes: From Failure to Sustainability, HangZhou, China 2018.

Impact of Initial In-Situ Stress Field on Soil Response During Cavity Expansion Using Discrete Element Simulation

Yang Dong[1(✉)], Behzad Fatahi[1], Hadi Khabbaz[1], and Jeff Hsi[2]

[1] School of Civil and Environmental Engineering, University of Technology Sydney (UTS), Ultimo, Australia
Yang.Dong@uts.edu.au
[2] EIC Activities Pty Ltd, Sydney, Australia

Abstract. Discrete element method (DEM) is gaining its popularity in investigating many complicated geotechnical related problems due to its discontinuous nature in simulating granular materials. Particularly when simulating the processes involving large deformation and displacement of soil (e.g. pile penetration), DEM demonstrates distinct advantages over other numerical solutions that may confront convergence problems. Despite the facts that DEM analysis has been conducted to study the mechanism of the cavity expansion, there is a very limited number of investigations conducted to study the effects of the initial stress field on the soil response. Hence, in this study, a three-dimensional numerical analysis has been conducted using PFC^{3D} to investigate the soil response under different initial stress conditions during cavity expansion. A large-scale model containing an adequate number of particles has been constructed to simulate the soil medium, in which, microscopic contact properties were calibrated against existing experimental data to mimic the realistic behaviour of a sandy soil. To examine the effects of the initial in-situ stresses, several cylindrical cavities were created and expanded gradually from an initial radius to a final radius, while stress and strain variations were monitored during the entire simulation. It should be noted that the internal cylinder boundary was loaded using a constant strain rate, while the outer boundary was controlled through a servo mechanism to maintain a constant external pressure adopting appropriate subroutines. The results obtained confirmed that the initial stress conditions have significant effects on the soil response during cavity expansion.

1 Introduction

The recent advancement in computational technology has facilitated the application of numerical analysis for many complex geotechnical problems. Discrete element method (DEM) has been extensively adopted due to its capability to incorporate the discontinuous nature of granular materials. For instance, installation of driven pile or self-boring pressure meter in cohesionless materials are often simulated using Discrete Element Method due to its merits in coping with problems involving large displacement (Falagush et al. 2015; Jiang et al. 2006). To interpret the installation mechanism

© Springer International Publishing AG, part of Springer Nature 2019
S. Shu et al. (eds.), *New Developments in Materials for Infrastructure Sustainability and the Contemporary Issues in Geo-environmental Engineering*, Sustainable Civil Infrastructures, https://doi.org/10.1007/978-3-319-95774-6_1

of driven piles and self-boring pressure meters, cylindrical cavity expansion modelling has been employed by many researchers (Carter et al. 1979; Fahey 1986; Gibson 1961). Analysing the mechanism of vertical drain installation required for ground stabilisation in saturated or partially saturated soils can be conducted using cavity expansion theory to estimate the smear zone characteristics since the smear zone permeability and over consolidation ratio can affect the ground improvement plan considerably (Azari et al. 2014; Ho and Fatahi 2015; Ho et al. 2015, 2016; Parsa-Pajouh et al. 2015). Extensive research has been conducted to investigate cavity expansion mechanism in sandy soils adopting different techniques such as analytical solutions and numerical analysis based on continuum method (Carter and Yeung 1985; Collins et al. 1992). However, closed form solutions and many constitutive models adopted in numerical simulations have been obtained from a number of idealised material models and therefore may not capture the realistic behaviour of soil to capture particularly elasto-plastic response (Carter et al. 1979). Thus, the discrete element method may be an alternative to investigate the mechanical behaviour of granular materials at the microscopic level due to its discontinuous nature (Cundall and Strack 1979). Despite the fact that there are studies conducted to investigate the soil behaviour during cavity expansion using DEM, there is a lack of studies on the effect of initial in-situ stresses on the predications. Therefore, the purpose of this paper is to investigate the effect of initial in-situ stress conditions on the soil response during the cylindrical cavity expansion in sandy soil using discrete element simulation.

2 Calibration Using Triaxial Test Results

Investigating the effect of initial stress state cannot be validated if the constitutive contact behaviour of the soil adopted in the simulation is not realistic. Hence, the contact model adopted in this study was calibrated to mimic realistic behaviour of sandy soil by matching the results obtained from the numerical simulations of triaxial compression tests to existing laboratory results. In discrete element modelling, selecting the most appropriate contact model is of paramount importance. Despite the fact that classical theories believe that it is mainly the sliding that plays the dominant role in controlling the strength and dilatancy of sandy soil, Oda and Kazama (1998) has observed that the effect of rolling is more prominent. Therefore, The Rolling Resistance contact mode, illustrated in Fig. 1, has been adopted due to its characteristics in capturing the rolling mechanism between sand particles, which is to incorporate a torque acting on the contacting pieces to counteract rolling motion (Iwashita and Oda 1998). Experiment adopted for calibration was conducted by Cornforth (1964) to obtain the drained strength of medium to fine grained sand under the plane strain condition. Considering the current computational power, it is not feasible to simulate the triaxial test following the true particle size distribution adopted by Cornforth (1964). Therefore, upscaling the grain size and hence reducing the number of particles becomes necessary. To ensure that simulation is computationally feasible and in order to conduct the calibration exercise efficiently, an upscaling factor can be back-calculated while ensuring that the void ratio remains same as the experimental measurement. By trial and error, an upscaling factor of 16.5 has been adopted to apply to the original particle size distribution uniformly so that the same porosity as reported in the experiment was achieved in the numerical simulation. Hence, the triaxial simulation comprised of 20,000 spherical particles, ranging from 1.25 to 24.50 mm after upscaling, and the setup of the triaxial test is shown in Fig. 2. Following the operational procedures given by Cornforth (1964), isotropic stresses were applied to consolidate the sample to its desired initial stress state. Once the initial condition was established, the lateral stresses in X direction were maintained while the facets in the intermediate direction (Y direction) were fixed to simulate the plane strain condition, similar to the laboratory condition, and then axial load was applied on top of the specimen (Z direction) at a constant strain rate. Stress and strain variations as well as volumetric changes were continuously recorded in order to match the experimental results. Calibrated results for stress-strain relationship as well as volumetric behaviour are plotted in Figs. 3 and 4, respectively. It is noteworthy to state that the numerical predictions have a good agreement with experimental data. The calibrated contact properties adopting rolling resistance contact model include the ratio between normal and shear stiffness, friction coefficient as well as rolling resistance coefficient, which are 1.75, 0.40 and 0.10, respectively.

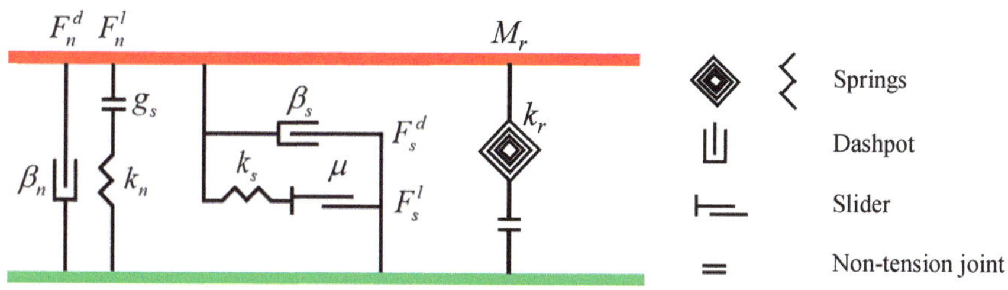

K_n, K_s and K_r are normal stiffness, shear stiffness and rolling stiffness, respectively
β_n and β_s are normal and shear dashpots, respectively
M_r is rolling resistance moment
F_n and F_s are normal force and shear force, respectively
μ is friction coefficient

Fig. 1. Rheological model for rolling resistance contact model

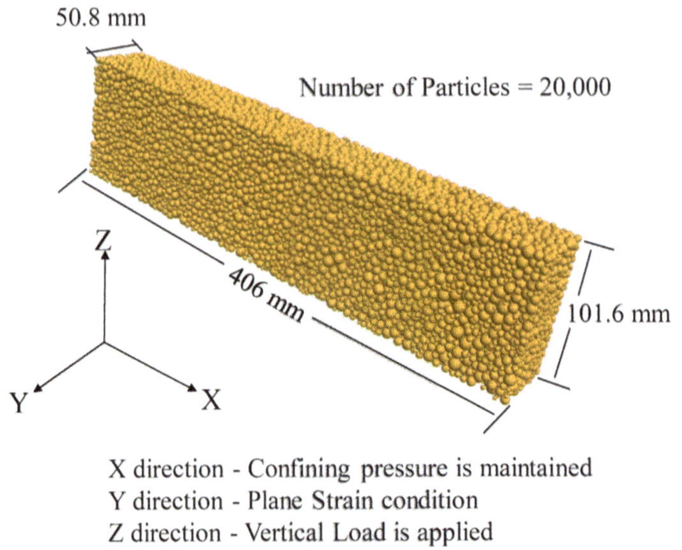

Fig. 2. Triaxial compressive test in plane strain condition setup in PFC^{3D}

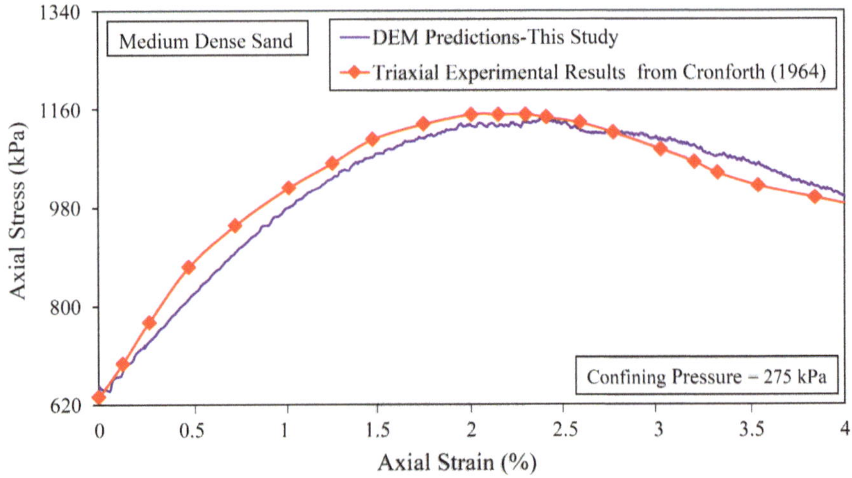

Fig. 3. Variations of axial stress with axial strain (DEM predictions versus experimental measurement)

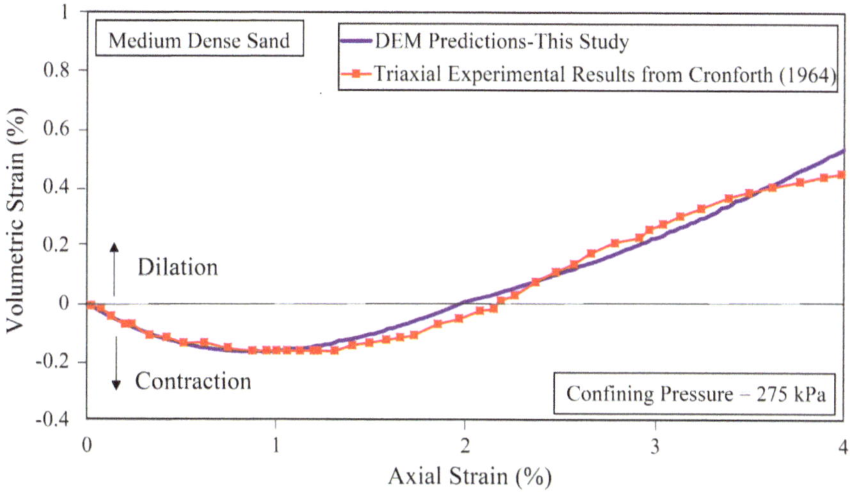

Fig. 4. Variations of volumetric strain with axial strain (DEM predictions versus experimental measurements)

3 Cavity Expansion Simulation

To study the effect of initial stress conditions on soil response subjected to cavity expansion, a large scale three dimensional numerical model simulating the cylindrical cavity expansion has been proposed. Considering the fact that many cavity expansions related problems are in plane strain condition (Alshibli and Sture 2000), and considering the axisymmetric geometry, only a quarter of geometry with plane strain condition has been simulated, as shown in Fig. 5. The model built contained 75,000 spherical particles following the particle size distribution adopted in the calibration and triaxial test in the laboratory. The model was generated using the radius expansion method, and the porosity and coordination numbers have been constantly monitored to ensure that the DEM sample has the identical properties to the soil in calibration exercise. The void ratio (or porosity) is measured during the entire process of the simulation adopting numerous prediction spheres, where the porosity is represented based on the ratio between volume of voids and the volume of the measurement sphere. The friction coefficient between particles was lower during the specimen generation stage to achieve a homogenous particle assembly. In this case, the void ratios measured adopting prediction spheres reveal very similar results regardless of their locations. Additionally, the void ratios reported in this paper are the average values obtained by those prediction spheres in discrete element simulation. Furthermore, the investigations were carried out for the points on the angular bisector (as illustrated in Fig. 5), where the boundary effects could influence the predictions to less extent. The external boundary was located 1 m away from the centre of the initial cavity (Fig. 5). The measured soil porosity was approximately 0.390, similar to the porosity of the medium dense sand specimen used for calibration and reported in Cornforth (1964). The ultimate objective of this paper is to investigate the effect of the initial stress field on the soil response during cavity expansion. Hence, four different initial stress fields ($\sigma_{r0}=\sigma_x=\sigma_y$) 62.5 kPa, 125 kPa, 250 kPa, and 500 kPa were adopted, respectively. In-situ stresses have been applied through the wall servo mechanism, ensuring that the radial stress (σ_{r_0}) is uniformed applied to the soil medium. After the initial conditions were satisfied, the cavity was expanded gradually from an initial cavity radius of 55 mm to a final cavity radius of 100 mm at a constant rate of 0.001 m/s to ensure the quasi-static loading, during which, the boundary pressure remained same as the initial stress. Cavity pressure is measured using coded subroutines by measuring the contact forces acting on the internal cavity wall and then divided by the corresponding contact area. Furthermore, stress and strain variations were continuously recorded using measurement spheres and subroutines at different radial distances from the cavity.

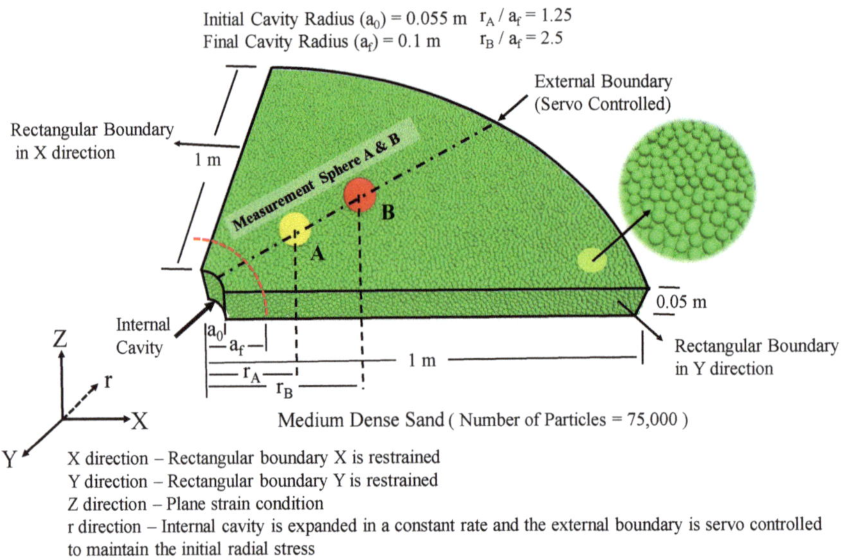

Fig. 5. Cavity expansion model setup for DEM simulation

4 Results and Discussion

Variations of internal cavity pressure (P) during the cavity expansion process has been plotted against the normalised cavity radius (a/a_0) in Fig. 6. Since the objective is to investigate the effects of in-situ stress field on the soil response, the cavity pressure has not been normalised. The results evidently reveal that an increase in the initial in-situ stress has contributed to escalation of the cavity pressure. For instance, the limit pressure (P_{max}) measured from the cavity expansion with σ_{r_0} = 125 kPa is approximately 750 kPa, while the limit pressure corresponding to σ_{r_0} = 500 kPa is P_{max} = 2500 kPa. In other words, the pressure required to expand an existing cavity in the deeper ground is larger than the corresponding pressure in shallower ground. Indeed, similar results were reported by Arroyo et al. (2011) when interpreting cone penetration test results. As indicated by Arroyo et al. (2011), measured cone resistance was enhanced from 0.5 to 2.0 MPa when the confining pressure increased from 60 kPa to 300 kPa, respectively, demonstrating that the cone resistance measured was highly correlated to the in-situ stress conditions. Hence, it can be readily confirmed that the limit pressure is largely dependent on the soil initial stress state, and such correlation may be used to evaluate the pile toe resistance during the pile driving process. Furthermore, comparisons were made on the deviatoric stress and the volumetric strain variations with the shear strain during the cavity expansion process. Prediction spheres A & B located along angular bisector (see Fig. 5) were used to continuously record the variations of deviatoric stress, shear strain as well as volumetric strain during the cavity expansion process.

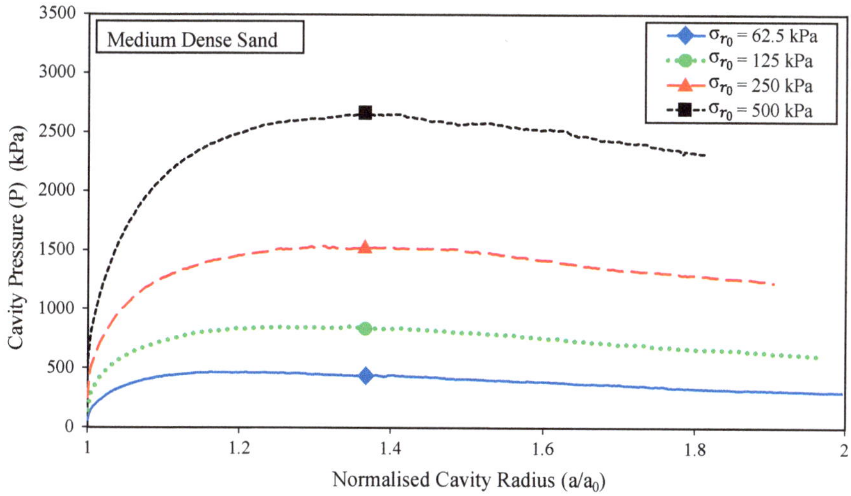

Fig. 6. Cavity pressure variations during the cavity expansion

Figures 7 and 9 present the results for $r/a_f = 1.25$ and $r/a_f = 2.5$, respectively (r being the radial distance from the centre of the cavity; a_f being the final cavity radius). It can be concluded that given the same deviatoric strain, the magnitude of the deviatoric stress q is highly dependent on the initial stress state. For example, as the in-situ horizontal stress increases from 250 kPa to 500 kPa, the maximum deviatoric stresses (q_{max}) measured at $r/a_f = 1.25$ and $r/a_f = 2.5$ increased by 400 kPa and 150 kPa, respectively. As shown in Figs. 8 and 10, with an increase in the initial in-situ stress, less significant volumetric strain changes are observed during the cavity expansion. This may explain the fact that the soil volume changes (expansion) during the pile installation are more evident near the ground surface, while the phenomenon becomes less pronounced at deeper soil layers, where initial in-situ stresses are larger. Additionally, referring to Figs. 8 and 10, volumetric strain variations obtained for different initial in-situ stresses showed a slight contraction during the initial stage of cavity expansion and then dilation as cavity expanded further, confirming the medium dense state of the in-situ soil. Furthermore, referring to Figs. 7 8, 9 and 10, comparisons can be made for the variations of the deviatoric stresses and volumetric strains at two different prediction locations (i.e. prediction spheres A & B shown in Fig. 5). It can be evidently concluded that the soil close to the cavity (predication sphere A) experienced higher shear stresses and deformations compared to the soil further away from the cavity (prediction sphere B). Moreover, it is projected that the measured deviatoric stresses and volumetric strains would eventually approach zero for the points located far away from the cavity centre (i.e. r = ∞).

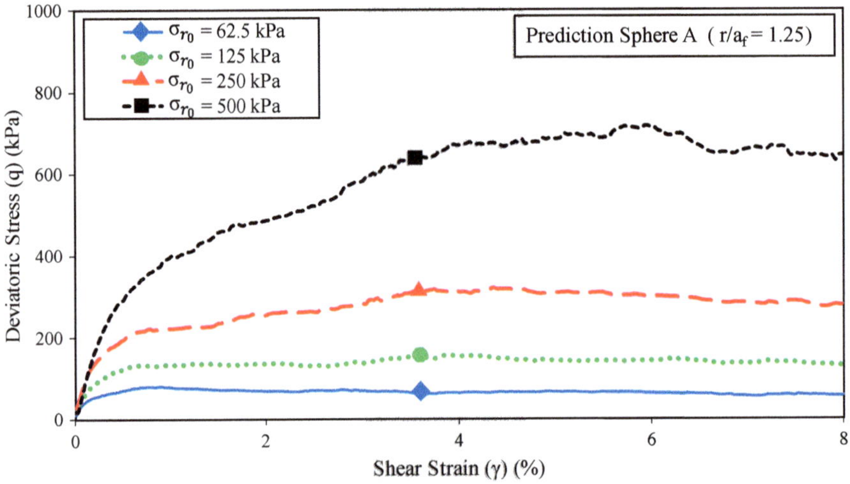

Fig. 7. Deviatoric stress—shear strain relationship at prediction sphere A

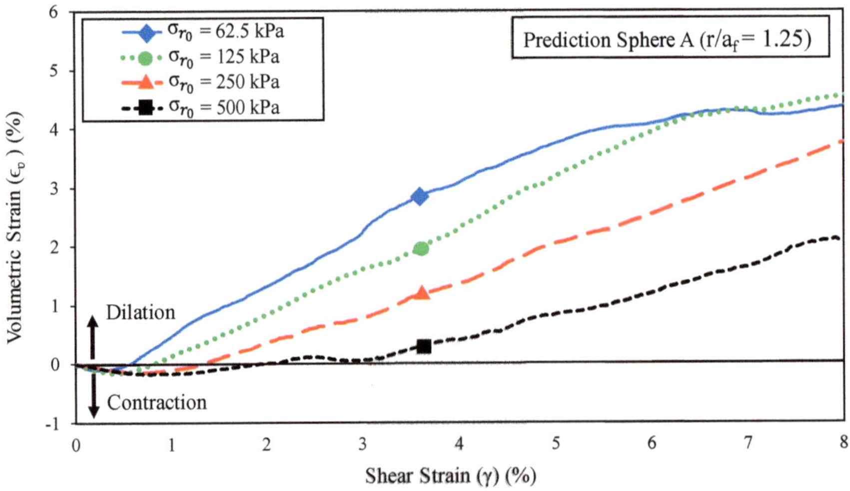

Fig. 8. Volumetric strain—shear strain relationship at prediction sphere A

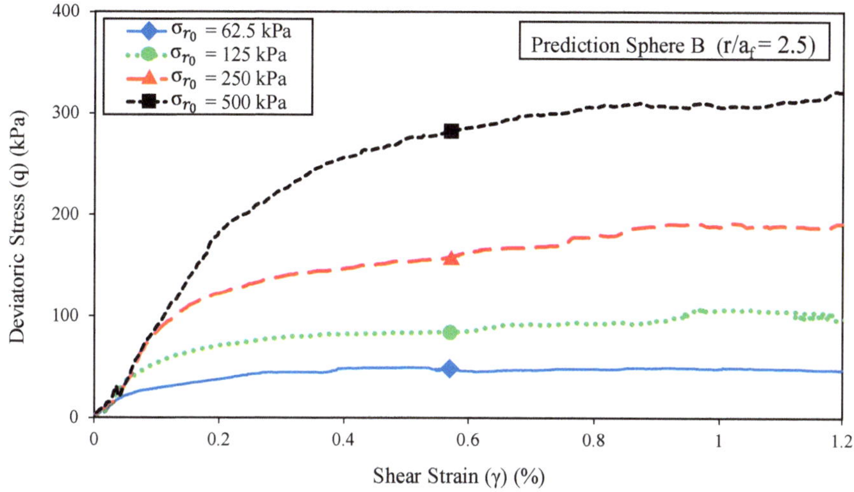

Fig. 9. Deviatoric stress—shear strain relationship at prediction sphere B

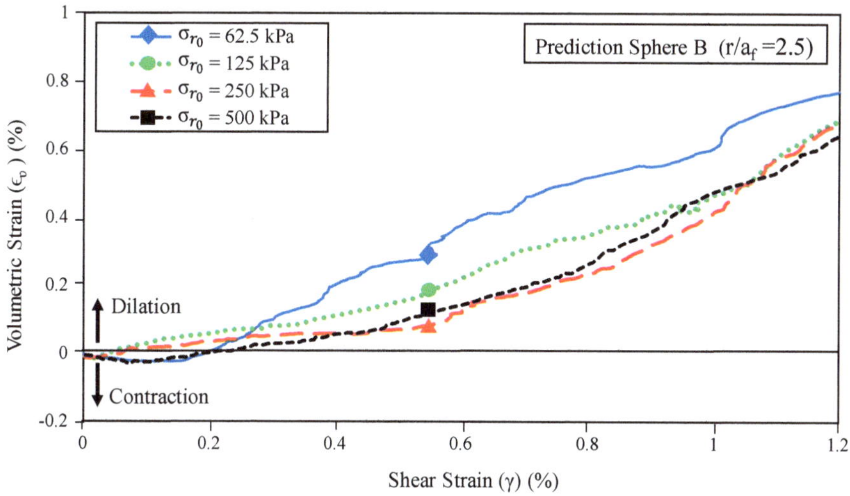

Fig. 10. Volumetric strain—shear strain relationship at prediction sphere B

5 Conclusions

In this paper, the effects of initial in-situ stresses on the soil response during cylindrical cavity expansion have been studied using a three-dimensional discrete element simulation. Comparisons were made for the variations of the cavity pressure, deviatoric stresses as well as the volumetric strains during the cavity expansions with the different initial in-situ stresses. Numerical results confirmed that the initial stress field plays a significant role in determining the soil response during the cavity expansion. Cavity pressures and deviatoric stresses showed a direct correlation with the value of the in-situ stresses, whereas the volumetric strain reduced as the in-situ stress increased. Furthermore, it was observed that soil close to the internal cavity experienced larger shear stresses and volumetric strains (dilation) compared to corresponding points locates further away from the cavity.

References

Azari, B., Fatahi, B., Khabbaz, H.: Assessment of the elastic-viscoplastic behavior of soft soils improved with vertical drains capturing reduced shear strength of a disturbed zone. Int. J. Geomech. **16**(1), B4014001 (2014)

Alshibli, K.A., Sture, S.: Shear band formation in plane strain experiments of sand. J. Geotech. Geoenvironmental Eng. **126**(6), 495–503 (2000)

Arroyo, M., Butlanska, J., Gens, A., Calvetti, F., Jamiolkowski, M.: Cone penetration tests in a virtual calibration chamber. Géotechnique **61**(6), 525–531 (2011)

Carter, J.P., Randolph, M., Wroth, C.: Stress and pore pressure changes in clay during and after the expansion of a cylindrical cavity. Int. J. Numer. Anal. Meth. Geomech. **3**(4), 305–322 (1979)

Carter, J.P., Yeung, S.: Analysis of cylindrical cavity expansion in a strain weakening material. Comput. Geotech. **1**(3), 161–180 (1985)

Collins, I., Pender, M., Yan, W.: Cavity expansion in sands under drained loading conditions. Int. J. Numer. Anal. Meth. Geomech. **16**(1), 3–23 (1992)

Cornforth, D.H.: Some experiments on the influence of strain conditions on the strength of sand. Geotechnique **14**(2), 143–167 (1964)

Cundall, P.A., Strack, O.D.: A discrete numerical model for granular assemblies. Geotechnique **29**(1), 47–65 (1979)

Fahey, M.: Expansion of a thick cylinder of sand: a laboratory simulation of the pressuremeter test. Geotechnique **36**(3), 397–424 (1986)

Falagush, O., McDowell, G.R., Yu, H.-S., de Bono, J.P.: Discrete element modelling and cavity expansion analysis of cone penetration testing. Granular Matt. **17**(4), 483–495 (2015)

Gibson, R.: In situ measurement of soil properties with the pressuremeter. Civil Eng. Pub. Wks. Rev. **56**(568), 615–618 (1961)

Ho, L., Fatahi, B.: Analytical solution for the two-dimensional plane strain consolidation of an unsaturated soil stratum subjected to time-dependent loading. Comput. Geotech. **67**, 1–16 (2015)

Ho, L., Fatahi, B., Khabbaz, H.: A closed form analytical solution for two-dimensional plane strain consolidation of unsaturated soil stratum. Int. J. Numer. Anal. Meth. Geomech. **39**(15), 1665–1692 (2015)

Ho, L., Fatahi, B., Khabbaz, H.: Analytical solution to axisymmetric consolidation in unsaturated soils with linearly depth-dependent initial conditions. Comput. Geotech. **74**, 102–121 (2016)

Iwashita, K., Oda, M.: Rolling resistance at contacts in simulation of shear band development by DEM. J. Eng. Mech. **124**(3), 285–292 (1998)

Jiang, M., Yu, H.S., Harris, D.: Discrete element modelling of deep penetration in granular soils. Int. J. Numer. Anal. Meth. Geomech. **30**(4), 335–361 (2006)

Oda, M., Kazama, H.: Microstructure of shear bands and its relation to the mechanisms of dilatancy and failure of dense granular soils. Geotechnique **48**(4), 465–481 (1998)

Parsa-Pajouh, A., Fatahi, B., Khabbaz, H.: Experimental and numerical investigations to evaluate two-dimensional modeling of vertical drain–assisted preloading. Int. J. Geomech. **16**(1), B4015003 (2015)

Equivalent FEM Meshes from Axisymmetric (AXID) to Three (3D) Dimensions Applied to Tunnels in Clay

Luisa N. Equihua-Anguiano[1(✉)], Iván Rubio-Saldaña[1], Marcos Orozco-Calderón[2], Eleazar Arreygue-Rocha[1], and Carlos Chávez-Negrete[1]

[1] UMSNH University, Gral Francisco J., Mugica S/N Ciudad Universitaria, 58030 Morelia, Mich, Mexico
nicteea@yahoo.com.mx, irs_mx02@hotmail.com, elearreygue@gmail.com, cachavez@umich.mx
[2] Pemex, Cd. Del Carmen Campeche, Mexico City, Mexico
marcos.orozco@pemex.com

Abstract. Efficient tunnel design involves the use of analytical methods and numerical analysis in three dimensions. This paper presents equivalent meshes in axisymmetric conditions (AXID) to three dimensions (3D) using the Finite Element Method (FEM), as well as a displacement nomograph to gauge values of the 3D displacements at a heading and at floor level of a horseshoe tunnel section, and according to a cylindrical symmetry numerical analysis (AXID) from a circular tunnel section. The software used was RS2® for the AXID analysis and RS3® for the 3D simulations. The constitutive model was the Mohr-Coulomb in elastic and anisotropic conditions. Soil parameters: cohesion c, friction angle ϕ, elastic modulus E, Poisson ratio ν and Unit weight γ correspond to endemic soil properties within the city of Morelia. For tunnel simulations circular geometry with $R = 5$ m of radius and an equivalent area from the circular section to build the horseshoe tunnel section were considered. A total length excavation $L = 100$ m considering 1 m of the thickness in the 3D mesh in the process of the numerical simulation was done. This study gives the possibility to obtain the maximum 3D displacement for a horseshoe tunnel, carrying out an axisymmetric analysis in RS2® or using the proposed displacement nomograph.

1 Introduction

Design methods for tunnel construction were initiated by Terzaghi (1942, 1946), Broms and Bennermark (1967) contributed with the study of vertical excavations and Peck (1969) established other concepts related to tunnel support and for the evaluation of the superficial settlements. Subsequently, criteria of tunnel analysis was developed to establish analytical and empirical methods, which focused in reviewing the face and the stability of the tunnel in general. Nowadays, tunnel design employs two (2D) and three (3D) dimensional software, based in numerical methods as the Finite Element

© Springer International Publishing AG, part of Springer Nature 2019
S. Shu et al. (eds.), *New Developments in Materials for Infrastructure Sustainability and the Contemporary Issues in Geo-environmental Engineering*, Sustainable Civil Infrastructures, https://doi.org/10.1007/978-3-319-95774-6_2

(FEM). Nevertheless, low computer resources and the design elevated requirements make 3D one of the not preferred methods for the construction industry. In practice, engineering tunnel design, takes advantage of two (2D) dimensional and axisymmetric (AXID)-FEM, due to the possibility to make faster and cheaper analysis. However, despite axisymmetric conditions (AXID) being a 3D analysis, it is not able to model different tunnel sections in a circular environment and neither is it possible to model asymmetric loads, since the simulation is rotationally symmetric about an axis.

In this way and to avoid long 3D calculations, there is a method reported in the bibliography that allows obtaining a 3D response from 2D FEM analysis (Vlachopoulos and Diederichs 2014). This method is based on the calculi of the plastic zones or plastic radius around the excavation site, obtained from a 2D FEM numerical model which is established under principles of the constitutive model of Hoek et al. (2002), applicable for rocks. Moreover, it is necessary to study other possibilities to interpolate results from 2D to 3D models in soils in order to obtain results quickly, due to the inherent complexity and evolution during construction of this kind of infrastructure.

Considering the previous statements, this article presents a study that allowed obtaining a nomograph to estimate the 3D elastic displacements for a horseshoe tunnel quickly. The presented graphs have also the advantage to achieve the 3D effect of the excavation length. It is also possible to calculate the AXID-3D relationship from the circular to the horseshoe tunnel displacements in elastic conditions.

2 Study Methodology and Tunnel Characteristics

First at all, equivalent meshes in axisymmetric and three dimensions (AXID-3D) using the finite element method (FEM) are presented. The purpose was to evaluate AXID and 3D displacements at the periphery of a circular tunnel and get the same results and both simulations. Then, with the equivalent meshes, it was possible to obtain a nomograph in 3D, modeling a typical horseshoe tunnel section, what is not possible to model in AXID conditions. Finally, the goal is to estimate the elastic displacements in a horseshoe tunnel section quickly, as well as the possibility of performing an AXID simulation to extrapolate its 3D maximal displacement, without using RS3® software.

The presented graphs show also the advantage to achieve the 3D effect of the excavation length. It is important to note, that support was not considered in FEM simulations. The considered depth for the tunnel was $H_o = 50$ m from the upper horizontal boundary to the center of the excavation (Fig. 1). Soil parameters of the study correspond endemic soil properties within the city of Morelia, Mexico. Software used in this study was RS2® for axisymetric conditions using a circular section and RS3® for three dimensions for circular and the horseshoe tunnel sections.

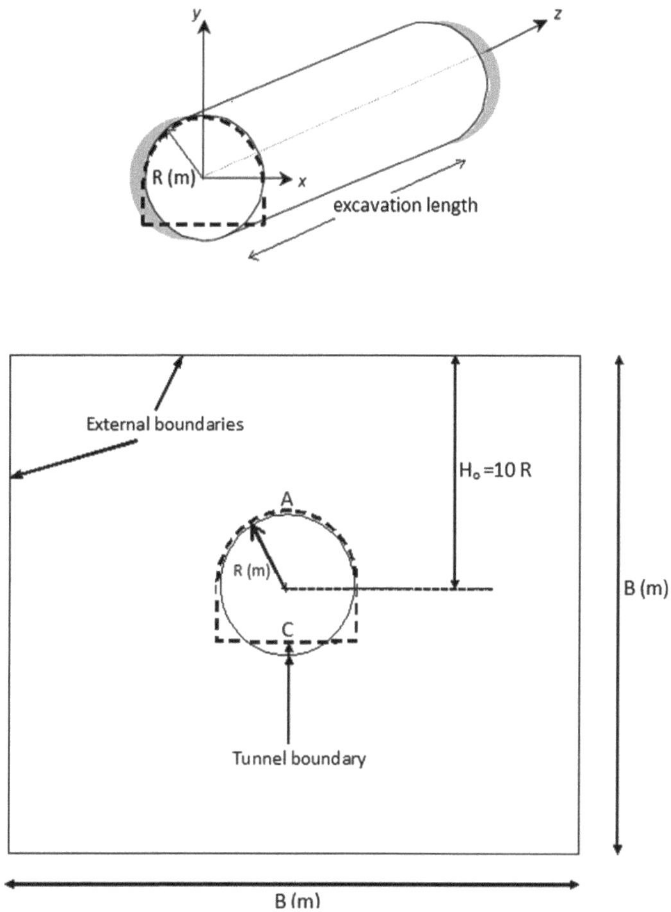

Fig. 1. Circular—Horseshoe geometries and 3D axes convention

3 Numerical Model Validation

Elastic theory is often used to obtain the tunnel radial response. First at all displacements in the tunnel periphery were obtained using Eq. (1) (Deere et al. 1969). These were compared with the displacements obtained in a 2D model (Equihua-Anguiano et al. 2017), considering elastic parameters (see Table 1).

Table 1. Soil parameters for the Mohr-Coulomb constitutive model

c kPa	ϕ (°)	γ (kN/m^3)	E (kN/m^2)	ν
12	22	17.4	15000	0.35

$$u = (\gamma\, H_o - p_a)\frac{(1+v)\,D}{2\,E} \tag{1}$$

where: u = elastic displacement in the tunnel periphery; γ = soil unit weight; H_o = depth from the ground surface to the center of the tunnel; p_a = soil radial stress in the tunnel periphery; v = Poisson ratio; D = tunnel diameter and E = soil elastic modulus.

Figure 2 shows comparisons among 2D-FEM displacements and the analytical response using Eq. (1). The radial stress p_a ($-y$ axis) in the tunnel periphery was varied starting from the "in situ" stress condition (γH_o = 870 kPa) until zero. Analytical displacements, which are represented in Figure with rhombuses and the 2D displacements in the points A an B of the tunnel periphery are very similar to each other. As a result, tunnel closure is the circumference with a decrease of the diameter (see Fig. 6). From these results AXID- and 3D- dimensions simulations were carried out, considering an excavation length L = 100 m, and to obtain the equivalent meshes in AXID- and 3D- presented in the next section and subsequently simulate the horseshoe tunnel section. In the same way Equihua-Anguiano (2017) obtained that, from equivalent meshes in 2D and 3D dimensions, displacements are the same for similar conditions as presented in the Fig. 2 and considering a thickness of 1 m in the 3D mesh to have similar conditions in 2D and 3D FEM simulations.

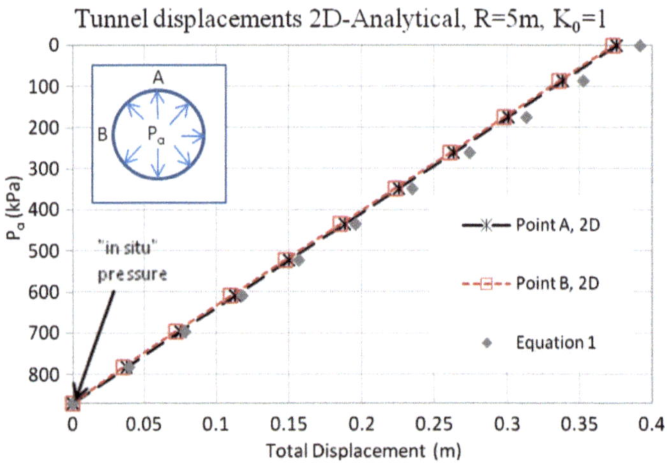

Fig. 2. Analytical-2D displacements versus soil radial stress in the tunnel periphery (p_a) in points A and B at the tunnel periphery.

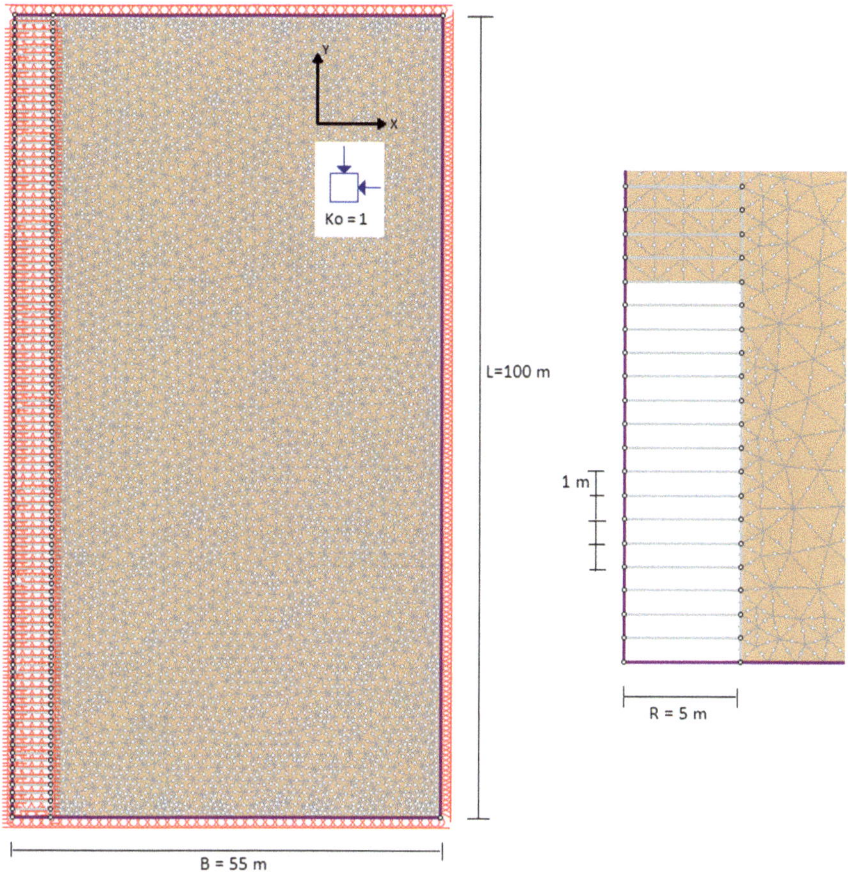

Fig. 3. Axisymmetric-AXID mesh, 55 × 00 m, $R = 5$ m, 5289 elements and 6 nodes tetrahedron, total excavation $L = 100$ m simulated each 1 m.

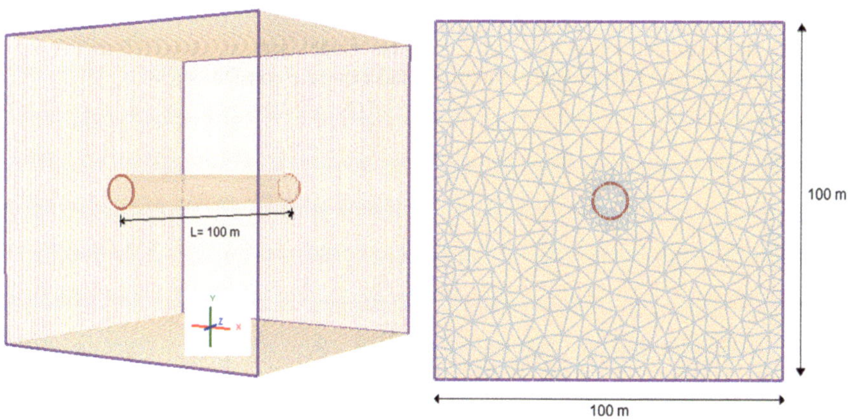

Fig. 4. Three dimensional-3D mesh, 100 × 100 × 100 m, $R = 5$ m, 5289 elements and 10 nodes tetrahedron, total excavation $L = 100$ m simulated each 1 m.

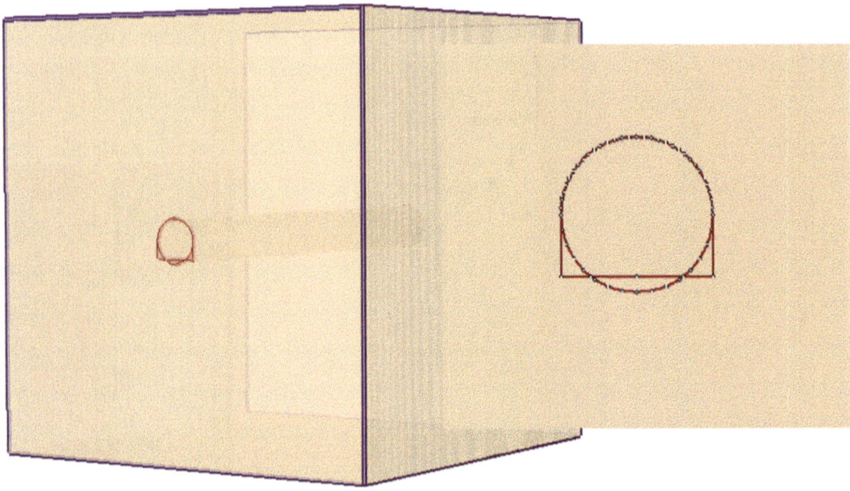

Fig. 5. Three dimensional-3D mesh, 100 × 100 × 100 m, horseshoe representation section, 5289 elements and 10 nodes tetrahedron, total excavation L = 100 m simulated each 1 m.

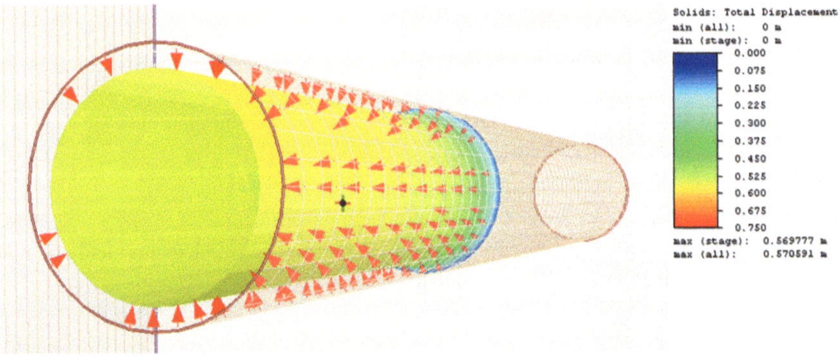

Fig. 6. 3D displacement kinematics for a circular tunnel section, A = 78.5 m^2, R = 5 m, L = 100 m and E = 5000 kPa, elastic conditions and isotropic constant stress.

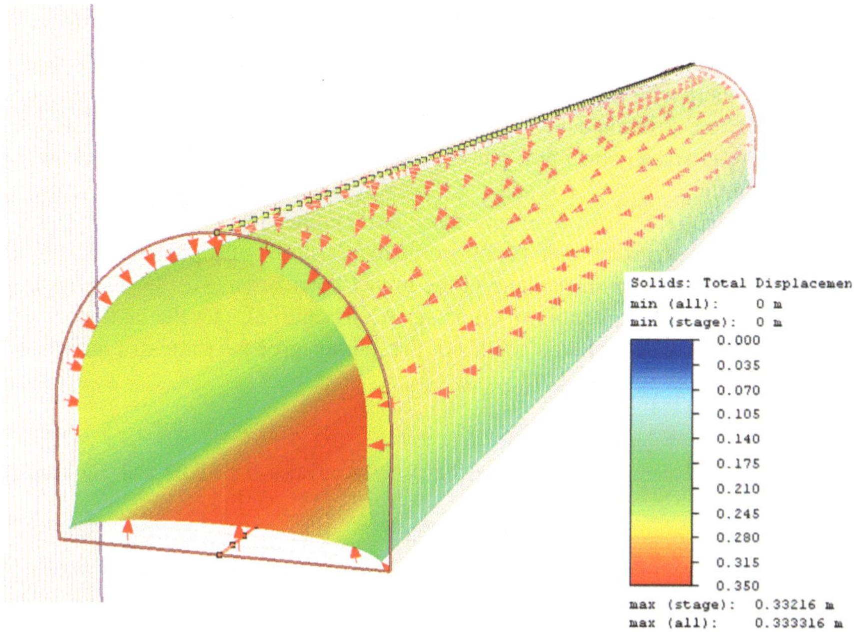

Fig. 7. 3D displacement kinematics, for a horseshoe tunnel section $A = 78.5\ \text{m}^2$, $L = 100$ m and $E = 5000$ kPa, elastic conditions and isotropic constant stress.

4 Numerical Modeling in Axisymmetric and 3D Conditions

Numerical analysis corresponds to the strain generated in the soil. Two points were analyzed: (A) at a heading and (C) in the floor of a tunnel horseshoe section as depicted in Fig. 1.

4.1 Soil Parameters

A Mohr-Coulomb constitutive model was used by considering elastic and isotropic conditions ($K_o = 1$). The geotechnical parameters, friction angle (ϕ), cohesion (c), and soil unit weight (γ) correspond to endemic soil properties within the city of Morelia, Mexico. Elastic modulus (E) value corresponds to a medium stiffness soil, this being the main limitation of results presented in this article, since according to the elastic theory displacements depend directly on this parameter. Table 1 shows the soil parameters. In the case of the Poisson ratio (v), value corresponds to a soil in unsaturated conditions. The friction angle (ϕ) and the cohesion (c), do not have an influence in the resulting displacements.

4.2 Equivalent FEM Meshes in AXI and 3D for a Circular Tunnel Section

Geometry of the equivalent meshes in AXID and 3D conditions for a circular tunnel with $R = 5$ m are presented in Figs. 3 and 4 respectively. Meshes dimensions and general FEM characteristics are present in Table 2. Figures 3 and 4 shows the dimensions in the *x*-, *y*- (AXID) and *z*- (3D) coordinates, considering more refinement in the tunnel area ($R = 5$ m) to ensure more accurate results. External boundary conditions were zero displacement in the vertical and horizontal planes and zero displacement in the perpendicular plane of the tunnel boundaries (Fig. 1). The mesh was extended $L = 100$ m in depth to compare AXID and 3D results and obtain the equivalent meshes. These characteristics of the meshes give the same displacements as shown in the results section.

Table 2. Meshes dimensions and general FEM characteristics for AXID and 3D numerical simulations

A (m^2)	Section tunnel geometry	External boundaries	Tunnel boundaries	AXID BxL (m)	3D B × B × L (m)
78.5 ($R = 5$ m)	Circular	Zero-displ. x–y	zero-displ. –y	55 × 100	100 × 100 × 100
				Graded mesh 6 noded triangles	Graded mesh 10 noded tetrahedron
78.5	Horseshoe	Zero-displ.– x, –y, –z	zero-displ. –z	–	100 × 100 × 100
					Graded mesh 10 noded tetrahedron

4.3 3D Mesh for a Horseshoe Tunnel Section

In the same way, a 3D mesh considering a horseshoe section was modeled, taking as a base the equivalent 3D mesh of the circular tunnel section. Horseshoe's geometry is standard for horseshoe tunnels cross sections (see Fig. 1). The considered area (*A*) is equivalent to the circular tunnel area resulting of the $R = 5$ m, so that value is $A = 78.5$ m^2 in both cases, circular and horseshoe sections. The horseshoe section in the mesh was placed according to the circular section as seen in Figs. 1 and 5. Point A is common in both tunnel sections and the C point corresponds only to the floor of the horseshoe tunnel. Table 2 shows geometry and mesh numerical characteristics. Horseshoe tunnel boundaries are the same that those used in the circular tunnel.

5 Results

Results obtained in the calibration of the equivalent meshes in AXID and 3D are presented in Fig. 8. It can be observed the numerical results for the circular tunnel's point A in AXID- with a continuous line and 3D- with triangles symbols respectively. The two curves (continuous line-AXID and triangles-3D) match fairly well among them for the case of the circular section. In this way, numerical mesh characteristics of the simulations are taken as the equivalent meshes. Results also show that there is a point of inflection where the excavation length cease to the influence development in the generated displacements in the tunnel (L = 12 m approximately) and a maximum value is presented.

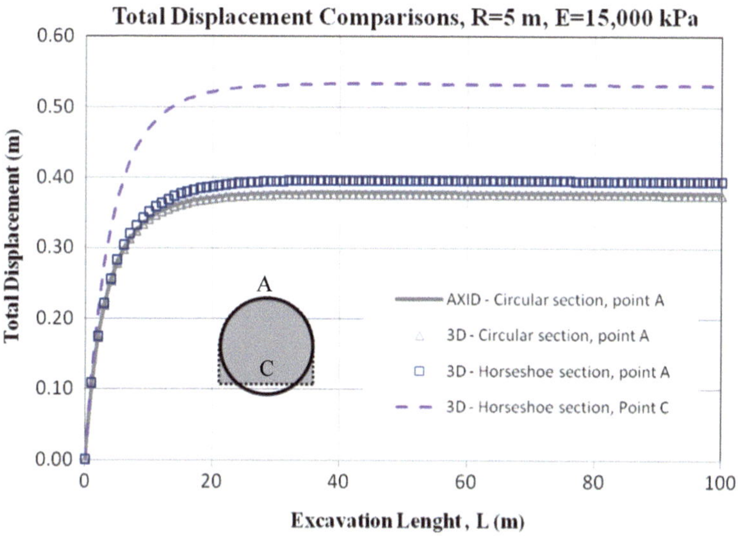

Fig. 8. Total Displacement versus Excavation length L, for a R = 5 m, E = 15,000 kPa and for the circular and horseshoe tunnel sections in AXID in 3D.

Results corresponding for the horseshoe tunnel section to the points A and C are presented in the same Fig. 8 with the void square symbols and the dashed line respectively. Displacements are bigger that those obtained for the circular tunnel. Displacements in the floor of the tunnel (point C) are the most unfavorable in respect to the circular tunnel section and in respect to the horseshoe's point A. Figure 7 shows the 3D displacement kinematics, for the horseshoe tunnel section and the greatest displacement is presented in the floor of the tunnel section. From Fig. 8 a displacement nomograph is deduced and presented in Fig. 9, to obtain this graph, total displacement curves were normalized respect to the soil specific weight γ, the depth from the ground surface to the center of the tunnel H_o and of the soil radial stress in the tunnel periphery p_a. It is important to note, that displacements using the nomograph, can be obtained for any properties soils (γ, ϕ and c), radial stress in the tunnel periphery and also for any depth from the ground surface to the center of the tunnel. The previous is based on the principle that displacements obtained with the elastic theory, depends only of the elastic soil modulus E. Limitations of results presented in this paper are the numerical simulations of only one elastic modulus corresponding to a medium soil consistency and one equivalent area for the horseshoes section (E = 15,000 kPa and A = 78.5 m^2).

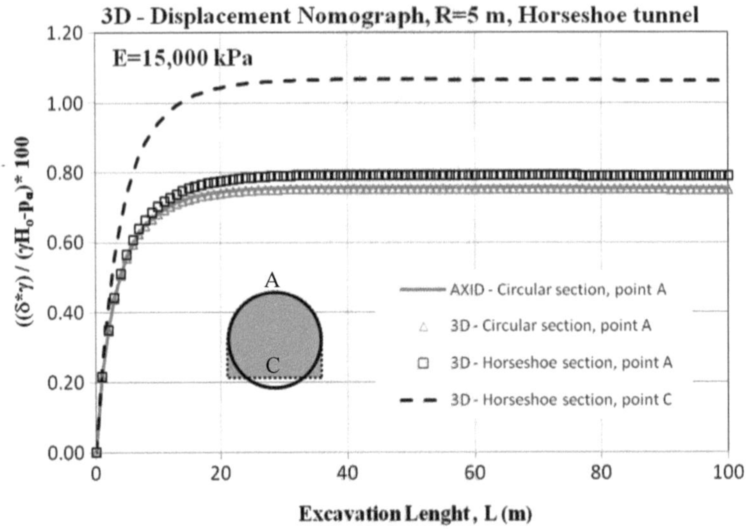

Fig. 9. AXID-3D Displacement nomograph for $R = 5$ m, $L = 100$ m and E = 15,000 kPa, normalized in respect to γ, H_o, and p_a.

Figure 10 shows the relation among displacements obtained from the horseshoe section in respect to the circular section. Results of the total displacements δ_i of the points A and C were normalized in respect to the maximum displacement of the circular tunnel $\delta_{maxcirc}$ curves, exposed that the maximal displacements in the point C are 40% bigger than those obtained from an AXID numerical simulation and 5% bigger in the case of the horseshoe section in point A. Then from an AXID simulation it is possible to interpolate this displacement augmenting the value obtained in this simulation in the mentioned proportion and obtaining maximal displacements developed in a horseshoe tunnel section.

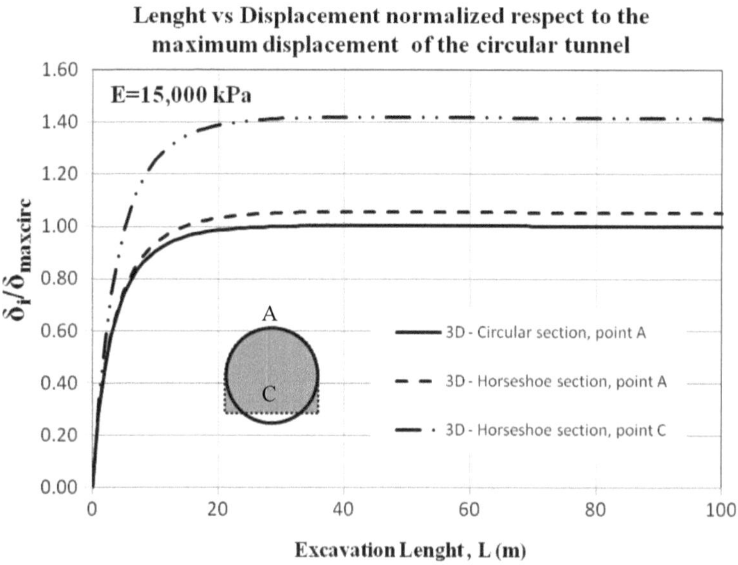

Fig. 10. Excavation length, L versus Total displacements δ normalized respect to the maximum displacement $\delta_{maxcirc}$ at the tunnel's point A, R = 5 m, K_0 = 1.

6 Conclusions

Tunnels are geotechnical facilities of great importance that require the development of tools that improve and make more efficient design in practice. In this article, an analysis using FEM was carried out using Rocscience®. Equivalent FEM meshes were presented from AXID to 3D for a circular tunnel in elastic and isotropic conditions. A good agreement between AXID and 3D results in terms of displacements was obtained, as well as with the analytical 2D solution provided by the theory of elasticity. These results allowed performing numerical simulations for a horseshoe equivalent tunnel section considering the obtained equivalent mesh for a circular tunnel. Results allow to compare the influence, in both forms of numerical simulations, from the obtained results; first, with the equivalent meshes and then, with a horseshoe tunnel section. Subsequently, from an AXID simulation it is possible to interpolate 3D displacement augmenting a value obtained from the AXID simulation.

Displacements obtained with the elastic theory, depends of the elastic modulus E, then in the results presented in this paper, nomograph only is applied to E = 15,000 kPa and a given an equivalent section area. It is necessary to obtain a nomograph for different soil stiffness (E variation) and other tunnel areas. In the same way, other horseshoe sections need to be simulated, in order to have results for more efficient tunnel sections.

Analyzed elastic and isotropic conditions are a limitation in this study, elasto-plastic analysis with water tables and anisotropic conditions are necessary, these will allow obtaining more real conclusions related to the behavior of the soil.

References

Broms, B.B., Bennermark, H.: Stability of clay at vertical opening. J. Soil Mech. Found. Div. ASCE **93**(1), 71–94 (1967)

Deere, D.U., Peck, R.B., Monsees, J.E., Schmidt, B.: Design of tunnel liners and support systems. Final Report, University of Illinois, Urbana, for office of High Speed Transportation, U.S. Department of Transportation, Washington, D.C., Contract No. 3-0152 (1969)

Equihua-Anguiano, L.N., Viveros-Viveros, F., Pérez-Cruz, J.R., Chávez-Negrete, C., Arreygue-Rocha, E., Orozco-Calderón, M.: Displacement nomograph from two (2D) to three (3D) dimensions applied to circular tunnels in clay using finite element. In: Proceedings of the 19th International Conference on Soil Mechanics and Geotechnical Engineering, Seoul, Korea, September (2017)

Hoek, E., Carranza-Torres, C., Corkum, B.: Hoek-Brown failure criterion-2002 edition. In: Proceedings of the 5th North American Rock Mechanics Symposium and the 17th Tunneling Association of Canada Conference: NARMS-TAC 2002, vol. 1, pp. 267–273, Toronto, Ontario, Canada (2002)

Peck, R.B.: Deep excavations and tunneling in soft ground. State of the Art Report. In: Proceedings of the 7th International Conference on Soil Mechanics and Foundations Engineering, pp. 225–325, Mexico (1969)

RS2 Version 9 Modeler, Version 9.020 64-bit, 2D finite element analysis for rock & soil, Copyright© 1990–2016, Rocscience Inc., Toronto, Ontario, Canada, www.rocscience.com

RS3 Version 1.017 64-bit, 3D finite element analysis for rock & soil, Copyright© 2013–2015, Rocscience Inc., Toronto, Ontario, Canada, www.rocscience.com

Vlachopoulos, N., Diederichs, M.: Appropriate uses and practical limitations of 2D numerical analysis of tunnels and tunnel support response. Geotech. Geolog. Eng. **32**(2), 469–488 (2014). Springer. https://doi.org/10.1007/s10706-014-9727-x

Terzaghi, K.: Shield tunnels of the Chicago subway. J. Boston Soc. Civil Eng. **29**, 163–210 (1942)

Terzaghi, K.: Rock defects and loads on tunnel support. In: Proctor, R.V., White, T.L. (eds.) Rock Tunneling With Steel Supports, pp. 1–15. Commercial Shearing and Stamping Co., Youngstown, Ohio, U.S.A., Section 1 (1946)

Earthquake-Induced Deformation of Breakwater on Liquefiable Soil with and Without Remediation: Case Study of Iran LNG Port

Yaser Jafarian[1], Mohsen Bagheri[2(✉)], and Mehdi khalili[3]

[1] International Institute of Earthquake Engineering and Seismology (IIEES), Tehran, Iran
yiafarianm@iiees.ac.ir
[2] Department of Civil Engineering, Babol University of Technology, Babol, Iran
m.bagheri@stu.nit.ac.ir; bagheri.m@hotmail.com
[3] Islamic Azad University, Science and Research Branch, Tehran, Iran
mkhalili.civil@gmail.com

Abstract. Seismic response of a breakwater founded on a thick liquefiable silt layer is evaluated through the numerical modeling. An economic design of breakwater rested on extremely thick sediments of loose or liquefiable soils may prefer no ground improvement. However, in a seismic environment, such breakwaters are potentially vulnerable to failure due to earthquake-induced pore pressure generation within the underlying deposits. Results of the present models are compared with the centrifuge data, as the verification stage of the numerical study. In Part I, a benchmark model without any liquefaction countermeasure has been considered. In Part II, additional models with remediation are analyzed and compared with no remediation case. The stone column method has been considered as the liquefaction countermeasure.

Keywords: Breakwater · Liquefiable soil · Stone columns · Deformation

1 Introduction

Breakwater is an artificial offshore structure which is built to protect port and harbor facilities from destructive effects of sea waves, currents, typhoons, and tsunamis by reflecting and dissipating their energies (Chaudharya et al. 2016). It sufficiently creates calm water for safe navigation, anchorage of vessels, loading and unloading cargoes, and other harbor activities. In the recent decades, more and more marine infrastructures such as breakwaters are either constructed or extended due to increase in global economic activities. The breakwaters are constructed on seabed, and maybe vulnerable to earthquake and tsunami. During the 2011 Tohoku-oki earthquake, the Pacific Coast of Tohoku and many coastal structures including breakwaters were damaged through earthquake and the subsequent tsunami (Hazarika et al. 2012, 2013, 2015; Hara et al. 2012.

There are some other breakwaters (e.g., Hattaro Breakwater in Hachinohe Port, Aomori Prefecture; Ryujin-zaki Breakwater at Miyako Port, Iwate Prefecture and

© Springer International Publishing AG, part of Springer Nature 2019
S. Shu et al. (eds.), *New Developments in Materials for Infrastructure Sustainability and the Contemporary Issues in Geo-environmental Engineering*, Sustainable Civil Infrastructures, https://doi.org/10.1007/978-3-319-95774-6_3

breakwater at Onagawa Port, Miyagi Prefecture) which also collapsed due to failure of their foundations during the 2011 earthquake and tsunami. Following the failure of breakwaters, tsunami could easily strike the coastal areas, and it leads to catastrophic losses for the structures and residents over there.

In seismic-prone regions, excess pore water pressure is generated by earthquakes and causes decrease in bearing capacity of the foundation soil. It may result in high deformation of mound, and heavy settlement of the breakwater. In addition, seismic inertia forces act on the breakwater.

To determine the effect of earthquake on the performance of breakwater, Hazarika et al. (2015) conducted a series of centrifuge model tests for the stability of breakwater against earthquake and tsunami, and found that earthquake had significant impacts on the stability of the breakwater which hit the breakwater before tsunami. Although extensive research has been done on the liquefaction occurrence in the marine environment, the numerical analysis of marine structures such as breakwaters that are relied on liquefiable soil ground is rarely found in the literature.

Although recent advances in physical model experiments (Chaudharya et al. 2017), field case histories (Hazarika et al. 2013, 2015; Hara et al. 2012), and computational modeling of liquefaction-induced ground deformation (Takahashi et al. 2011; Elsafti and Oumeraci 2016; Verhaeghe et al. 2014; Cihan et al. 2012; Yc and Jeng 2013, Jafarian 2010; Moghadam et al. 2017; Asgari et al. 2014; Shooshpasha and Bagheri 2014) are quite promising, some challenges remain on this critical problem. This study aims to gain insight into the influence of liquefiable soil on the breakwater settlements and lateral displacements. Details of the numerical modeling are given in the paper and the results of the analyses for the breakwater with and without stone column are compared and discussed.

2 Iran LNG Port

The Iran LNG plant is located on the north-western banks of the Persian Gulf, in Tonbak, 50 km northwest of Asaluyeh and 15 km Southeast of Kangan-Pars 2 (Tonbak) (Moghadam et al. 2017). The breakwater of LNG Port is located on the southwest of the Kangan port. This breakwater is laid on a thick layer of silt with liquefaction capacity.

Currently, it has been constructed in Pars Energy Zone of Assaluyeh in Tonbak. According to the original design, there was possibility of failure due to liquefaction of the silt layer. Therefore, seismic deformation analysis was studied to determine the potential of failure in this structure due to earthquake. The geotechnical analysis conducted on the site revealed that the breakwater might be considerably damaged due to the design earthquake. Hence, the LNG facilities were decided to be kept away from the breakwater to decline the importance grade of the structure.

In this study, the liquefaction-induced damage to breakwater is evaluated using nonlinear effective stress analysis performed by FLAC (Itasca), and failure mechanisms of LNG breakwater are discussed as well. Since the obtained level of deformations did not meet the allowable damages criteria for the breakwater, the soil improvement against liquefaction is considered.

In the present study, the critical cross section of the breakwater located in LNG harbor was evaluated in two modes including the breakwater prior to ground improvement and the one in the presence of a stone column with a diameter of 80 cm and center to center distance of 240 cm.

3 Numerical Simulation

3.1 Model Geometry and Soil Properties

The finite difference program (FLAC) was used for the numerical analyses of the breakwater. The purpose of the numerical analyses was to confirm the effectiveness of the soil reinforcing technique, and to determine the behavior of the reinforcement-soil-breakwater system during earthquake loadings.

The seabed (foundation soil) was made of loose non-plastic silt (D_r = 30%), and modeled as a cohesionless and homogeneous material by the Mohr-Coulomb failure criterion. The soil profile involves a thick layer of non-plastic silt. The thickness of the foundation silt was equal to 12.5 m based on the geotechnical site investigation. After construction of the model, the level of water was increased up to above the seabed. The seabed and mound were made fully saturated. The material properties of the foundation soil are shown in Table 1.

The numerical model of breakwater and the foundation soils are shown in Fig. 1. The Finn model was used to simulate seismic behavior of the liquefiable silt layer. This constitutive model is a coupled effective stress model that works with combining Mohr-Coulomb criterion and volume change-based pore water pressure buildup. All displacements were restrained at the bottom of the model. The model was solved for static equilibrium before applying seismic loadings.

3.2 The Equivalence of Three-Dimensional Stone Columns in Plane Strain Case

In this research, the three dimensional method of stone columns were represented as (two-dimensional) plane strain method, validated by Tan et al. (2008). From the two simplified methods suggested by Tan et al. (2008), the first one tends to match soil permeability with analytical relations, while in the second; the width of the stone column could be obtained by equalizing the cross section of the stone column.

In the second method, an alternative geometrical transformation tends to be based on equalizing the column drainage capacity in both axisymmetric and plane-strain conditions, whose concept was proposed in a vertical drain study by Indraratna and Redana (1997) to convert vertical drain system into equivalent plane-strain drain walls. This method would preserve cross sectional areas of the stone column and the surrounding soil for the same total area in both conditions. The stone column width is determined by the following relationship based on the equivalence of area replacement ratio (Fig. 2):

Table 1. Physical and mechanical parameters of breakwater and its underlying soil

Material	Normalized standard penetration test $N_{1(60)}$ (blow)	Porosity (n)	Permeability K (m/s)	Poisson's ratio (ν)	Dilation angle (ψ)	Friction angle (φ)	Maximum Shear modulus G_{max} (kPa)	Unit dry weight (ϒdry) (kg/m3)
Armor	–	0.2	1×10^{-2}	0.27	11	44	$13000\frac{(2.17-e)^2}{1+e}(\bar{\sigma}_0)^{0.55}$	18
Filter	–	0.2	1×10^{-3}	0.27	10	41	$8400\frac{(2.17-e)^2}{1+e}(\bar{\sigma}_0)^{0.6}$	17
Core (Gravel)	–	0.3	1×10^{-4}	0.27	0.3	38	$8400\frac{(2.17-e)^2}{1+e}(\bar{\sigma}_0)^{0.6}$	16
Foundation (Liquefiable soil)	8	0.44	1×10^{-7}	0.35	0	28	$8400\frac{(2.17-e)^2}{1+e}(\bar{\sigma}_0)^{0.6}$	14.6
Foundation (Gravel)	–	0.26	1×10^{-6}	0.27	10	40	$8400\frac{(2.17-e)^2}{1+e}(\bar{\sigma}_0)^{0.6}$	20.4
Stone column	–	0.3	1×10^{-3}	0.27	10	41	$8400\frac{(2.17-e)^2}{1+e}(\bar{\sigma}_0)^{0.6}$	17

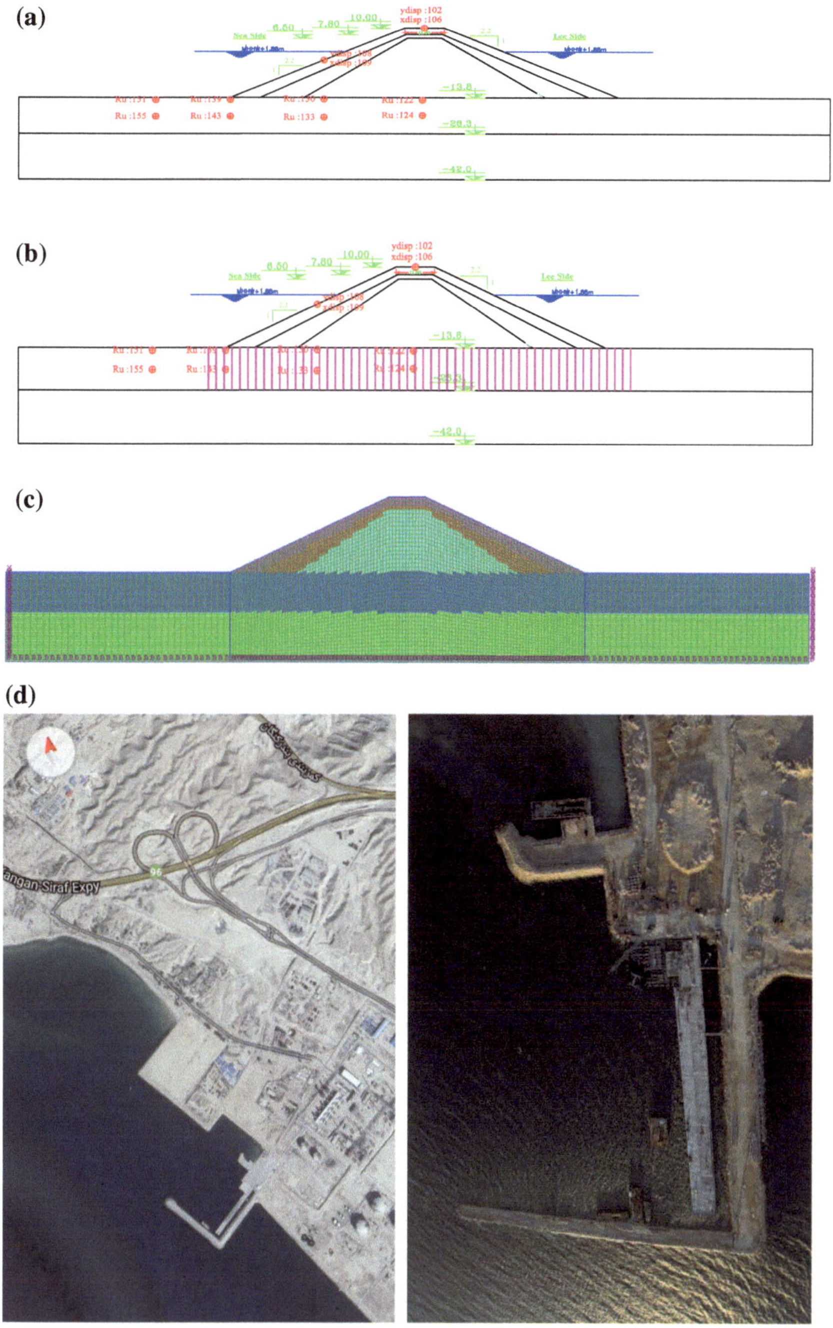

Fig. 1. Model geometry, (a) and (b) breakwater with and without stone columns and (c) its meshed model and (d) general layout of Iran LNG plant.

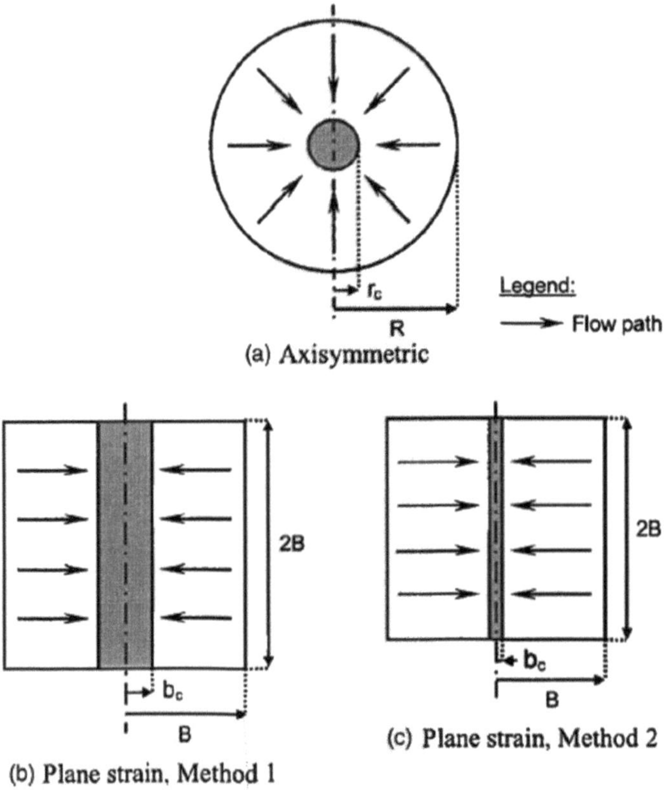

Fig. 2. Cross-sections of unit-cell stone column and plane-strain conversions Tan et al. (2008)

$$b_c = B\frac{r_c^2}{R} \tag{1}$$

where b_c, B, R and r_c stand for plane-strain column or wall's width, the equivalent plane-strain width, the radius of drainage zone, the equal axisymmetric column diameter, respectively.

Numerical simulations were performed on (i) breakwater on unreinforced foundation and (ii) breakwater foundation reinforced by stone columns (reinforced foundation), shown in Fig. 1. Each model was subjected to ground motion events obtained from scaling of the El Centro (1940) earthquake record (Fig. 3). Each model is subjected to a ground motion record that is selected from the PEER Strong Motion Database (Table 2) (Elsafti and Oumeraci 2016).

Comparison of excess pore water pressure ratio, $r_u = \frac{\Delta u}{\sigma_0^{\prime}}$ horizontal and vertical deformations at the crest and the slope of the breakwater was made during the earthquake to determine the impact of stone columns. In order to improve liquefiable soil with relative density Dr = 30%, Stone column with diameter of 80 cm (D = 80 cm) and distance between two columns (240 cm center to center (S = 240 cm)).

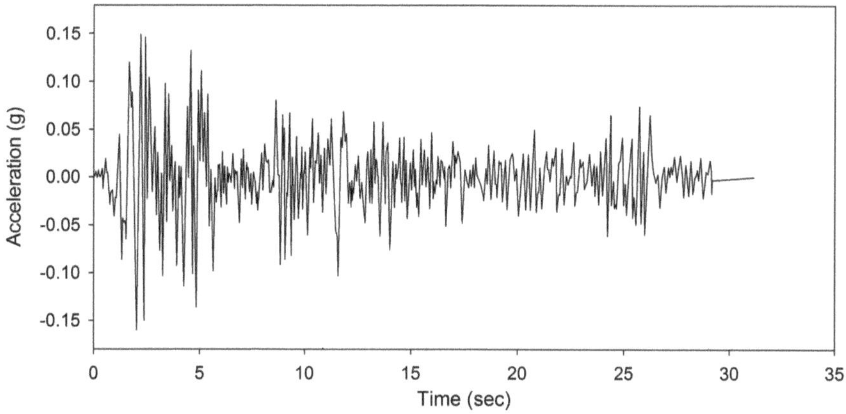

Fig. 3. EL Centro earthquake accelerograms with maximum acceleration (PGA) of 0.15 g

Table 2. Characteristic of input motion wave

Earthquake motion parameters	El Centro (USA)
Date of occurrence	1940
Magnitude of earthquake, M_w	6.9
Maximum horizontal acceleration, (g)	0.349
Predominant period, T_p (s)	0.56
Significant duration, D_{5-95} (sec)	24.58
Time of MHA (t_p (s))	4.1
PGV/PGA (s)	0.102
Arias intensity (m/s)	1.758
SIR (m/sec/sec)	0.117
Energy flux (J.m^{-2}.s^{-1}) PGA = 0.35g(J.m^{-2}.s^{-1})	2144.177
Type	Far field
Hypocentral distance (km)	15.69

$SIR = I_{a(5-75)}/D_{(5-75)}$

Obtained from Pacific Earthquake Research Center (PEER 2015)

4 Validation of the Numerical Model

The explicit finite difference code FLAC (Itasca) has been employed to perform fully coupled nonlinear dynamic analysis on models of embankment dam founded on shallow layers of liquefiable foundation (Adalier and Sharp 2002a, b). to assess the seismic stability of earth dam. A fully coupled analysis is one in which calculation of displacements is carried out by solving a coupled system of equations that includes the motion equation and the diffusion equation. In the code used herein, saturated soil is treated as a two-phase material using Biot coupled equations for the soil and water phases.

In the static analysis, the earth dam was under gravity loading only; the base boundary was fixed in all directions and the side boundaries were fixed in the x direction. Furthermore, in order for pore pressure not to interact with mechanical volume changes, the water bulk modulus was assumed zero.

Following the model presented by Martin et al. (1975) and Byrne (1991) proposed the following empirical equation which relates the increment of volume decrease, $\Delta\varepsilon_{vd}, \gamma$ to the cyclic shear-strain amplitude:

$$\frac{\Delta\varepsilon_{vd}}{\gamma} = c_1 \exp\left(-c_2 \frac{\varepsilon_{vd}}{\gamma}\right) \tag{2}$$

where c_1 and c_2 are constants with different interpretations (Itasca). The above mentioned formulation is available in FLAC as a built-in model and was adopted in this study along with a bilinear, elastic-perfectly plastic stress-strain relationship and Rayleigh damping. Therefore, during dynamic analysis, as effective stresses decrease with pore pressure increment, the soil begins to yield and increments of permanent deformation are accumulated.

The approach described above has been verified and calibrated in the literature through analysis of well-documented centrifuge model tests (Adalier and Sharp 2002a, b) and was adopted in this study for the dynamic analysis of the model. In the centrifuge model study which used in the current study for the verification of numerical model, the liquefaction countermeasure study (Adalier and Sharp 2002a, b) was conducted at a 100 g gravitational acceleration. Under this gravity field, the centrifuge models (Fig. 4) simulated a prototype earth dam of 10 m in height and 39.5 m in width, resting on a sand foundation deposit of 9 m in thickness. The earth dam core was composed of Kaolin clay compacted at about 33% water content, with a dry unit weight of 13.4 kN/m3 and an estimated unconfined shear strength (s_u) of 16–18 kPa. Clean Nevada No. 120 sand at a relative density (D_r) of about 70% was used to construct the embankment slopes. The same sand was also used as the foundation material, at a Dr of about 35% (for the nondensified zone) and 70% (for the densified zone). The foundation layer was saturated with a fluid at a prototype permeability of about 1.3×10^{-4} m/s (D_r = 35%) and 1.0×10^{-4} m/s (D_r = 70%), within the range of fine sand Lambe and Whitman (1969). Water was used as the reservoir fluid (Fig. 4), resulting in a prototype permeability of about 6×10^{-3} m/s within the embankment (in the range of coarse sand) (Table 3).

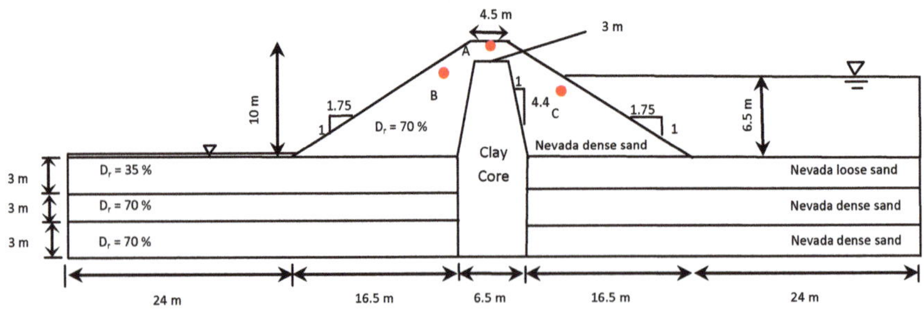

Fig. 4. Model configuration and dimension of earth embankment and liquefiable foundation system in centrifuge test conducted by Adalier and Sharp (2002a, b)

Table 3. Properties for foundation and embankment soils in the centrifuge test (Adalier and Sharp 2002a, b)

		Mass density (kg/m3)	Maximum Shear modulus (kPa) G_{max}	Poisson-s ratio	Cohesion (kPa)	Friction angle (degrees)	Porosity	Hydraulic conductivity (m/s)	N1(60) (blows)
Foundation	Nevada loose sand	1600	$6908\frac{(2.17-e)^2}{1+e}(\bar{\sigma}_0)^{0.5}$	0.45	0	31	0.424	1.3e-5	7
	Nevada dense sand	1800	$6908\frac{(2.17-e)^2}{1+e}(\bar{\sigma}_0)^{0.5}$	0.45	0	35	0.373	1e-5	20
Embankment	Core	1500	$3230\frac{(2.97-e)^2}{1+e}(OCR)^k(\bar{\sigma}_0)^{0.5}$	0.45	20	15	0.35	1e-9	–
	shell	1800	$6908\frac{(2.17-e)^2}{1+e}(\bar{\sigma}_0)^{0.5}$	0.45	0	35	0.424	1.3e-5	20

Figure 4 represents details of the experimental configuration. All models were subjected to similar lateral excitation of about 30 cycles, 0.2 g peak amplitude, and 1.5 Hz dominant frequency (Fig. 5). Results of the validation model as shown in Fig. 6 in terms of vertical displacement of the embankment confirm successful applicability of the numerical model in obtaining seismic performance of breakwater founded on layers of liquefiable foundation.

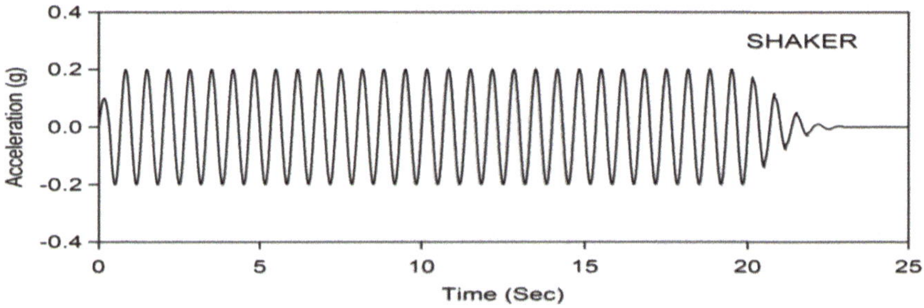

Fig. 5. Recorded input acceleration (g) in the centrifuge model (Adalier and Sharp 2002a, b)

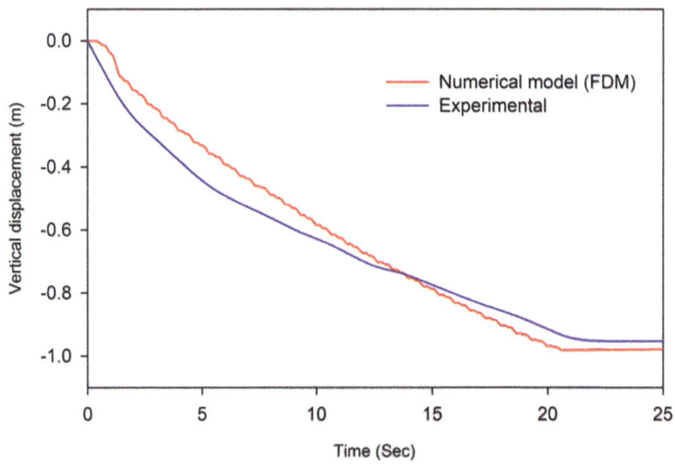

Fig. 6. Comparing the vertical displacement of embankment crest in the centrifuge model test (Adalier and Sharp 2002a, b) and the numerical model (FDM).

5 Breakwater on Unreinforced Versus Reinforced Ground

5.1 Breakwater on Unreinforced Ground

Figures 7 and 8 obviously illustrates how the lack of stone columns would cause ground longer to incur substantial reduction in resistance after strong earthquake shaking. Due to the weight of the LNG breakwater, horizontal and vertical displacements increased, and a differential settlement, which can be the greatest threat to breakwaters. Figures 7 and 8 depict excess pore water pressure at different depths of soil under the breakwater. Comparing the pore water pressure ratio (r_u) at different parts of the breakwater to the free field obviously shows how the increased overburden induced by weight of the breakwater can significantly increase effective stress and decrease pore pressure ratio. However, the soil susceptible to liquefaction under the breakwater not improving even at low elevations caused settlement, and differential and significant horizontal displacement due to increasing shear stress, which is considered the greatest threat to earth structure, resulting in significant damage.

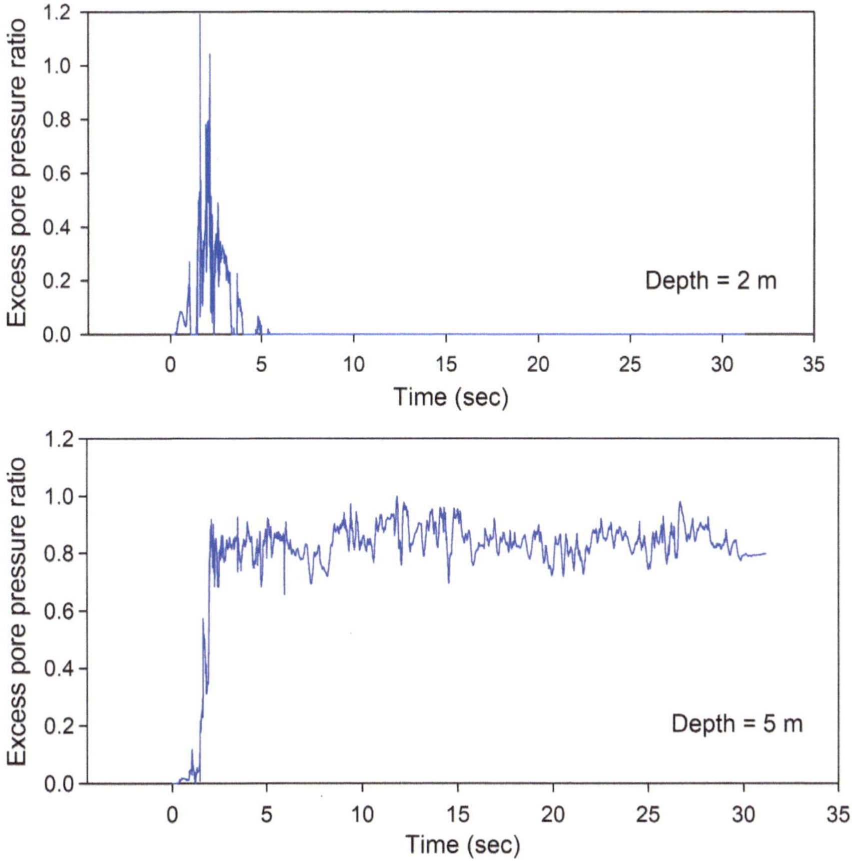

Fig. 7. Changes of pore water pressure ratio in terms of time in free field in the points (151) and (155) at a depth of 2 and 5 m under the EL Centro 0.15 g record

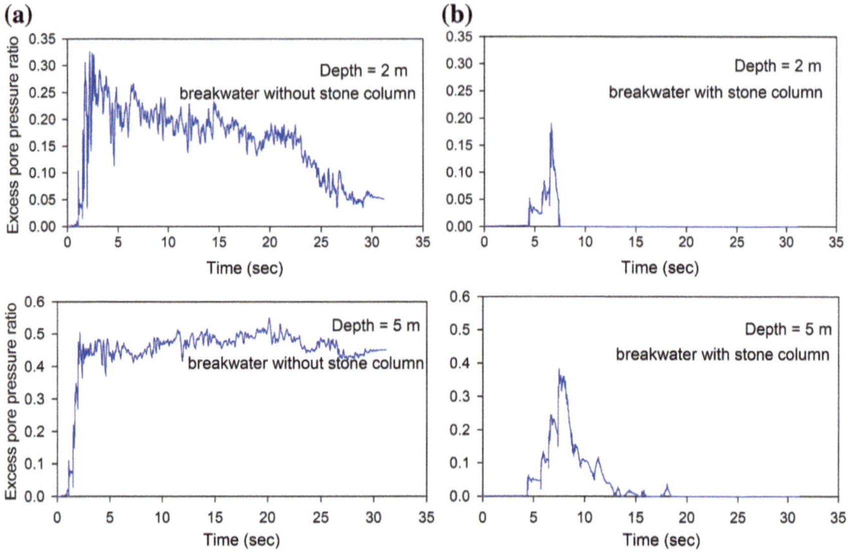

Fig. 8. Changes of pore water pressure ratio in terms of time in the middle of the model in the points (122) and (124) at a depth of 2 and 5 m below the breakwater with or without stone column under the EL Centro 0.15 g record

As can be seen in Fig. 9, the liquefiable soil not improving under the breakwater, even at low overburden pressure, caused significant differential settlement and horizontal displacement, which could damage the breakwater.

5.2 Breakwater on Stone-Columned Ground

Effectiveness of the reinforcing technique was determined in terms of reduction in settlement and horizontal displacement of the breakwater, and comparisons were made between the reinforced and unreinforced ground of the breakwater. Figure 8 clearly illustrates how stone columns caused rapid depreciation of excess pore water pressure, applied in a short time after the implied earthquake. This means that soil could restore a major part of its resistance quickly. The ratio of pore water pressure in the middle of the breakwater was approximately under 0.5, whereas after a strong shock significantly reduced.

The comparison shows that stone columns could improve drainage performance in surface depths, but their efficiency reduced by increasing the length of stone columns. It also shows that the breakwater could incur much less vertical and horizontal displacement than before.

Figure 9 shows that stone columns could reduce the changes in vertical and horizontal displacement of breakwater crest by approximately 80% and 70%, respectively, which could be more by increasing number, diameter, and hardness of the stone columns. Figure 8 also indicates that even with stone columns, there would be a significant difference in the vertical displacement of the crest and slope of the breakwater. As clearly seen, the breakwater incurred less horizontal and vertical

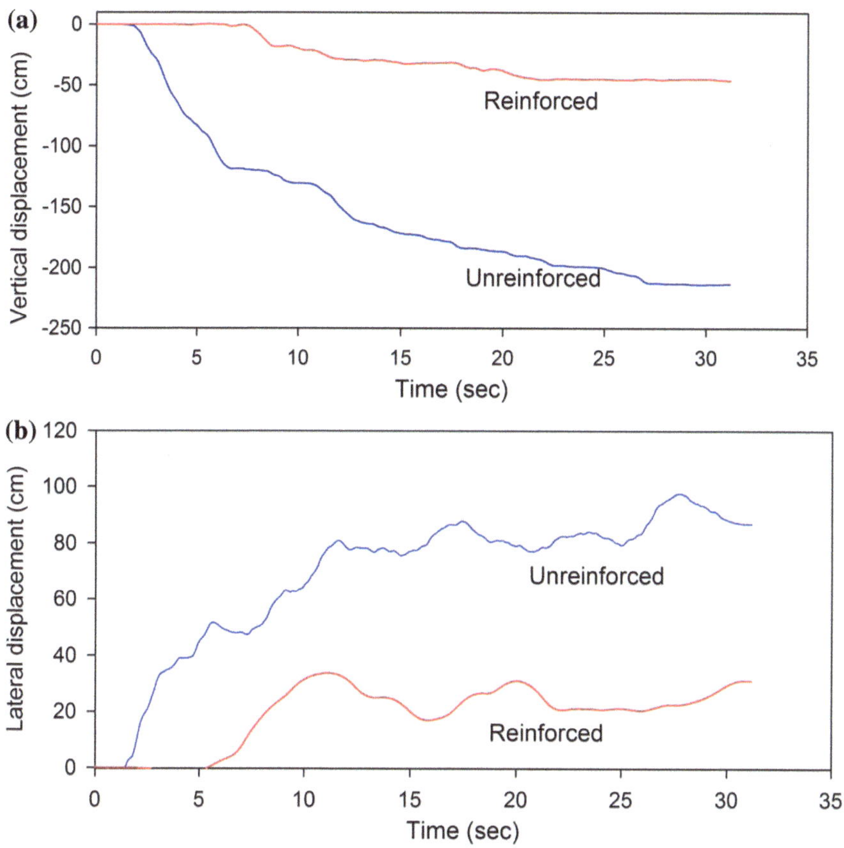

Fig. 9. (a) Vertical displacement, and (b) horizontal displacement of breakwater crest (point 102) with and without stone column under EL Centro 0.15 g record

displacement, and differential displacement significantly dropped as well. Changes in vertical and horizontal deformations in the presence of stone columns under the breakwater showed that the breakwater suffered a homogenous settlement that could decrease to a greater extent by increasing the diameter and hardness of stone columns, resulting in smaller damage to the breakwater.

Comparing Figs. 7 and 8 indicates that the amount of pore pressure was significantly less than unity, which possibly meant that because of the stone column, and increasing effective stress due to the weight of the structure on the soil, the soil resistance could improve.

6 Conclusion

In this study, seismic deformation of a breakwater rested on liquefiable silt layer prior and after soil improvement was addressed through the numerical analyses. Based on two-dimensional numerical analyses conducted in this study, the predicted results were in good agreement with the results of a centrifuge model, previously conducted on the same problem. Thus, the numerical model could be used with acceptable accuracy for assessment of breakwater seismic behavior. The silt layer was then improved by stone columns in the numerical model to prevent liquefaction triggering. The following findings can be concluded from the numerical results of seabed improvement:

1. The excess pore water pressure of the soils around the stone columns significantly declined depending on the diameter and permeability of the columns compared with the free environment.
2. Although weight of breakwater increased the effective stress and reduced the ratio of excess pore water pressure (r_u), the simultaneous effect of increasing vertical and horizontal displacements beneath the breakwater could cause considerable damages when liquefaction of the seabed is not prevented.
3. Stone columns showed good performance in shallow depths; however, its effect was negligible in deeper zones.
4. The results clearly showed that the longer the time of seabed liquefaction, the more the possibility of breakwater damage due to significant decrease of soil resistance and the weight of the breakwater, and asymmetric settlement and reversal, which is the greatest threat to breakwaters.
5. The presence of stone columns could reduce vertical and horizontal deformations of breakwater crest by approximately 80% and 70%, respectively compared with the no remediated case. Much more decrease could be expected by increasing the number, the diameter, and the hardness of the stone columns.
6. The results also indicated that stone columns could rapidly dissipate excess pore water pressure shortly after the main earthquake strikes. It means that soil was able to quickly restore a great part of its resistance.

References

Adalier, K., Sharp, M. K.: Dynamic centrifuge modeling of earth dams on liquefiable ground. In: Proceedings of 3rd U.S.–Japan Workshop on Advanced Research on Earthquake Engineering for Dams, San Diego (2002a)

Adalier, K., Sharp, M. K.: Embankment dam on liquefiable foundation—Dynamic behavior and densification remediation. J. Geotech. Geoenvironmental Eng. **130**,(11) (2002b). (November 1, 2004)

Asgari, A., Golshani, A., Bagheri, M.: Numerical evaluation of seismic response of shallow foundation on loose silt and silty sand. J. Earth Syst. Sci. **123**(2), 365–379 (2014)

Byrne, P.M.: A cyclic shear-volume coupling and pore pressure model for sand; In: Second International Conference on Recent Advances in Geotechnical Earthquake Engineering and Soil Dynamics, vol. 2, pp. 47–55. St. Louis, Missouri (1991)

Chaudharya, B., Hazarikab, H., Nishimurab, K.: Effects of duration and acceleration level of earthquake ground motion on the behavior of unreinforced and reinforced breakwater foundation. Soil Dyn. Earthquake Eng. **98**, 24–37 (2017)

Cihan, K., Yuksel, Y., Berilgen, M., Ozkan, Cevik E.: Behavior of homogenous rubble mound breakwaters materials under cyclic loads. Soil Dyn. Earthquake Eng. **34**, 1–10 (2012)

Elsafti, H., Oumeraci, H.: A numerical hydro-geotechnical model for marine gravity structures. Comput. Geotech. **79**, 105–129 (2016)

Hazarika, H., Kataoka, S., Kasama, K., Kaneko, K., Suetsugu, D.: Composite ground disasters caused by the earthquake and tsunami in Aomori, Iwate Prefecture, Northern. Geotech. Eng. J. **7**(1), 13–23 (2012). [Special Issue on 2011 Great East Japan Earthquake, in Japanese

Hazarika, H., Kasama, K., Suetsugu, D., Kataoka, S., Yasufuku, N.: Damage to geotechnical structures in waterfront areas of northern Tohoku due to the March 11, 2011 tsunami disaster. Indian Geotech. J. **43**(2), 137–152 (2013)

Hazarika, H., Chaudhary, B., Nishimura, K., Kasama, K., Noda, T., Yamada, S.: Physical and numerical modeling of resilient breakwater foundation subjected to earthquake and tsunami. In: Proceedings of the 50th Indian geotechnical conference. Pune (India) (2015)

Hara, T., Okawara, M., Osumi, T., Yamanaka, M., Ishihara, Y., Tsunekawa, Y.: Damages in south-central coastal area of Iwate prefecture in 2011 Off the Pacific Coast of Tohoku Earthquake. Jpn. Geotech. J. **7**(1), 25–36 (2012)

Indraratna, B., Redana, I.W.: Plane-strain modeling of smear effects associated with vertical drains. J. Geotech. Geoenviron. Eng. **5**, 474–478 (1997)

Indraratna, B., Redana, I.W.: Numerical modeling of vertical drains with smear and well resistance installed in soft clay. Can. Geotech. J. **37**(1), 132–145 (2000)

Itasca, F.: Fast Lagrangian Analysis of Continua, Version 5.0 User's Guide, Itasca Consulting Group, Inc., Thrasher Square East, 708 p (2002)

Jafarian, Y., Alielahi, H., Sadeghi Abdollahi, A., Vakili, A.: Seismic numerical simulation of breakwater on a liquefiable layer: IRAN LNG port. Electr. J. Geotech. Eng. 1–11 (2010)

Ye, J., Gang, W.: Seismic dynamics of offshore breakwater on liquefiable seabed foundation. Soil Dyn. Earthquake Eng. **76**, 86–99 (2015)

Kubilay, C., Yalcin, Y., Mehmet, B., Esin, O.C.: Behavior of homogenous rubble mound breakwaters materials under cyclic loads. Soil Dyn. Earthquake Eng. **34**, 1–10 (2012)

Lambe, T.W., Whitman, R.V.: Soil Mechanics. Wiley, N.Y. (1969)

Martin, G.R., Finn, W.D.L., Seed, H.B.: Fundamentals of liquefaction under cyclic loading. J. Geotech. ASCE **101**(GT5), 423–438 (1975)

Moghadam, M.R., Alielahi, H., Sadeghi, Abdollahi A.: Numerical evaluation of liquefaction-induced damages in composite breakwaters and its application for performance-based improvement design. Mar. Georesour. Geotechnol. **35**(3), 376–396 (2017)

Pacific Earthquake Engineering Research Center (PEER): PEER Ground Motion Database. Univ. of California, Berkeley, CA (2012)

Shooshpasha, I., Bagheri, M.: The effects of surcharge on liquefaction resistance of silty sand, Arabian J. Geosci. **7**(1), 1029–1035 (2014). (March 2014)

Tan, S.A., Tjahyono, S., Oo, K.K.: Simplified Plane-Strain Modeling of Stone-Column Reinforced Ground. J. Geotech. Geoenvironmental Eng. ASCE **134**(2), 185–194 (2008)

Takahashi, S., Kuriyama, Y., Tomita, T., Kawai, Y., Arikawa, T., Tatsumi, D., Negi, T.: Urgent survey for 2011 Great East Japan earthquake and tsunami disaster in ports and coasts part I (tsunami). Technical note 1231. Japan: Port and Airport Research Institute, pp. 1–9 (2011)

Verhaeghe, H., Vos, L.D., Boone, E., Goemaere, J.: Using field data to improve the settlement prediction model of a breakwater on soft soil. J. Waterway Port Coast. Ocean Eng. **140**(2) March 1, 2014

Ye, J.H., Jeng, D.S.: Three-dimensional dynamic transient response of a poro-elastic unsaturated seabed and a rubble mound breakwater due to seismic loading. Soil Dyn. Earthquake Eng.g **44**, 14–26 (2013)

Effect of the Seismic Vulnerability of Water Pipelines on the Collapsible Soils of the North of Chile

Yolanda Alberto[1,3(✉)] and Juan Carlos de la Llera Martin[2,3]

[1] The University of Tokyo, Tokyo, Japan
h-yolanda@t-adm.t.u-tokyo.ac.jp
[2] School of Engineering, Pontifical Catholic University of Chile, Santiago, Chile
[3] National Research Center for Integrated Natural Disaster Management (CIGIDEN), Santiago, Chile

Abstract. During the recent seismic events in Chile (2010 Maule Earthquake, 2014 Arica Earthquake and 2015 Illapel Earthquake), the water pipelines have been subjected to ground motion and ground failure that have affected their performance and increased the leakage ratio of the system. In arid areas, as the North of Chile, this has led to the saturation of saline soils that eventually have collapsed and affected more than 2,000 houses. This paper presents a framework of seismic damage to pipelines and reports the results of a geotechnical experimental program on the collapsible soils found in the North of Chile and subjected to wetting due to pipelines leakage. A review of the water system in the northern area is provided along with the damage experienced after the 2014 Earthquake. An experimental program was conducted and the soil was characterized to determine its physical and mechanical properties, including consolidation and collapsibility tests. The risk associated with leakage of water pipelines is evaluated in the context of increasing the vulnerability of houses founded on collapsible soils. Conclusions provide recommendations to mitigate and prevent this issue, regarding foundation design and seismic retrofit of underground pipelines.

1 Introduction

The recent earthquakes in Chile (2010 Maule, 2014 Pisagua, 2015 Illapel) have caused a significant amount of damage on the potable water system. Several failures have been reported by the companies that have led to large increase in leakage rate. This issue is usually overcome after later repairs and pipe replacement. In the North of Chile, however, the nature of soils transform this into a serious problem. These saline soils undergo considerable reduction of shear strength and deformation modulus due to wetting (Kannan 1999).

Given the arid weather that characterizes the Atacama Desert, the soils have high contents of salt which causes them to be prone to collapse when in contact with water. Geologically there are three conditions that determined the characteristics of the saline soils in this area. First, the arid climate allows the accumulation and storage of nitrate

© Springer International Publishing AG, part of Springer Nature 2019
S. Shu et al. (eds.), *New Developments in Materials for Infrastructure Sustainability and the Contemporary Issues in Geo-environmental Engineering*, Sustainable Civil Infrastructures, https://doi.org/10.1007/978-3-319-95774-6_4

deposits, second, during the late Tertiary and early Quaternary periods the deposits started to grow gradually in that area and third, there was a scarcity of micro-organisms and plants requiring nitrate from the soil, which eventually contributed to the buildup of these deposits (Ericksen 1981). These soils have loose structures composed of silt to fine-sand-size particles. It has been observed that the collapse potential of saline soils increases when there is a rapid increase of an acting pressure at which filtration of water takes place (Baghabra Al-Amoudi and Abduljauwad 1995). Saturation of these soils destroys the bonds and cementation, and leads to the change of soil structure, with or without loading. Nevertheless, the formation of sinkholes is associated with anthropogenic causes.

In Alto Hospicio, a city located in the region of Tarapaca near the Atacama Desert (Fig. 1), there have been several occurrences of sinkholes. This city has increased its population 16 times in the past 20 years, rising the need for more sanitary infrastructure. On the other hand, the recent seismic events have triggered large amounts of leakage of the buried pipelines used for water supply. A geotechnical campaign was conducted to identify the mechanical properties of the saline soils of Alto Hospicio. The risk associated due to the increase in seismic-induced breaks and leaks of the pipelines, will be explained.

2 Geotechnical Conditions in Alto Hospicio

2.1 Field Testing

The stratigraphy of Alto Hospicio was obtained by Sernageomin (2014) near the sands of the Dragon Mountain (Table 1). The interbedded lens of salt and gravel within layers of weakly cemented brown sand are commonly found in the city. Sinkholes usually appear when there is a limestone layer in the general formation. In some cases, sand ravels into the void spaces and creates weak spots in the sand cover. The sand slowly filters into the subsurface of these depressions for a long time. Then, the cover finally collapses on top of the permeable layers (Brinkmann et al. 2008).

2.2 Laboratory Testing

For this testing program, samples were retrieved from sinkhole locations of Alto Hospicio. The tests conducted allowed the definition of physical and mechanical properties, including: grain size distribution, Atterberg limits, water content, density of solids, dry density, direct shear test and collapsibility. Table 2 shows a summary of the physical parameters.

Fig. 1. Location of Alto Hospicio

Table 1. Alto Hospicio stratigraphy

Geographic coordinates	WGS 84 coordinates	Depth (m)	Thickness (m)	Description
S 20°15′ 21.6″ W 70°05′ 50.3″	E: 385.399 N: 7.759.809	0.00–0.10	0.10	Brown silty sand, loose, well graded, homogeneous, dry
		0.10–0.40	0.30	Light brown medium to coarse sand, stiff due to salt cementation, homogeneous, low water content, non-plastic, well graded
		0.40–0.48	0.08	White salt lens crystalized
		0.48–1.55	1.07	Brown coarse sand, well graded, stiff due to salt cementation, low water content, non-plastic, interbedded salt lens between 1.50 and 1.55 m
		1.55–3.00	1.45	Brown coarse sand, stiff due to salt cementation, homogeneous, well graded, medium to low water content
S 20°17′ 31.0″ W 70°05′ 31.4″	E: 385.971 N: 7.755.833	0.00–0.10	0.10	Brown silty sand, medium dense, homogeneous, dry, well graded, low to non-plastic fines
		0.10–0.15	0.05	White salt lens crystalized
		0.15–1.80	1.65	Brown coarse sand, well graded, stiff due to salt cementation, heterogeneous, dry, non-plastic. Gravels of about 2 inches. Rock at the end of the bore.

Table 2. Physical parameters of the samples

Sample	Liquid limit, LL	Plastic limit, PL	Plasticity index, PI	Apparent density, ρ (kg/m^3)	Water content, w (%)	Specific gravity, G_s
Sample I	18	16	2	1940	10.0	2.82
Sample II	23	20	3	1940	12.2	2.82

The grain size distribution of four different samples exhibits the homogeneity of the soil in the city (Fig. 2). The coefficient of uniformity (C_u) is about 60 and the coefficient of curvature (C_c) is 1.5. The soil is well graded and is classified as silty sand (SM) by the Unified Soil Classification System (USCS).

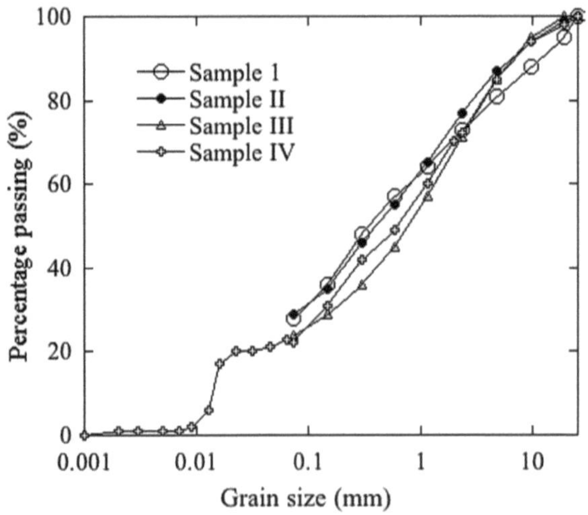

Fig. 2. Grain size distribution of the three samples obtained with the sieving method

2.2.1 Direct Shear Tests

The direct shear tests were run according to the ASTM D3080 in drained and saturated samples that were remolded with the collected material. The direct shear box has a contact area of 60 by 60 mm. Table 3 shows the conditions of the tests and the range of vertical effective stress, σ'_v, used.

Table 3. Direct shear test conditions

Sample	Depth (m)	Conditions	Vertical effective stress σ'_v (kg/cm^2)	Initial dry density, ρ (kg/m^3)	Water content (%)
Sample I	3.00	Cut, undisturbed sample	0.5, 1.0, 2.0	1940	8.6

The horizontal stress-deformation history for each normal stress, is shown in Fig. 3 (a). These shear stress-displacement curves exhibit a well-defined peak shear strength. In Fig. 3(b), for the three normal stresses, sand has a dilating behavior. After reaching the peak of the interface shear strength, the vertical displacement progress of curves for 0.5 and 1 kg/cm^2 have a difference of values of about 20%, while for the 1 and 2 kg/cm^2 curves, there is a difference of about 40%.

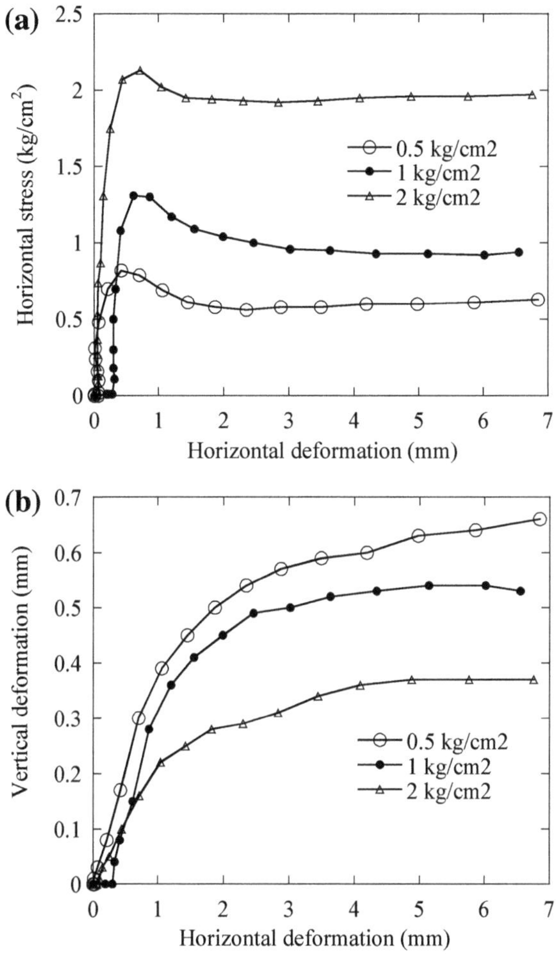

Fig. 3. (a) Stress-deformation relationship and (b) vertical displacement-shear displacement

For each normal pressure σ'_v (0.5, 1, 2 kg/cm^2), the corresponding peak shear stresses were 0.82, 1.31 and 2.13 kg/cm^2 (Fig. 4), while the residual shear strength values are 0.63, 0.96 and 1.97 kg/cm^2. The Mohr Coulomb shear strength parameters are cohesion, c = 0.41 kg/cm^2 and direct shear friction angle, ϕ_{DS} = 40.8°. These values can be corrected for triaxial conditions and then used for design in numerical models.

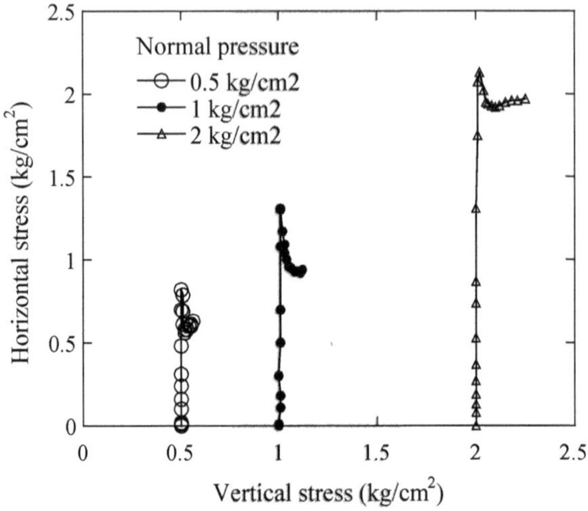

Fig. 4. Peak shear strength for each normal stress

2.2.2 Consolidation Test

Given the collapse potential of soils, a series of consolidation tests was necessary to obtain the deformability parameters of the saline soils. The parameters of the two samples are shown in Table 4. Two different dry densities were used, ρ_d = 1.50 g/cm^3 and ρ_d = 1.94 g/cm^3, to consider both loose and medium dense states. The consolidation curves for both dry densities are displayed in Fig. 5. The preconsolidation pressure is 1.2 kg/cm^2.

The time-settlement behavior for each loading stage is seen in Fig. 6. For the loads of 2, 4 and 8 kg there is a sudden deformation of the sample, indicating that the degree of saturation achieved in combination with the time of load application, led to more deformation.

2.2.3 Determination of Salt Content

The amount of soluble salts was determined through aqueous leaching (chlorides and sulfates). The criteria considered for the determination of collapsible soils is that if the soluble salt content is less than 2%, the soil is not collapsible; if the amount of soluble salt is more than 2%, the soil is potentially collapsible and it is recommended to adopt design and construction measures to avoid and diminish the effects of the soil on the foundations and the undesirable settlement.

Table 4. Sample initial and final conditions

Sample	Initial height (mm)	Final height (mm)	Area (cm^2)	Initial moisture content (%)	Final moisture content (%)	Initial void ratio, e_0	Initial dry density (%)	Initial saturation (%)	Final saturation
1	16.08	21.12	20	11.02	17.20	0.78	1.49	37.51	34.12
2	16.08	23.22	20	10.36	12.95	0.37	1.94	74.33	35.12

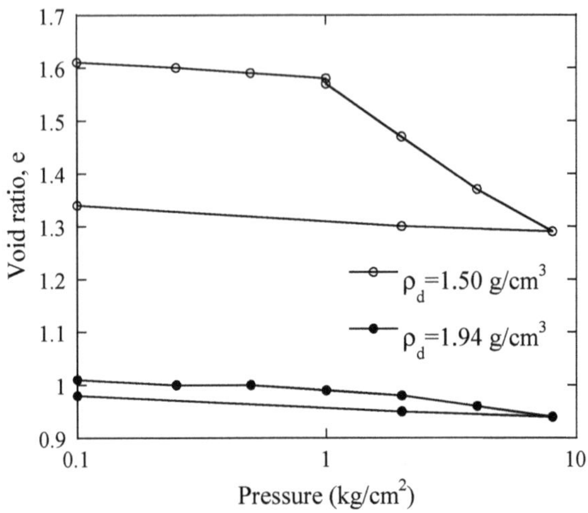

Fig. 5. Consolidation curves

The amount of soluble salt obtained from the samples collected was 9.2%. This value compares to those obtained by Sernageomin (2014) and shown in Table 5. This indicated that these soils are highly susceptible to collapse and should have special considerations for foundation design. The use of deep foundations or mixed solutions can reduce the amount of differential settlement.

3 Seismic Analysis of the Water Pipelines

3.1 Model for Simulation of Pipeline Leakage and Breakage

Pipeline damage in hydraulic simulations is classified as breaks and leaks. There are different ways to account for these failures in the models, for instance, the breakage is simulated disconnecting the original pipeline completely and opening the broken ends into the atmosphere, and a leak is simulated as an orifice in the pipe wall.

It is considered that earthquake damage on pipelines can be classified as 80% leaks and 20% breaks (NIBS 2002). As water systems tend to have leakage rates of 5 to 20% in normal conditions, during a seismic simulation, the energy loss due to leakage is considered a minor loss. The leak simulation in hydraulic network analysis is derived from the hydraulic head loss and is expressed in Eq. (1):

$$Q = \left(\frac{2g}{\gamma_w}\right)^{0.5} A_1 p_1^{0.5} \tag{1}$$

Where Q is the water loss from leaks, A_1 is the cross-sectional area of the orifice in the pipeline, p_1 is the initial internal pressure, γ_w is the unit weight of water and g is the gravitational acceleration.

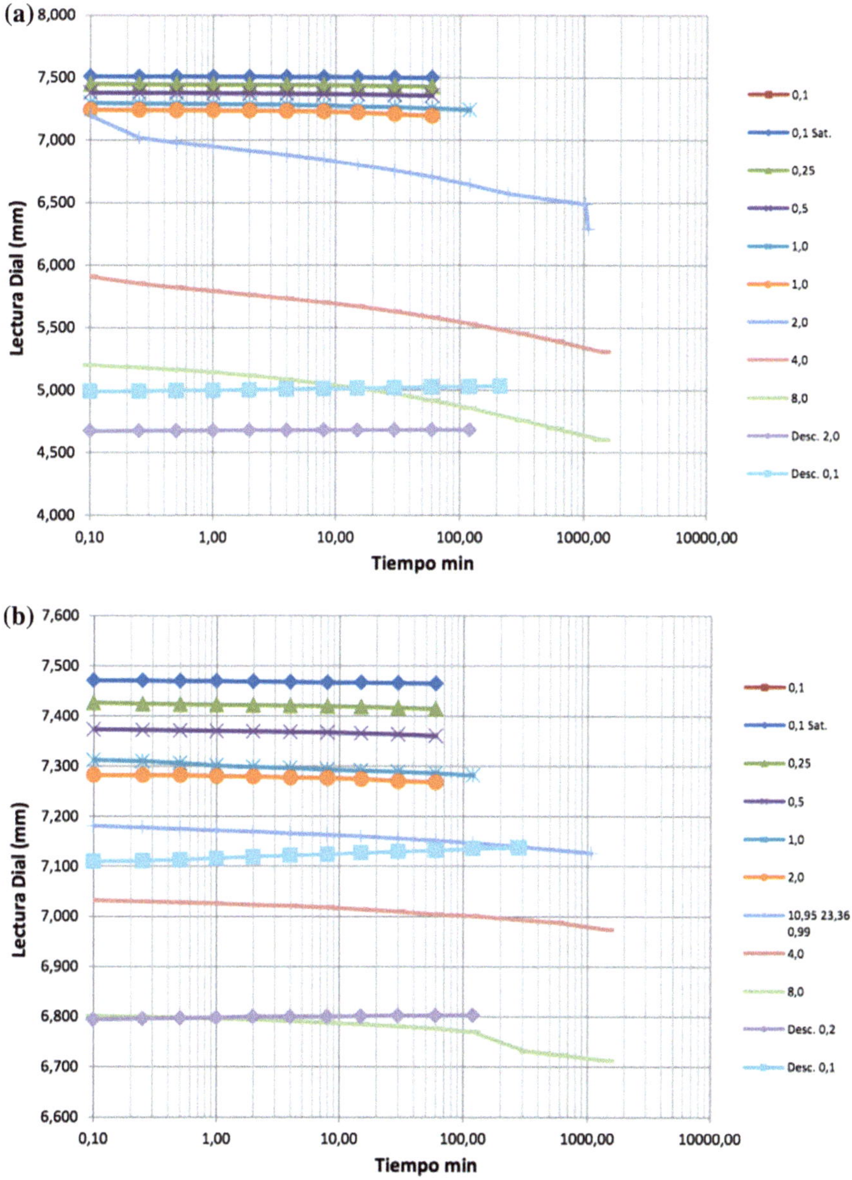

Fig. 6. Time-settlement behavior (a) sample with $\rho_d = 1.49$ g/cm^3 and (b) sample with $\rho_d = 1.94$ g/cm^3

Table 5. Soluble salt content – Alto Hospicio

Depth (m)	Total soluble salt (%)	Chlorides (%)	Sulfates (%)
0.00–0.50	7.19	3.65	0.34
0.50–1.00	3.80	1.94	0.15
1.00–1.50	5.51	2.80	0.24
1.50–2.00	4.89	2.42	0.24
2.00–2.50	3.30	1.53	0.16
2.50–3.00	4.82	2.49	0.17

The different leakage scenarios can be classified into (Shi and O'Rourke 2008):

1. Annular disengagement, joint looseness of segmented pipelines due to axial pullout caused by the seismic loading.
2. Round crack, the circumferential cracking of pipe barrel or joint under the effects of bending or the combination of bending and tensile forces.
3. Longitudinal crack, the cracking of the pipe barrel or seam along the pipe longitudinal direction caused by the loading or high internal pressures during earthquakes.
4. Local loss of pipe wall, caused by the loss of a small portion of pipe wall, which deteriorates by corrosion under the seismic loading.
5. Local tear of a pipe wall, commonly caused by a rupture in the bell casing of wrinkled welded slip joints caused by compressive forces.

From these scenarios, the longitudinal crack causes the biggest water loss for small diameters (300–600 mm), and the round crack, for large dimeter pipes (1500–1800 mm).

3.2 Simulation of Damage

When conducting reliability analyses, the result is usually the probability that a pipeline system complies with one or more performance criteria (O'Rourke et al. 1995). To account only for pipeline failure, the simulation can be done in either deterministic approach, adding the concise damage scenarios, or probabilistic, randomly generating the distributed pipeline breaks and leakages in the system according to pipeline repair rate, length, and conditional probability of pipe break given that damage occurs. The type of leakage is also implemented probabilistically. This process is carried out in three steps, generating pipe damage, deciding on damage states (leak or break), and determining leak type.

According to the damage observed in the 1995 Kobe Earthquake, for brittle pipes 20% of the damage was considered breakage, while for ductile pipes there was no breakage. Using this as a hypothesis, for brittle pipes, probabilities larger than 0.2 are considered to be leakage and for ductile pipes, all probabilities are for leakage (Javanbarg and Takada 2009).

3.3 Water Damage After the 2014 Pisagua Earthquake

A $M_w = 8.2$ earthquake struck off the coast of Chile on April 1st, 2014. Immediately after the earthquake, there was a stoppage of the water service, as part of the emergency protocol. After this step, a survey of the damage was conducted by Aguas del Altiplano, the company in charge of water distribution in Alto Hospicio.

One of the main struggles was the lack of electric supply and the access limitations caused by the tsunami alert. In Arica, it was decided to lower the pressure of distribution to protect the distribution networks.

To guarantee the water service, distribution points were established and tanker trucks were habilitated to serve the population. On April 3rd, the distribution of water with tanker trucks started since 08:30 local time. The dialysis, policlinics and Hospital centers received their supply by means of tanker trucks. In total in Alto Hospicio, there were 6 tanker trucks for the population, 5 distribution points and 4 portable tanks.

After the monitoring, the restoration of Alto Hospicio continued. The most affected area was El Boro, which accounted for 20% of the total customers, while the rest of the city tried to reach 80% of the performance on April 3rd.

Again, the lack of electrical energy delayed the logistics for restoration of the service. In the case of Iquique and Alto Hospicio, the restitution of the electric service in the production facilities and the repair of a large-diameter pipelines increased the likelihood of repairs. Aguas del Altiplano implemented a call-center, webpages and Twitter to let the users report failures.

Aguas del Altiplano continued with the repairs after the earthquake. However, as of March of 2015, the break of water matrices in the area of La Pampa continued to create sinkholes.

4 Sinkhole Formation and Wetting Collapse Mechanism

The formation of a sinkhole can be summarized in four steps (Fig. 7). The initial stage, where the void forms due to the filtration of water into the saline stratum. At the critical stage, a shear zone develops between the void and the surface, making the cover soil fall over the void and generating slope instability on the walls of the sinkhole (Drumm et al. 1990). One of the main factors influencing the distribution of sinkhole collapses is the thickness of overlying soil cover and the cone of groundwater depression (Gongyu and Wanfang 1999). Once the sinkhole is formed, the process for it to become bigger depends on the surrounding slip surfaces and the mechanical properties of soil and load conditions. This process can happen in a matter of minutes or hours. Since the size of the initial void is unknown, it is difficult to predict where the sinkholes will appear unless some geophysical tools are used to study the underground. In the area of Alto Hospicio, there have been different openings in residential areas that take a long time to be evaluated and fixed by the corresponding authorities. For this reason, people start filling the sinkholes with debris which actually causes more stress in the surrounding area, increasing the chances of a bigger opening.

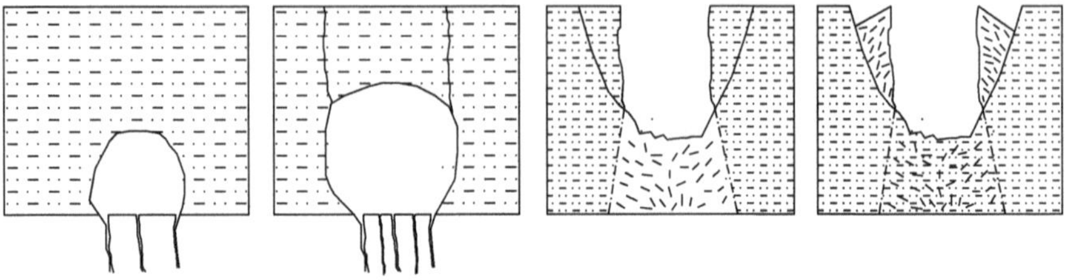

Fig. 7. Development of a sinkhole (Drumm et al. 1990)

On other instances, the opening is fixed in one house, but internally has expanded to the neighboring areas and after the repair, another sinkhole appears. This situation should be evaluated with the aim of geophysical tools. Sometimes, the right combination of cavity diameter and vertical stress, circumferential stresses may be sufficient to maintain soil stability (Drumm et al. 1990). To assess that stability, the vertical stress required to cause failure can be estimated if the failure is idealized as an elastic-plastic model, in which the behavior is represented as elastic until the ultimate or yield strength is reached. At the yield strength, the soil is said to be at critical state and yields continuously at constant stress until the stress state is changed.

Other option is to consider the soil as rigid-perfectly plastic, in which case, the plastic deformation is large with respect to the elastic deformation period. If the failure is assumed to constitute a cylindrical void, extending the full height of the residual soil, surface loads are applied over an area that is large with respect to the cavity diameter. Only the stress distribution within the soil is considered, since the strength and stiffness of the bedrock are assumed to be large with respect to that of the soil.

4.1 Wetting-Induced Collapse Mechanism

There are several theories to explain the wetting-induced collapse mechanism of saline soils such as loss of capillary tension, shortage of clays, and the porous fabric of saline soils (Li et al. 2016). The solution of soluble salts that act as bonds in these unsaturated soils can be applicable to Alto Hospicio regarding the collapse mechanism given the large amount of soluble salt found in these soils (>9%). In terms of the measurable change of shear strength and volume change, the saturated condition of these kind of soils is the most critical state, although even before the full saturation state, soil can undergo large volume changes. It has been seen that full collapse can occur above 65–75% of saturation (Li et al. 2016), therefore, after pipeline breaks, soil can achieve this degree of saturation and be in risk of collapse.

A numerical study conducted by Hosseinalipour and Aghakhani (2011) showed that the leakage velocity initially depends on the diameter of the pipe. As the time increases, the effect of the diameter disappears, which could indicate that even a small pipe break or leakage, could cause potential damage with time.

One important factor to consider is that steady state conditions lead to stability of the void, while sinkhole formation is associated to dynamic or short duration loading that can increase pore pressure (Tharp 1999), therefore the occurrence of a water matrix break or seismic-induced breaks of pipelines, would cause a decrease in the shear strength of the void.

5 Conclusions

The saline soils of Alto Hospicio have posed a significant risk for the population due to the poor maintenance of the sanitary systems. The increase of leaks and underground water due to the recent earthquakes in the northern area of Chile have aggravated the problem and in some cases, exposed the formation of already existing sinkholes. In order to protect the population and their properties, is necessary that the organizations responsible of the sanitary system run continuous monitoring of the system and carry out maintenance.

Due to the seismic conditions in the area, a seismic analysis of the water network should be carried out to identify the most vulnerable spots of the network, realize soil improvement and improve the pipelines condition.

Cone penetrometer, ground penetrating radar, seismic refraction or resistivity surveys can be used to identify existing sinkholes and do the appropriate urban planning for relocation, demolition and improvement.

Acknowledgments. The authors would like to thank the National Research Center for Integrated Management of Natural Disasters for the funding provided for the geotechnical campaign. The authors would like to express their gratitude to Carlos Javier Castro Escalante and to the Municipal Office of Alto Hospicio for the support and information provided for this research.

References

Baghabra Al-Amoudi, O.S., Abduljauwad, S.N.: Compressibility and collapse characteristics of arid saline sabkha soils. Eng. Geol. **39**(3–4), 185–202 (1995)

Brinkmann, R., Parise, M., Dye, D.: Sinkhole distribution in a rapidly developing urban environment: Hillsborough County, Tampa Bay area, Florida. Eng. Geol. **99**(3–4), 169–184 (2008)

Drumm, E.C., Kane, W.F., Yoon, C.J.: Application of limit plasticity to the stability of sinkholes. Eng. Geol. **29**(3), 213–225 (1990)

Ericksen, G. E.: Geology and origin of the Chilean nitrate deposits. Geological Survey Professional Paper, 1188, 42 (1981)

Gongyu, L., Wanfang, Z.: Sinkholes in karst mining areas in China and some methods of prevention. Eng. Geol. **52**(1–2), 45–50 (1999)

Hosseinalipour, S.M., Aghakhani, H.: Numerical simulation of the flow of leaking buried pipe in an unsaturated porous media. Int. J. Math. Comput. Phys. Electr. Comput. Eng. **5**(6), 632–637 (2011)

Javanbarg, M.B., Takada, S.: Seismic reliability assessment of water supply systems. In: 10th International Conference on Structural Safety and Reliability (March 2013), 8 (2009)

Kannan, R.C.: Designing foundations around sinkholes. Eng. Geol. **52**(1–2), 75–82 (1999)
Li, P., Vanapalli, S., Li, T.: Review of collapse triggering mechanism of collapsible soils due to wetting. J. Rock Mech. Geotech. Eng. **8**(2), 256–274 (2016)
NIBS: A guide to using HAZUS for mitigation. National Institute of Building Sciences, 51, Apr 2002
Sernageomin: Geologic information of Alto Hospicio (in Spanish), 10 (2014)
Shi, P., O'Rourke, T.: Seismic response modeling of water supply systems. MCEER-08-0016 (2008)
Tharp, T.M.: Mechanics of upward propagation of cover-collapse sinkholes. Eng. Geol. **52**(1–2), 23–33 (1999)

Experimental Study on the Dynamic Response of Saturated Sandy Soil with Different Clayey Particle Content and Skeleton Sand Size

Guocai Wang(✉), Hao Wang, Yong Zhang, Xiaobing Xu, and Qianqian Liu

Institute of Geotechnical Engineering, Zhejiang University of Technology, Hangzhou, China
wgc@zjut.edu.cn, liuqianqian267@163.com

Abstract. The sandy soil in nature is not pure, but contains a certain content of clayey particles, which shows different characteristics under dynamic loadings. Using cyclic simple shear system, a series of tests were conducted on the saturated sandy soil to reveal the effect of fine particle content and skeleton sand size on the dynamic response of sandy soil. The experimental results show that SSR (ratio of shear stress amplitude), vibration frequency, fine particle content, and size of skeleton sand have important influence on the dynamic response of sandy soil. There exists a sharply increasing inflection point in the P-P peak shear strain curves. The relationship between the position of inflection point and the values of SSR is negatively correlated, but shows positive correlation in the first several vibration frequencies. The vibration frequency is an important factor affecting the damage of samples. There exists a critical vibration frequency and the soil samples will be destroyed if the vibration frequency exceeds the critical value. With the increasing of clayey particle content, the inflection point becomes more and more ambiguous. If the content of clayey particle exceeds 40%, the inflection point will not emerge obviously in the curves. The change law of the dynamic response and dynamic shear stress of sandy soil is different for soils with different skeleton sand size and different clayey particle content.

Keywords: Clayey particle content · Amplitude of shear stress
Skeleton sand size · Critical vibration frequency · Test

1 Introduction

It is well known that sandy soil is easy to liquefy when subjected to dynamic loadings. Among the reasons of foundation damage, sand liquefaction is one of the most important factors that needs to be considered seriously. The whole or local liquefaction of sand can cause serious damage to buildings. Most of soil liquefaction occurs in sand, silt and silty soil. Evans and Zhou (1995) performed a series of undrained cyclic triaxial tests on sand-gravel composite specimens with different gravel contents and found that the liquefaction resistance of sand-gravel composites might increase significantly with the increasing of gravel content. Puri et al. (1996) and Baziar et al. (2016) pointed out that filling sands and filling sedimentary soils are more prone to slip. At present, the

© Springer International Publishing AG, part of Springer Nature 2019
S. Shu et al. (eds.), *New Developments in Materials for Infrastructure Sustainability and the Contemporary Issues in Geo-environmental Engineering*, Sustainable Civil Infrastructures, https://doi.org/10.1007/978-3-319-95774-6_5

Canadian Ottawa sand, Japan Toyoura sand or Fujian standard sand are often used as raw materials to carry out experimental investigation on the dynamic response of sandy soil (Amini 2002; Bouckovalas et al. 2003; Lee and Seed 1967; Silver and Seed 1971). However, these kinds of sandy soils are uniform, stable and pure, which is significantly different from natural sandy soil. Therefore, the experimental results using pure sand to actual non-pure sand projects will inevitably cause some differences or even errors. Based on this reason, it is necessary to study the influence of different particle content and skeleton sand size on the dynamic properties of sand soil, which will be closer to real soil conditions and can guide the engineering practice.

The effect of fine particles on the dynamic response and liquefaction of sand has been studied by various scholars for many years. For example, Lee and Seed (1967) used cyclic loading triaxial system and investigated the liquefaction of saturated sand, where several important factors and different conditions (initial liquefaction, partial liquefaction, complete liquefaction and failure) were considered. Vaid et al. (1990) and Boulanger and Idriss (2006) thought that the liquefaction resistance of sand was negatively correlated with the fine particle content, which was opposite to the conclusions obtained by Amini (2002). Guo and Prakash (1999) found that there was no definite criterion to evaluate the liquefaction potential of silt and silt-clay mixtures and that there was confusion on the influence of clay content, plasticity index and void ratio. The effects of soil fabric, ageing and other factors on the dynamic response of silt and silt-clay mixtures are still unclear. Bouchovalas et al. (2003) found that the initial confining pressure had a significant effect on the relationship between the fine particle content and liquefaction resistance of sand using sand test of fine grain content less than 30%. When the initial confining pressure is less than 60 kPa, the liquefaction resistance of sand will increase with the increase of fine particle content. If the initial confining pressure is over 60 kPa, it increases with the decrease of fine particles. Xenaki and Athanasopoulos (2003) thought that if the porosity of saturated sand was appropriate, the minimum value of liquefaction resistance would occur if the fine particle content was about 44%. Cetin and Bilge (2012) used cyclic simple shear and triaxial test, and semi-empirical probabilistic models to assess cyclic large strain and induced excess pore-water pressure responses of fully saturated clean sands. It is necessary to point out that, the conclusions of these studies are not the same, and even contradictory to some extent. It is generally accepted that there exists critical fine particle content where the liquefaction potential of sand reaches its maximum.

Silver and Seed (1971) used cyclic load simple shear tests and investigated the dynamic response of soil, where the change law and factors affecting the dynamic modulus and hysteretic damping were analyzed using different testing equipment. Seed et al. (1985) clarified the meaning of the values of standard penetration resistance used in correlations of field observations of soil liquefaction using SPT tests to eliminate the variable energy delivery efficiency associated with the rope and pulley technique. Evans (1992) investigated the water-content redistribution and specimen density caused by membrane compliance during undrained cyclic loading. Fragaszy et al. (1992) proposed a new method to evaluate static strength and deformation behavior of gravelly soils containing particles too large to be tested using conventional equipment, which was based on the idea that large particles floating in a matrix of finer-grained material do not significantly affect the strength and deformation characteristics of the media. Zlatovic and Ishihara (1997) thought that the sandy soil containing some silty content was more prone to static liquefaction than pure sandy soil. Youd et al. (2001) developed a methodology termed the "simplified procedure" for evaluating liquefaction resistance of soils, which later became a standard of practice used by many countries of the world. Prakash and Puri (2010) used cyclic triaxial tests on reconstituted soil samples and investigated the liquefaction behavior of silts and silt clay mixers over a range of plasticity index values, where the results were compared with those of sand.

In order to further reveal the effect of fine particle content and skeleton sand size on the dynamic response of sandy soil, a series of tests are conducted on the saturated sandy soil using Geocomp Shear Trac-DSS II simple cyclic shear apparatus. The influence of fine particle size on the dynamic performance of sand is studied in detail. The experimental results show that SSR (ratio of shear stress amplitude), vibration frequency, fine particle content, and skeleton sand size have important influence on the dynamic response of sandy soil. There exists a sharply increasing inflection point in the shear strain curves of P-P peak values of samples. The change law of the dynamic response of sandy soil is different for different skeleton sand.

2 Experimental Scheme

A series of tests are undertaken to study the dynamic response of saturated sandy soil under different water wave loading and the effects of shear stress amplitude ratio, clayey particle content and skeleton sand size on the dynamic response are investigated. The clayey particle content used in the test is chosen to be 10, 20, 30 and 40%, respectively. The skeleton sand is divided into three kinds: fine sand (d_{max} < 0.25 mm), medium sand (0.25 mm < d_{max} < 0.5 mm) and coarse sand (0.5 mm < d_{max} < 1.0 mm). The ratio of shear stress amplitude is chosen to be 0.08, 0.09, 0.1, 0.11, 0.12, 0.13, 0.14, 0.15, 0.16 and 0.17, respectively. The test scheme is shown in Table 1.

Table 1. Experimental scheme

Type of skeleton sand	Clayey particle content/%	Shear stress amplitude ratio
Fine sand	10	0.08, 0.09, 0.10
	20	0.10, 0.15, 0.20
	30	0.15, 0.16, 0.17
	40	0.15, 0.16, 0.17
Medium sand	10	0.08, 0.09, 0.12
	20	0.09, 0.10, 0.12
	30	0.10, 0.12, 0.15
	40	0.12, 0.13, 0.15
Coarse sand	10	0.07, 0.08, 0.09
	20	0.10, 0.11, 0.12
	30	0.10, 0.11, 0.12
	40	0.12, 0.14, 0.16

3 Sample Preparation and Testing

The test equipment used includes testing apparatus and sample making apparatus. The sample preparation process includes sample making, saturation and consolidation.

(1) Sample making
The remodeling samples are used in the experiment, where the Fujian standard sand is used as raw material. After classification, the sand is divided into three kinds: fine sand, medium sand, and coarse sand. The size of the sample is: diameter d = 63.5 mm, height h = 25.4 mm. During the sample preparation, four points need to pay attention:
(a) All the water needed should not be added in one time, but in several times by spraying to ensure the water evenly distributed during stirring the sand and clay mixture.
(b) Vaseline must be coated into the inner side of the ring knife, otherwise it will be very difficult to push out the sample from the ring knife. This is due to that the sand is rough and the friction between the sand and inner of ring knife is large.
(c) The soil should be added in three or four times. The surface of each layer should be scraped using scraper to ensure each layer closely connected to the underlying layer.
(d) The hammer should not be carried too high to apply excessive weight, as long as the strike force can make the amount of soil weighted just putting into the ring knife.

(2) Saturation
The suction saturation method is used in the saturation of sand sample. The samples are loaded into the stacked saturator where permeable stone, filter paper, ring knife with sample, filter paper and permeable stone are stacked in order until the saturator is full. Tighten the screws of saturator. Finally, put the saturator into the vacuum saturation cylinder for pumping. The pumping time should not be less than 1 h. After finishing pumping, inject the air-free water into the vacuum saturated cylinder. Pay attention that no air enters the cylinder during injecting air-free water. The samples are put at rest for not less than 12 h after injecting water so as to ensure the samples are saturated fully.

(3) Consolidation
Place the sample into the shear box after saturation is finished and the drainage consolidation is taken by applying preset consolidation pressure. The consolidation time of sand is relatively shorter compared with that of clayey soil. The consolidation time can be chosen as 30–60 min to saturate the sand completely.

(4) Cyclic shear
According to the test scheme, preset the amplitude ratio of cyclic shear stress, the vibration period, the maximum strain, the vibration number, and the number of points collected for each cycle. Because the soil sample used in the test is sand, it is easy to reach the preset failure limit. Therefore, 2000 of vibration number and 10% strain of the p-p shear strain are set as test termination conditions.

4 Results and Discussions

In the test process, it will take longer time to finish the consolidation of silty soil and cohesive soil than sandy soil. At the same time, it needs large number of soil samples to undertake, and the using time of experimental equipment is limited. Therefore, the saturated sandy soil with different clayey particle content was chosen in the test to save experimental time and to consider more factors affecting the experimental results as possible. The effects of amplitude ratio of shear stress, clayey particle content and skeleton size on the dynamic characteristics of sandy soil subjected to wave loading were simulated in the test. In the experiment, different clayey particle content (10, 20, 30 and 40%), different size of sand (fine sand, medium sand and coarse sand) and different shear stress amplitude ratio (0.08, 0.09, 0.1, 0.11, 0.12, 0.13, 0.14, 0.15, 0.16 and 0.17) are considered, respectively. The experimental results are shown in Figs. 1, 2 and 3.

Figure 1 is the relation curves between the vibration number and cumulative deformation of fine sand under clayey particle content of 10, 20, 30 and 40% and different cyclic shear stress ratio. From the figure, it can be seen directly that cyclic shear stress amplitude ratio is an important factor for the final failure of the sample. When SSR is 0.05, the samples did not destroy till the end of the test. All the samples were destroyed if SSR is equal 0.08, 0.09 or 0.1, respectively. There exists an inflection point of sharp deformation increase in the curves of the development of P-P peak shear strain of the sample, except in the curve for SSR = 0.05. The vibration number corresponding to the inflection point decreases gradually with the increase of SSR, i.e., the larger the value of SSR, the earlier the time corresponding to the P-P peak shear strain sharply increased and the earlier the sample be destroyed. As shown in the figure that the value of SSR also determines the stable value of the P-P peak shear strain at the first several vibration number. The larger the value of SSR, the greater the stable value will be until the inflection point appears and the deformation increase sharply. The value of P-P peak shear strain is almost at the stable value.

From the figure, it can also be seen that the inflection point appears at the vibration number N = 35 and the sample is destroyed if N = 45, when SSR is 0.1. When SSR = 0.09, the inflection point appears at the vibration number N = 50 and the sample will be destroyed if N = 63. When SSR = 0.08, the inflection point appears at N = 233 and the sample will be destroyed if N = 243. The P-P peak shear strain of each group before the appearance of inflection point was stable, and almost all of the samples after the inflection point were destroyed in about 10 vibration number. This indicates that the destruction of the sample is connected not only to the value of SSR, but also to the vibration number. When the vibration number reaches the critical vibration number under certain conditions, the P-P peak shear strain will increase sharply, and will be destroyed in a few vibrations. For this test samples, the critical vibration number is 35, 50 and 233 for SSR = 0.1, 0.09 and 0.08, respectively.

Figures 2 and 3 are the relationship curves between P-P peak shear strain and cyclic vibration number of saturated medium sand and coarse sand under the consolidation pressure of 200 kPa, and clayey particle content of 10, 20, 30 and 40% for different cyclic shear stress amplitude ratio, respectively.

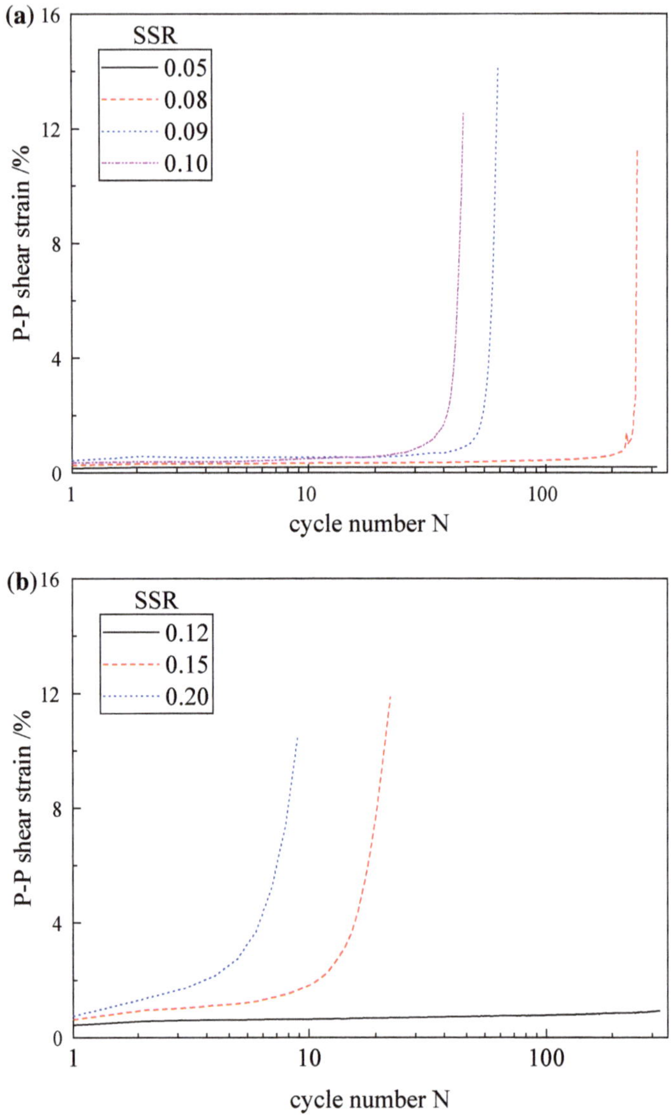

Fig. 1. The relationship of P-P peak shear strain versus vibration number of saturated fine sand under different clayey particle content: **a** 10%; **b** 20%; **c** 30%; **d** 40%

From Figs. 2a and 3a, it can be seen that: although the particle size of skeleton sand is different, the changeable trend of P-P peak shear strain of saturated sand is similar as long as the clayey particle content is still less than 10%. Under different value of SSR, there will exist a clear and obvious inflection point. The appearance of the inflection point is consistent with the previous rule: the bigger the value of SSR, the earlier the inflection point appears, indicating that the earlier appearance of critical vibration

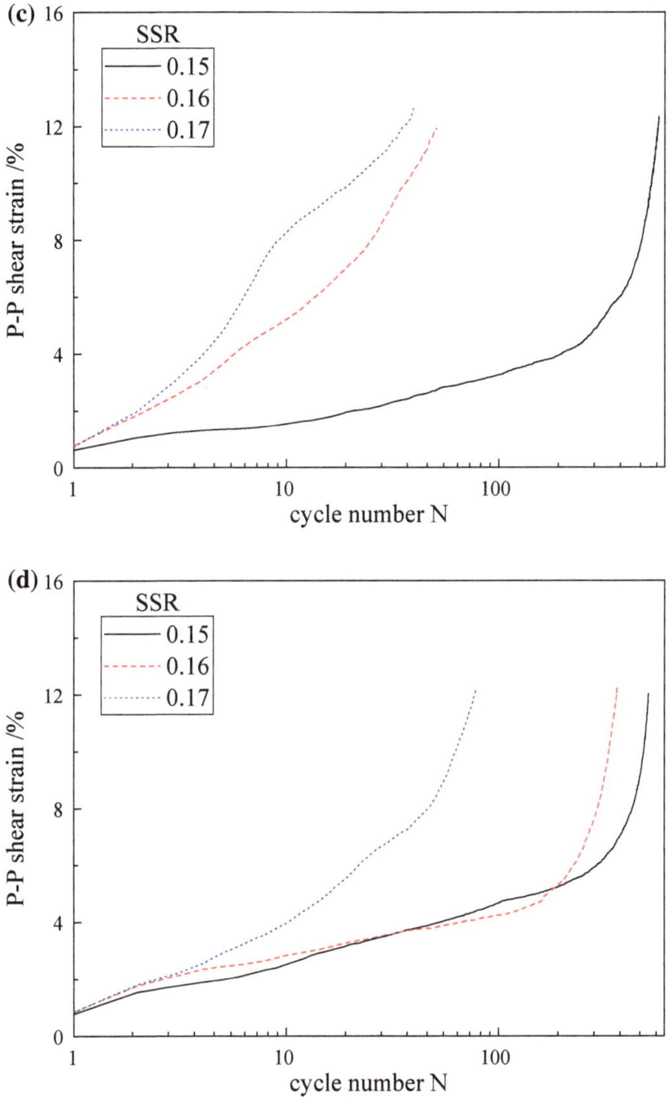

Fig. 1. (*continued*)

number. Compared with Figs. 1a, d, 2a, d and 3a, d respectively, it can be seen that the change law of P-P peak shear strain and appearance of inflection point of the sample are different for different clayey particle content. With the increase of clayey particle content, the appearance of inflection point, i.e., critical vibration number, is fuzzier until there is no obvious inflection point appeared when the final clayey particle content reaches 40%, which indicates that the clayey particle content is an important factor to affect the dynamic response of sandy soil.

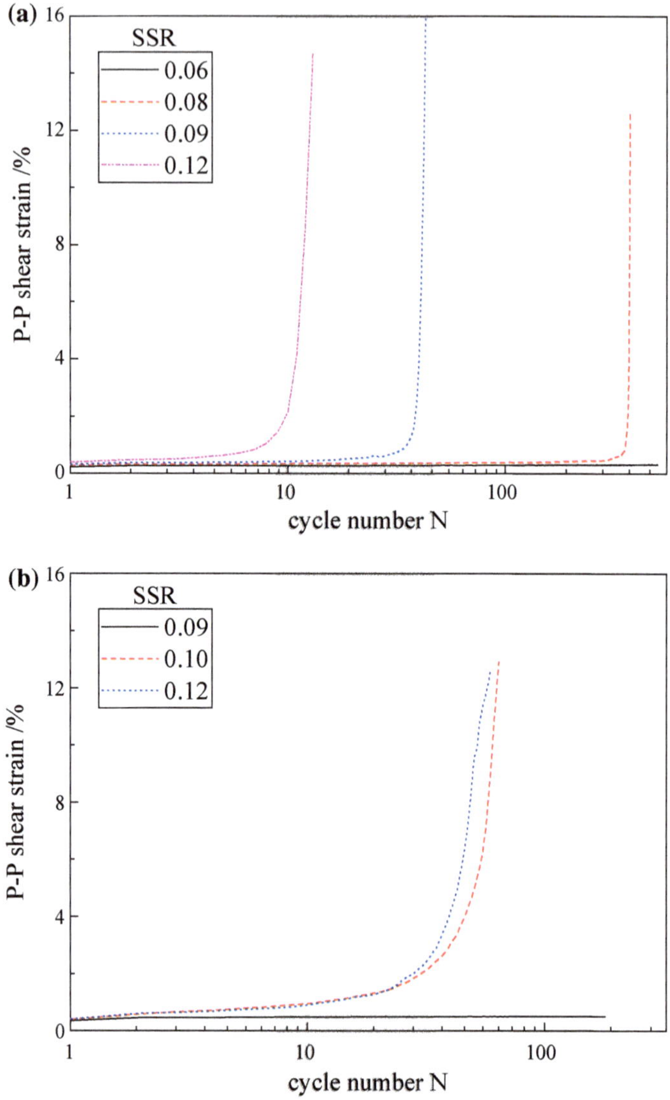

Fig. 2. The relationship of P-P peak shear strain versus vibration number of saturated medium sand under different clayey particle content: **a** 10%; **b** 20%; **c** 30%; **d** 40%

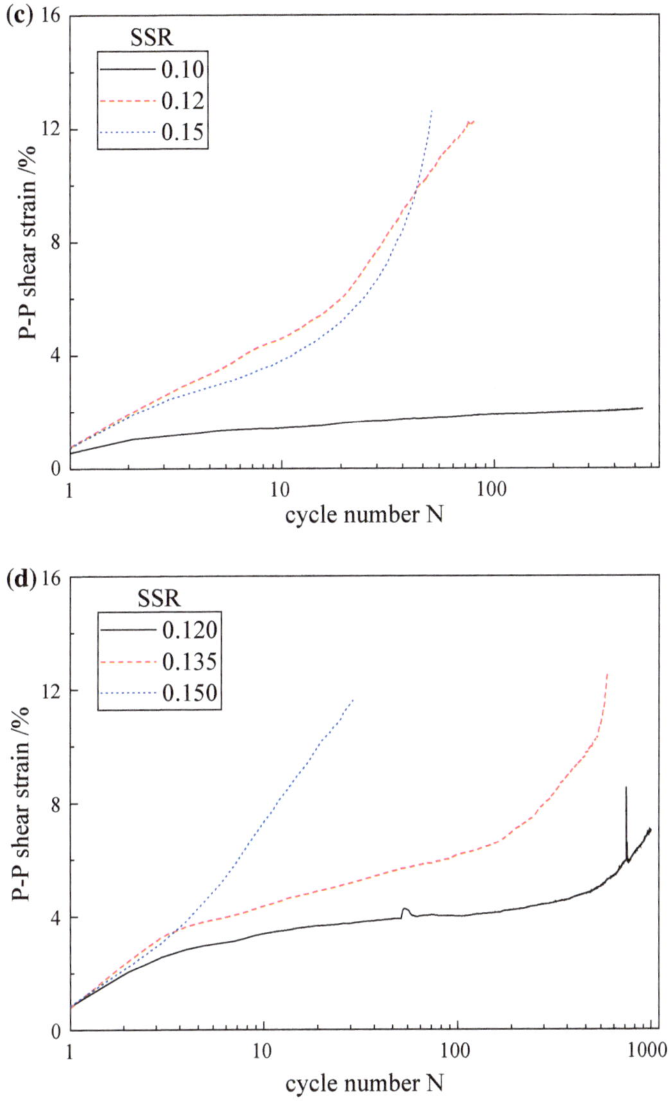

Fig. 2. *(continued)*

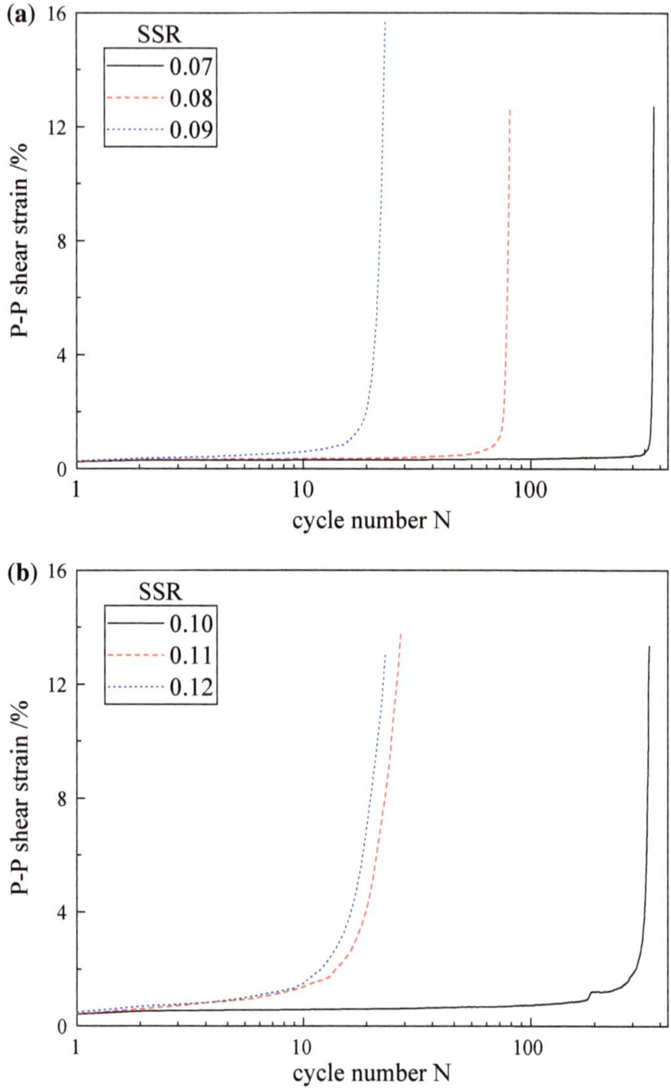

Fig. 3. The relationship of P-P peak shear strain versus vibration number of saturated coarse sand under different clayey particle content: **a** 10%; **b** 20%; **c** 30%; **d** 40%

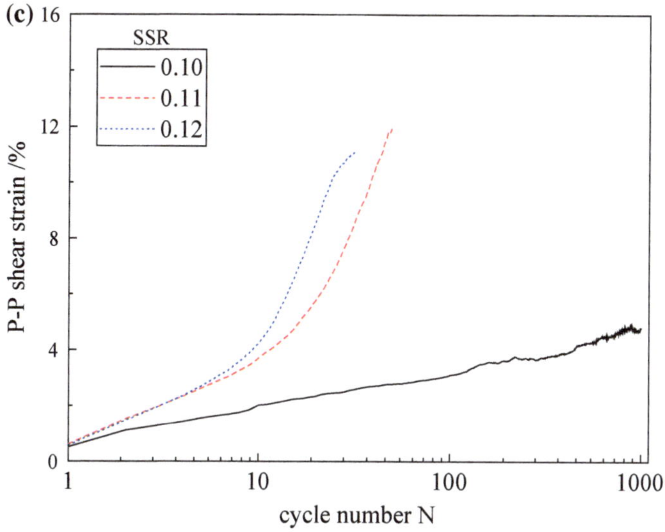

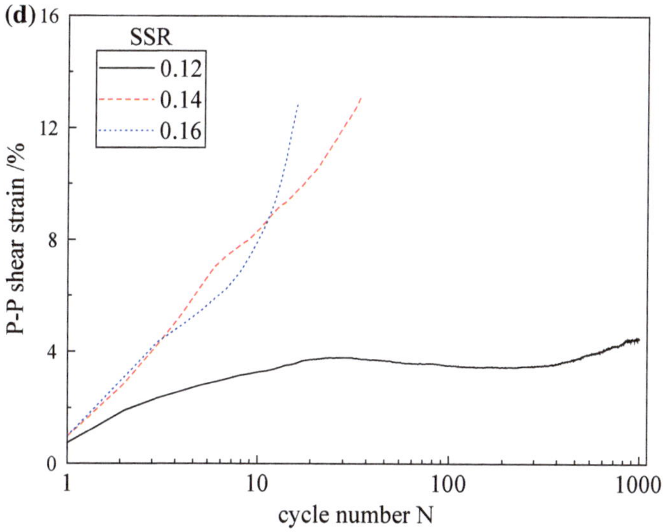

Fig. 3. (*continued*)

5 Conclusions

In order to reveal the dynamic response of seabed soil under wave load and its influence factors, a series of tests were undertaken to investigate on the saturated sandy soil with different clayey particle content. The effects of cyclic shear stress amplitude, clayey particle content and particle size of skeleton sand on the dynamic characteristics of soil were studied emphatically. Based on the experimental results, the following conclusions can be drawn as:

(1) The shear stress amplitude is an important factor to determine whether the sample is destroyed ultimately or not. There has an inflection point in the P-P peak shear strain curves of the sample, where the deformation increases sharply. The position corresponding to the number of vibration of the inflection point appeared is closely related to the value of SSR. With the increase of SSR, the position of the inflection point is decreasing, i.e., the bigger the value of SSR, the earlier the P-P peak shear strain of the sample will appear and the earlier sample will be destroyed accordingly.
(2) The value of the ratio of shear strain amplitude SSR also determines the magnitude of the stability value of P-P peak shear strain achieving the steady stage in the first several vibration numbers. The larger the SSR is, the greater the stability value in the steady stage will be. The P-P peak shear strain of the samples fluctuates around the stability value until the inflection point appeared and the deformation begins to increase sharply.
(3) The destruction of the sample is related not only to the value of SSR, but also to the number of vibration. There exists a critical vibration number. When the vibration number reaches the critical vibration number under certain conditions, the P-P peak value begins to increase sharply, and the sample will be destroyed within a few vibrations.
(4) The clayey particle content is an important factor affecting the dynamic characteristics of sand. With the increase of clayey particle content, the existing of the inflection point, i.e., critical vibration number becomes more and more ambiguous. There will be no obvious inflection point appeared if the clayey particle content is over 40%.
(5) The particle size of the skeleton sand has much influence on the dynamic response of soil. The difference in particle size of skeleton sand leads to different porosity between soil particles. Fine sand has less pores, medium sand has more ones, while coarse sand has the largest number of pores. With the incorporation of clayey particle into the pores of sand, it plays a role of "bearing" and will improve the dynamic strength of soil samples to a certain extent. With the increase of clayey particle content to a certain extent, the "bearing" role will disappear gradually, and eventually becomes "ball" role. At this moment, the characteristics of the soil changes completely, making the dynamic strength of the soil not to improve, but to decrease. The different skeleton sand size causes different interconversion of "bearing" and "ball" roles.

Acknowledgements. This study is partially supported by the National Natural Science Foundation of China (Grant No. 51278466, 41302229), which is gratefully acknowledged.

References

Amini, F.: Closure of "liquefaction testing of stratified silty sands". J. Geotech. Geoenviron. Eng. **128**(1), 96–96 (2002)

Baziar, M.H., Moghadam, M.R., Choo, Y.W., et al.: Tunnel flexibility effect on the ground surface acceleration response. Earthq. Eng. Eng. Vibr. **15**(3), 457–476 (2016)

Bouckovalas, G.D., Andrianopoulos, K.I., Papadimitriou, A.G.: A critical state interpretation for the cyclic liquefaction resistance of silty sands. Soil Dyn. Earthq. Eng. **23**(2), 115–125 (2003)

Boulanger, R.W., Idriss, I.M.: Liquefaction susceptibility criteria for silts and clays. J. Geotech. Geoenviron. Eng. **132**(11), 1413–1426 (2006)

Cetin, K.O., Bilge, H.T.: Cyclic large strain and induced pore pressure models for saturated clean sands. J. Geotech. Geoenviron. Eng. **138**(3), 309–323 (2012)

Evans, M.D., Zhou, S.: Liquefaction behavior of sand-gravel composites. J. Geotech. Eng. **121** (3), 287–298 (1995)

Evans, M.D.: Density changes during undrained loading-membrane compliance. J. Geotech. Eng. **118**(12), 1924–1936 (1992)

Fragaszy, R.J., Su, J., Siddiqi, F.H., Ho, C.L.: Modeling strength of sandy gravel. J. Geotech. Eng. **118**(6), 920–935 (1992)

Guo, T., Prakash, S.: Liquefaction of silts and silt-clay mixtures. J. Geotech. Geoenviron. Eng. **125**(8), 706–710 (1999)

Lee, K.L., Seed, H.B.: Cyclic stress conditions causing liquefaction of sand. J. Soil Mech. Found. Eng. Div. **93**(1), 47–70 (1967)

Prakash, S., Puri, V.K. Recent advances in liquefaction of fine grained soils. In: Fifth International Conference on Recent Advances in Geotechnical Earthquake Engineering and Soil Dynamics and Symposium in Honor of Professor IM Idriss, San Diego, California (2010)

Puri, V.K., Das, B.M., Prakash, S.: Liquefaction of silty soils. Int. J. Offshore Polar Eng. **6**(4), 308–312 (1996)

Seed, H.B., Tokimatsu, K., Harder, L.F., Chung, R.M.: Influence of SPT procedures in soil liquefaction resistance evaluations. J. Geotech. Eng. **111**(12), 1425–1445 (1985)

Silver, M.L., Seed, H.B.: Deformation characteristics of sands under cyclic loading. J. Soil Mech. Found. Div. **97**(8), 1081–1098 (1971)

Vaid, Y.P., Fisher, J.M., Kuerbis, R.H., et al.: Particle gradation and liquefaction. J. Geotech. Eng. **116**(4), 698–703 (1990)

Xenaki, V.C., Athanasopoulos, G.A.: Liquefaction resistance of sand-silt mixtures: an experimental investigation of the effect of fines. Soil Dyn. Earthq. Eng. **23**(3), 1–12 (2003)

Youd, T.L., Idriss, I.M., Andrus, R.D., et al.: Liquefaction resistance of soils: summary report from the 1996 NCEER and 1998 NCEER/NSF workshops on evaluation of liquefaction resistance of soils. J. Geotech. Geoenviron. Eng. **127**(10), 817–833 (2001)

Zlatovic, S., Ishihara, K.: Normalized behavior of very loose non-plastic soils: effects of fabric. J. Jpn. Geotech. Soc. **37**(4), 47–56 (1997)

Model-Scale Study on the Effect of Cyclic Loading on Pile Lateral Bearing Capacity at Different Directions

Dong Su[1(✉)], Junjie Huang[1], Bin Liu[1], W. M. Yan[2], and Xiaohua Bao[1]

[1] College of Civil Engineering, Shenzhen University, Shenzhen, China
sudong@szu.edu.cn
[2] Department of Civil and Environmental Engineering, The University of Auckland, Auckland, New Zealand

Abstract. Foundation piles are extensively used to support the lateral loads exerted on structures. In earthquake-active areas, piles in service usually experience cyclic lateral loadings induced by earthquakes. Many previous studies mainly focused on the pile lateral bearing capacity along the same horizontal direction that the cyclic loadings are applied. This paper reports two series of pile model tests conducted in soft and medium-stiff clay utilizing a computer-controlled biaxial motion platform. In these tests, the piles were loaded monotonically along different horizontal directions, after being cyclically loaded along a specified horizontal direction and allowed to return to the original position. The results indicate that the post-cyclic lateral resistance of pile depends on the intersection angle between the directions of cyclic loading and monotonic loading. The degree of reduction in the bearing capacity is a function of the shear strength of soil as well as the intersection angle.

1 Introduction

Foundation piles are widely used as deep foundations in onshore, coastal and offshore structures to resist lateral loads. The loads are triggered by waves and earthquakes under some circumstances, and are cyclic in nature. In the last several decades, large numbers of experimental, theoretical, and numerical researches have been performed to explore the behaviors of pile-soil systems under lateral loadings. Experimental studies include field tests (e.g., Dan et al. 1987; Rollins et al. 1998, 2006), 1 g model tests (e.g., Rao et al. 1998; Mayoral et al. 2005; Su 2012; Su et al. 2014; Su and Zhou 2016), and centrifuge model tests (e.g., Dyson and Randolph 2001; Zhang et al. 2011). Meanwhile, extensive finite element analyses have also been developed to quantify the behavior of piles subjected to lateral loadings (e.g., Rajashree and Sundaravadivelu 1996; Su and Li 2013). Based on the results from these studies, it was found that the degradation of pile resistance under lateral cyclic loading depends on the type of soil, overconsolidation ratio of clay or relative density of sand, loading frequency, number of loading cycles, and the cyclic displacement or loading level.

© Springer International Publishing AG, part of Springer Nature 2019
S. Shu et al. (eds.), *New Developments in Materials for Infrastructure Sustainability and the Contemporary Issues in Geo-environmental Engineering*, Sustainable Civil Infrastructures, https://doi.org/10.1007/978-3-319-95774-6_6

In earthquake-active areas, piles in service usually experience cyclic lateral loadings induced by earthquakes. Most of the previous studies focused only on the post-cyclic lateral resistance along the same direction to the cyclic loadings. In this study, two series of pile model tests have been conducted in soft and medium-stiff clay utilizing a computer-controlled biaxial motion platform. In these tests, the piles were loaded monotonically along different horizontal directions, after being cyclically loaded along a specified horizontal direction and allowed to return to the original position. The results shed light on the effects of the intersection angle between the directions of cyclic loading and monotonic loading and the shear strength of clay on the degradation of the lateral loading capacity.

2 Description of Experimental Setup

This research was carried out at the Geotechnical Center of Shenzhen University, by use of a computer-controlled biaxial motion platform. During the tests, the slide unit of the platform was connected to the model pile head. By controlling independently or simultaneously the two servo motors of the platform, the slide unit was moved independently, simultaneously or simultaneously along two horizontal directions (hereafter referred as x- and y-directions) that are orthogonal to each other. By this means, the pile head can be loaded along any predesigned displacement paths.

The model piles used for the tests were aluminum tubes. For the tests in the soft clay, the tube is 1 mm in thickness and 22 mm in outer diameter. For the tests in the medium-stiff clay, the tube is 2 mm in thickness and 22 mm in outer diameter. Conical aluminum plugs were used to form closed pile ends. The piles were embedded 600 mm into the clay while loaded at a point 250 mm above the clay surface. 16 pairs of half-bridge strain gauges were attached at eight different levels along the pile shafts, with eight pairs along the x-direction and the other eight pairs along the y-direction. The strain gauges were protected by a layer of epoxy coating, and then calibrated from moment measurement. The horizontal loads applied to the pile head were analyzed from the moments measured by the strain gauges on the piles at the level of the clay surface.

In the tests, the soil was contained in a cylindrical box. The box has an inner diameter of 500 mm and a height of 700 mm, and was constructed of stainless steel. Soft and medium-stiff clay ground models were used in the tests, both of which were made from kaolin powder. Table 1 summarizes some of the intrinsic properties of the soils. The soft clay model was constructed as follows. First, a batch of dry kaolin powder was poured into the container. A hand-held mixer was used to stir kaolin powder with sufficient water being introduced gradually. After powder and water were thoroughly mixed, the surface of the clay was scratched before adding another batch of kaolin powder and water. These procedures were repeated five times until the clay sample reach a height of about 650 mm. Then a rigid loading plate was placed on the surface of the mixture and the load was applied to consolidate the clay sample. The excess moisture inside the soft clay sample can be drained from the holes of the loading plate and be removed. The consolidation process was terminated when the height of clay sample reduced to 600 mm. The medium-stiff clay model was prepared by the compaction method. The kaolin powder was first mixed with a specified amount of water so that the water content of the soil was about 30%. After the mixture was stirred thoroughly, it was compacted by a hand-held proctor hammer layer by layer. The interlayer surfaces were scratched as well to avoid layering phenomena. The shear strengths of the soft and medium-stiff clay samples were measured via the portable vane shear test, and the results are also summarized in Table 1.

Table 1. Properties of kaolin clay

Parameter	Value
Specific gravity	2.68
Liquid limit (%)	51.9
Plastic limit (%)	28.5
Plasticity index (%)	23.4
Water content of soft clay (%)	94
Water content of medium-stiff clay (%)	30
Vane shear strength of soft clay (kPa)	6
Vane shear strength of medium-stiff clay (kPa)	60

The piles were loaded by controlling the displacement of the pile heads. As illustrated in Fig. 1, the tests consist of two loading stages, i.e., the cyclic loading followed by the monotonic loading. In each test, the pile head was first cyclically moved along x direction for 20 cycles, with a displacement amplitude of 10 mm, and then was monotonically moved along a specified horizontal direction. The directions for the monotonic loading are different in different tests, and these paths are labeled as P1, P2, P3, P4 and P5, respectively. The intersection angles between the direction of cyclic loading and the directions of different monotonic paths are summarized in Table 2. As a benchmark for the sake of comparison, a monotonic test without any cyclical-loading history was conducted. The loading path is labeled as P0. To investigate the influence of the shear strength of clay, this series of tests were performed in both soft and medium-stiff clay following the same displacement paths.

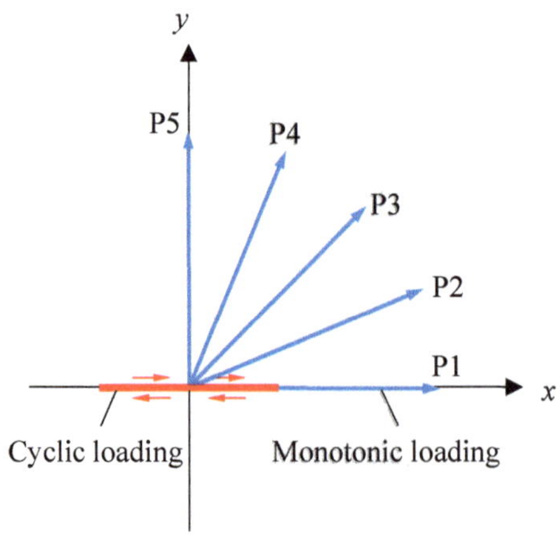

Fig. 1. Displacement paths at the pile head

Table 2. Summary of displacement paths

Displacement path	The intersection angle between the direction of cyclic loading and the direction of monotonic loading (°)
P1	0
P2	22.5
P3	45
P4	67.5
P5	90

3 Experimental Observations in the Soft Clay

Figure 2 shows the test results of the cyclic loading prior to the monotonic loading along P1. It is seen that the relationship between the force and the displacement was extremely nonlinear during the initial loading phase. The maximum lateral resistance was 81 N in the first cycle, but it dropped significantly in later cycles. The maximum force was 40.4 N and 39.8 N in the 9th and 10th cycles, respectively. The difference between them is less than 2%. After the 10th cycle, the change in the maximum force became negligible, and the behavior of the pile-clay system tended to be stable. The responses of the cyclic loadings in other tests are almost the same, which allows for comparison of the subsequent monotonic tests along different directions.

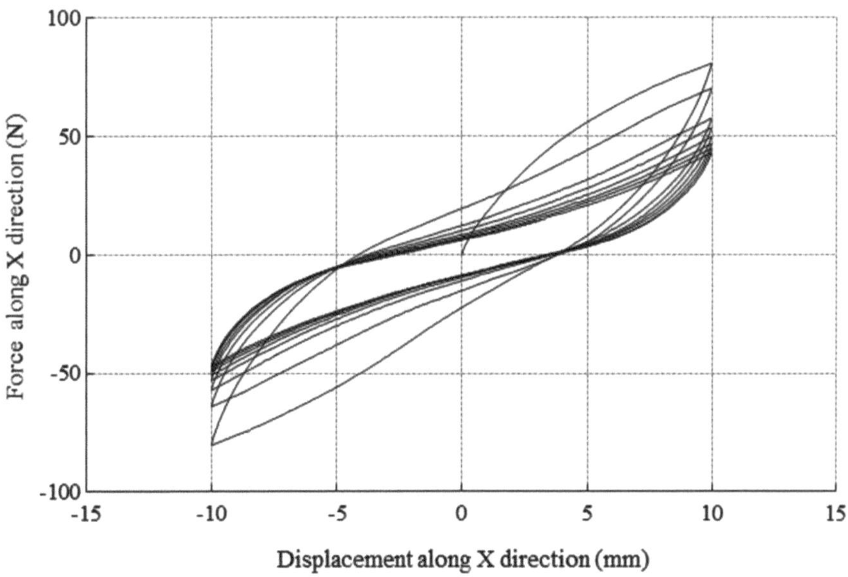

Fig. 2. Force-displacement relationship for the cyclic test in the soft clay

Figure 3 presents the force-displacement curves of the monotonic loadings along P1 ~ P5. For the purse of comparison, the result of the benchmark test along P0 is also plotted. To explore the influence of the intersection angle between the direction of the cyclic loading and the direction of the monotonic loading on the pile lateral bearing capacity, the resistance at the maximum displacement of each path (from P0 to P5) is summarized in Table 3. It can be seen from Fig. 3 that the features of the force-displacement curves under P1 ~ P5 are noticeably different from that under P0. The curves under P1 ~ P5 are not starting from the origin due to the residual forces when the pile returned to the original position after cyclic loading. The increment rate of force at low-level displacements under P1 ~ P5 is much smaller than that under P0, probably due to the disturbance of soil around the pile induced by the cyclic loadings. The lateral bearing capacity of the pile under P1 ~ P5 is also significantly lower than that under P0. As summarized in Table 3, the ratio of the resistance under P1 ~ P5 to that under P0 is between 61.4% and 71.5%. The impact of cyclic loading depends on the intersection angle between the direction of the cyclic loading and the direction of monotonic loading. The most severe degradation occurs when the monotonic-loading direction is the same as the cyclic-loading direction (i.e., the intersection angle is 0°). In this test, the history of cyclic loading caused a reduction of 38.6% in the resistance. As the angle becomes larger, the degradation becomes less significant. When the angle is 90°, the reduction in the pile lateral resistance is 28.5%.

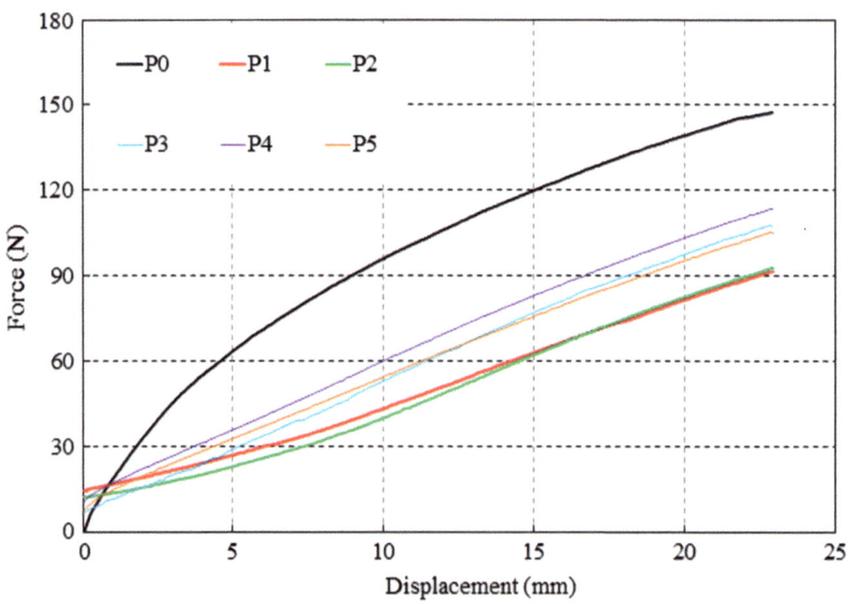

Fig. 3. Results of the monotonic loading tests in the soft clay

Table 3. Summary of test results in the soft clay

Displacement path	Resistance at maximum displacement (N)	Ratio (%)
P0	147.6	100
P1	90.7	61.4
P2	92.4	68.6
P3	108.1	73.2
P4	106.3	72.0
P5	105.6	71.5

After the test, it was observed that a gap developed around the pile, as shown in Fig. 4. The gap resulted from cyclic remolding of clay around the pile during the stage of cyclic loading of pile. It is probably the gap around the pile that caused the significant reduction in the monotonic lateral resistance of pile. The gap was formed mainly on the path of cyclic loading, and this explains that the impact of the cyclic loading on the monotonic pile resistance along the same direction is more significant than that on the resistance along the orthogonal direction.

Fig. 4. Development of a gap around the pile in the soft clay

4 Experimental Observations in the Medium-Stiff Clay

Six model pile tests were conducted in the medium-stiff clay with the same parameter setting in the soft clay. The results are presented in Figs. 5 and 6 and Table 4. Figure 5 shows that the maximum lateral resistance during the cyclic loading was 259 N, which appeared in the first cycle, and no significant reduction was noticed in later cycles. The force-displacement curves after the 5th cycle almost formed closed loops. This phenomenon indicates that the pile responses in the medium-stiff clay tended to stabilize after the 5th loading cycle. Cyclic degradation of the lateral pile resistance is much less severe in the medium-stiff clay than in the soft clay.

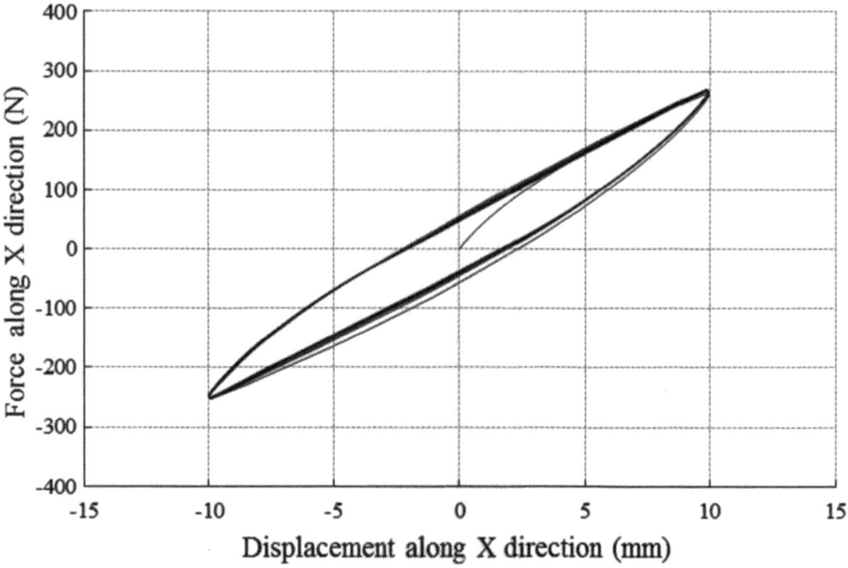

Fig. 5. Force-displacement relationship for the cyclic test in the medium-stiff clay

The force-displacement relationships obtained from the monotonic loading under P0～P5 in the medium-stiff clay are compared in Fig. 6. The curves under different paths are closer in this series of tests, in comparison with the results in the soft clay. To evaluate the influence of the intersection angle between the cyclic-loading direction and the monotonic-loading direction on the pile lateral bearing capacity, the resistance at maximum displacement under each displacement path (from P0 to P5) in the medium-stiff clay was obtained and compared in Table 4. The results indicate that the impact of the intersection angle on the degree of resistance degradation in the medium-stiff clay is similar to that in the soft clay. The most severe degradation occurred when the monotonic loading was applied along the direction of cyclic loading, while the least severe degradation occurred when the monotonic loading was applied

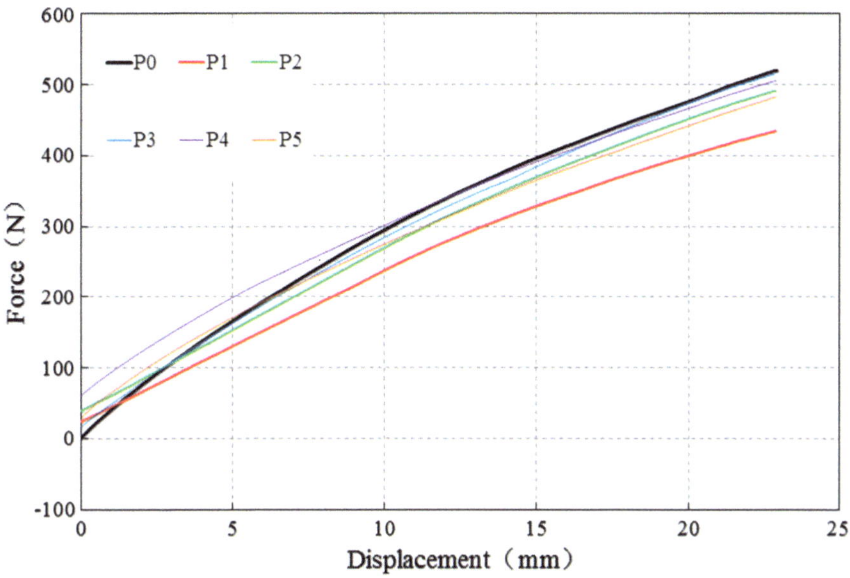

Fig. 6. Results of the monotonic loading tests in the medium-stiff clay

Table 4. Summary of test results in the medium-stiff clay

Displacement path	Resistance at maximum displacement (N)	Ratio (%)
P0	518.7	100
P1	434.0	83.7
P2	490.6	94.6
P3	517.5	99.8
P4	504.7	97.3
P5	482.1	92.9

along the direction perpendicular to it. But the degree of reduction is much less significant in the medium-stiff clay than in the soft clay. For instance, the reduction ratio under P1 is 16.3% in the medium-stiff clay, in comparison with 38.6% in the soft clay, while that under P5 is 7.1% in the medium-stiff clay, in comparison with 28.5% in the soft clay. Under the same path, the reduction ratio in the medium-stiff clay is about 20% less than that in the soft clay.

Under the cyclic loading, the medium-stiff clay surrounding the model pile was disturbed and the plastic strain accumulated as the number of cycle increased. But no obvious gap around the pile was observed during the tests, which is different from the phenomena observed in the tests of soft clay. This should be the main reason that the impact of the cyclic loading on the post-cyclic lateral resistance in the medium-stiff clay is much weaker than that in the soft clay.

5 Conclusions

In this study, two series of pile model tests were conducted in the soft and medium-stiff clay. The piles were loaded monotonically along different horizontal directions, after being cyclically loaded along a specified horizontal direction. The results indicate that the post-cyclic lateral resistance of pile depends on the intersection angle between the directions of cyclic loading and monotonic loading. The degree of reduction decreases with an increase of the intersection angle. The impact of cyclic loading on the resistance degradation is much more significant in the soft clay than in the medium-stiff clay. The reduction ratio is 16.3% in the medium-stiff clay, in comparison with 38.6% in the soft clay when the post-cyclic loading direction is along the direction of cyclic loading. The reduction ratio is 7.1% in the medium-stiff clay, in comparison with 28.5% in the soft clay when the post-cyclic loading direction is orthogonal to the direction of cyclic loading. In view of these results, influences of both the shear strength of soil and the intersection angle should be taken into account when quantifying the cyclic degradation effect.

Acknowledgments. The authors acknowledge the financial support provided by the National Natural Science Foundation of China under Grant No. 51478273.

References

Dan, A.B., et al.: Cyclic lateral loading of a large-scale pile group. J. Geotech. Eng. (1987). https://doi.org/10.1061/(ASCE)0733-9410(1987)113:11(1326)

Dyson, G.J., Randolph, M.F.: Monotonic lateral loading of piles in calcareous sand. J. Geotech. Geoenviron. Eng. (2001). https://doi.org/10.1061/(ASCE)1090-0241(2001)127:4(346)

Mayoral, J.M., et al.: Determination of multidirectional p-y curves for soft clays. Geotech. Test. J. (2005). https://doi.org/10.1520/GTJ11697

Rajashree, S.S., Sundaravadivelu, R.: Degradation model for one-way cyclic lateral load on piles in the soft clay. Comput. Geotech. (1996). https://doi.org/10.1016/s0266-352x(96)00008-0

Rao, S.N., et al.: Influence of rigidity on laterally loaded pile groups in marine clay. J. Geotech. Geoenviron. Eng. (1998). https://doi.org/10.1061/(ASCE)1090-0241(1998)124:6(542)

Rollins, K.M., et al.: Lateral load behavior of full-scale pile group in clay. J. Geotech. Eng. (1998). https://doi.org/10.1061/(ASCE)1090-0241(1998)124:6(468)

Rollins, K.M., et al.: Pile spacing effects on lateral pile group behavior: load tests. J. Geotech. Geoenviron. Eng. (2006). https://doi.org/10.1061/(ASCE)1090-0241(2006)132:10(1262)

Su, D.: Resistance of short, stiff piles to multidirectional lateral loadings. Geotech. Test. J. (2012). https://doi.org/10.1520/GTJ103840

Su, D., Li, J.H.: Three-dimensional finite element study of a single pile response to multidirectional lateral loadings incorporating the simplified state-dependent dilatancy model. Comput. Geotech. (2013). https://doi.org/10.1016/j.compgeo.2013.01.007

Su, D., et al.: Cyclic degradation of a multidirectionally laterally loaded rigid single pile model in compacted clay. J. Geotech. Geoenviron. Eng. (2014). https://doi.org/10.1061/(ASCE)GT.1943-5606.0001084

Su, D., Zhou, Y.G.: Effect of loading direction on the response of laterally loaded pile groups in sand. Int. J. Geomech. (2016). https://doi.org/10.1061/(ASCE)GM.1943-5622.0000544

Zhang, C., et al.: Centrifuge modeling of the cyclic lateral response of a rigid pile in soft clay. J. Geotech. Geoenviron. Eng. (2011). https://doi.org/10.1061/(ASCE)GT.1943-5606.0000482

Evaluation of Concrete Bored Piles Behaviour in Saturated Loose and Dense Sand During the Static Load Testing

Mehdi Aghayarzadeh[1(✉)], Hadi Khabbaz[2], and Behzad Fatahi[2]

[1] University of Technology Sydney (UTS), Sydney, NSW, Australia
mehdi.aghayarzadeh@student.uts.edu.au
[2] Geotechnical Engineering, University of Technology Sydney (UTS), Sydney, NSW, Australia

Abstract. Diversity of construction methods is available for the installation of piles; and generally piles are classified into two main groups in terms of their installation method, including non-displacement piles and displacement piles. However, non-displacement piles such as bored piles or drilled shafts cause less disruption to adjacent soil compared to displacement piles. Knowing the behaviour of piles, the load-displacement curve has an important role during the pile load testing. In this study, firstly, the background and concepts of three different constitutive soil models including Mohr-Coulomb, Hardening Soil and Hypoplastic models are discussed. Secondly, using these soil models and finite element software PLAXIS, the load carrying capacity of a single bored pile in saturated dense and loose sand are evaluated. Finally, different factors affecting load-displacement curve including Hypoplastic model parameters are assessed. The most important aim of this study is comparing the capability of different soil models to capture the large strain behaviour observed during the static load testing. The results revealed that all three soil models indicate a reasonable correlation to each other, except when the "Intergranular Strain" concept, defined in Hypoplastic model, is activated. This paper can be useful for the practicing engineers to identify the effect of different soil models in numerical simulation of static pile load testing.

Keywords: Pile load testing · Hypoplastic soil model · Solid pile
Load-displacement curve

1 Introduction

As explained by Yu and Yang (2012) the initial and primary consideration in pile designing is the evaluation of the foundation materials and the selection of the best substratum as the pile foundation support. In certain conditions comprising of cohesive subsoils, the pile lengths will be necessary to minimize settlement of the foundations rather than to develop load capacity. The proper selection of a type of pile for a given foundation should be made on the basis of a comparative study of cost, permanency, stability under vertical and a horizontal loading, long-term settlement and method of pile installation. As stated by Grand (1970), among different pile types, the concrete

© Springer International Publishing AG, part of Springer Nature 2019
S. Shu et al. (eds.), *New Developments in Materials for Infrastructure Sustainability and the Contemporary Issues in Geo-environmental Engineering*, Sustainable Civil Infrastructures, https://doi.org/10.1007/978-3-319-95774-6_7

cast-in place piles, and more particularly the prestressed concrete piles, can sustain high bending stresses and are frequently used in viaducts and trestle type of structures with the pile extending above ground or channel bottom level.

According to Dung (2009), the most important issue in FEM programs is capturing the soil behaviour, which changes continuously during the pile installation. Hence, it is vital to have an appropriate model that can simulate the real soil behaviour, which alters especially in driven piles. As explained in this study, Mohr-Coulomb model has a number of limitations to embrace the real soil behaviour, while the Hardening Soil model can be considered as an advanced model to simulate the soil behaviour. However, some important features are still not captured in this model. The findings of this study indicate that currently the Hypoplastic model seems to be the most suitable model, especially for driven piles, because of introducing the void ratio as a state variable and the intergranular strain concept.

In this study, using PLAXIS 2D version 2017, the behaviour of concrete bored pile in terms of the load-displacement curve resulted from static load testing is evaluated. In numerical modelling three constitutive soil models, introduced above, are used and the capabilities of each model during the simulation of static load testing are assessed.

2 Constitutive Models

In this section, the constitutive models used in the analyses are briefly introduced. In numerical simulations, the pile behaviour is assumed to be linear-elastic, while the soil behaviour is described by three different models including the Mohr-Coulomb, Hardening Soil and Hypoplastic (incorporating intergranular strain concept) models.

2.1 Mohr-Coulomb Model (MC Model)

According to Dung (2009), the Mohr-Coulomb model is a simple and well-known linear elastic perfectly plastic model, which can be used as a first estimation of soil behaviour. The linear elastic part of the Mohr-Coulomb model is based on Hooke's law of isotropic elasticity, while the perfectly plastic part is based on the Mohr-Coulomb failure criterion. Plasticity comprises the development of irreversible strains and in order to assess whether or not plasticity occurs in a calculation, a yield function "f" is introduced as a function of stress and strain. The two plastic model parameters appearing in the yield functions are the well-known friction angle ϕ and the cohesion c. The plasticity occurs when $f = 0$ and the full MC model yield condition includes six yield functions which represents a fixed hexagonal cone in principal stress space. For stress states represented by points within the yield surface, the behaviour is purely elastic and all strains are reversible.

As explained by PLAXIS (2017) and Wehnert and Vermeer (2004), in addition to the yield functions, six plastic potential functions are defined for this model. These functions contain a third plasticity parameter, called the dilatancy angle ψ. This parameter is required to model the plastic volumetric strain developed during plastic shearing. Clays (regardless of overconsolidated layers) are characterized by a very low amount of dilation ($\psi = 0$). As for sands, the angle of dilation depends on the angle of internal friction. For non-cohesive soils (sand, gravel) with the angle of internal friction $\phi > 30$ the value of dilation angle can be estimated as:

$$\psi = \phi - 30 \tag{1}$$

The linear elastic perfectly-plastic Mohr-Coulomb model requires a total of five parameters, which can be obtained from basic tests on soils samples. It is worth mentioning that according to PLAXIS (2017), in triaxial testing of soil samples the initial slope of stress-strain curve (tangent modulus) is usually indicated as E_0 and the secant modulus at 50% strength is referred as E_{50}. For materials with a large linear elastic range it is realistic to use E_0 but for loading of soils one generally uses E_{50}. The Mohr-Coulomb stress-strain curve obtained from triaxial test is shown in Fig. 1.

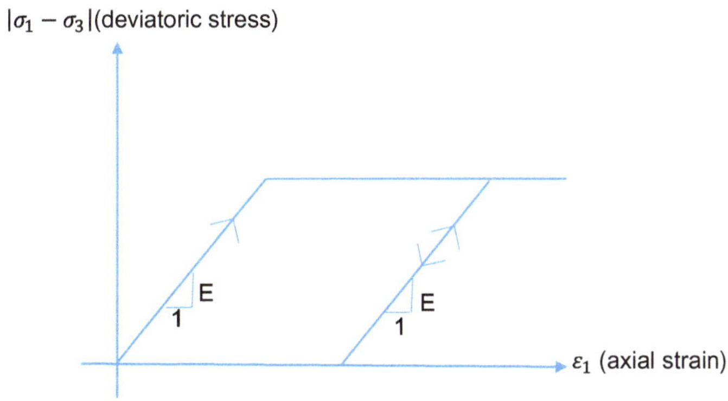

Fig. 1. Basic principle of MC model (after Wehnert and Vermeer 2004)

As mentioned by Dung (2009), although this model describes quite well the failure behaviour of soil, it consists of many limitations for the real soil behaviour. Firstly, the linear elasticity occurs until failure, hence it includes neither stress-dependency nor stress-path dependency of stiffness. Furthermore, care must be taken in undrained condition due to unrealistic stress path. Finally, this model does not include either hardening or softening rules.

2.2 Hardening Soil Model (HS Model)

According to PLAXIS (2017) the Hardening Soil model is an advanced model for simulating the behaviour of different types of soil, both soft soils and stiff soils. This model can consider two main types of hardening, including shear hardening and compression hardening. The advantage of the Hardening Soil model compared to the Mohr-Coulomb model is using of a hyperbolic stress-strain curve instead of a bi-linear curve and also the control of stress level dependency. When using the Mohr-Coulomb model, the user has to select a fixed value of Young's modulus, while for real soils this stiffness depends on the stress level. However, Hardening Soil model in calculations takes into account the stress-path dependent stiffness. The basic idea for the formulation of the Hardening Soil model is the hyperbolic relationship between the vertical strain, ε_1, and the deviatoric stress, q, in primary triaxial loading. This relationship is plotted in Fig. 2.

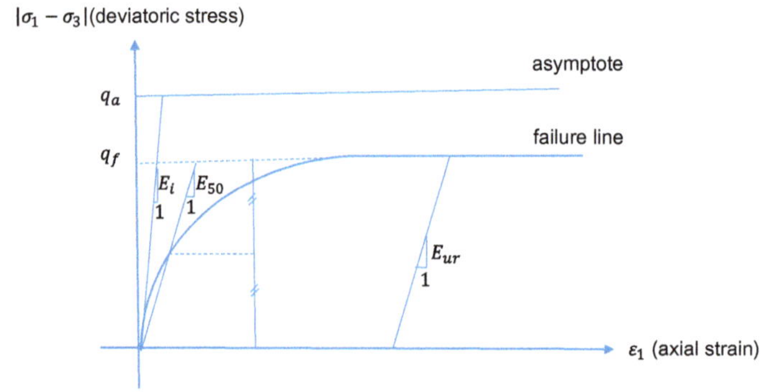

Fig. 2. Hyperbolic stress-strain relation in primary loading for a standard drained triaxial test (after PLAXIS 2017)

As explained by Wehnert and Vermeer (2004), the stress dependency stiffness for the HS model is described by E_{oed} and E_{50} and E_{ur}. E_{oed} is the oedometer stiffness being defined as a tangent stiffness modulus. The other two stiffness parameters E_{50} and E_{ur} are related to the standard drained triaxial tests. E_{50} is the confining stress dependent stiffness modulus for primary loading and E_{ur} the confining stress dependent Young 's modulus for unloading and reloading. In this model, as soon as deviatoric stress (q) in the standard drained triaxial test becomes equal to the ultimate deviatoric stress (q_f), the failure criterion is satisfied and perfectly plastic yielding occurs as described by the Mohr-Coulomb model. That is why, some parameters of the Hardening Soil model including cohesion, the friction angle and the dilation angle are the same as the Mohr-Coulomb model.

Based on Schanz et al. (1997) study, it can be mentioned that in practice, $E_{ur}^{ref} = 3E_{50}^{ref}$ is considered and for engineering practice $E_{oed}^{ref} = E_{50}^{ref}$ can be assumed.

As mentioned by Dung (2009), besides the advantages of stress-dependency and memory of pre-consolidation stress, the HS model still has a number of limitations. Firstly, this model does not capture the softening behaviour caused by dilatancy and de-bonding effect. In the HS model, the dilatancy continues to infinity, unless the dilatancy cut-off option is used. Moreover, the HS model does not take into account the increase of stiffness at a small strain in comparison with reduced stiffness at a large strain level. Lastly, this model also has a shortcoming in the region of small stress cycles due to an excessive accumulation of deformation in cyclic loading.

2.3 Hypoplastic Model (HP Model)

As explained by Mašın (2010), Hypoplasticity is a particular class of incrementally nonlinear constitutive models, developed specifically to predict the behaviour of soils.

The basic structure of the hypoplastic models has been developed during 1990's at the University of Karlsruhe. In hypoplasticity, contrary to elasto-plasticity, the strain rate is not divided into elastic and plastic parts, and the models do not use the notions of the yield surface and the plastic potential surface. However, the models are capable of predicting the important features of the soil behaviour, such as the critical state, dependency of the peak strength on soil density, non-linear behaviour in the small and large strain ranges, dependency of the soil stiffness on the loading direction and so on.

The first version of the hypoplastic constitutive law was proposed by Kolymbas (1985), describing the stress-strain behaviour of granular materials in a rate form. It has been developed from the stress rate, written as a function of the stress tensor and the strain rate. Later on, the void ratio was added as an additional state variable by Gudehus (1996). The general form of the hypoplastic model is:

$$\dot{T} = F(T, e, D) \tag{2}$$

where, $\dot{T}$ represents the objective stress rate tensor as a function of the current void ratio, e, the Couchy granular stress tensor, T, and the stretching tensor of the granular skeleton, D. The model proposed by Gudehus (1996) was later refined by von Wolffersdorff (1996) to incorporate Matsuoka-Nakai critical state stress condition. This model, is nowadays considered as a standard hypoplastic model for granular materials, and this version is also implemented in finite element software.

As explained by Elmi Anaraki (2008), the outstanding aspect of this model is its simplicity. As mentioned earlier, this model recognizes inelastic deformation without using any additional notions such as yield surface or plastic potential. The hypoplastic models can predict successfully the soil behaviour in the medium to large strain range. However, in the small strain range and upon cyclic loading they fail in predicting the high quasi-elastic soil stiffness. In fact, small deformation due to cyclic loading and creep are not included, leading to an excessive accumulation of deformations under stress cycles. To overcome this problem, Niemunis and Herle (1997), proposed an extension of the hypoplastic equation by considering an additional state variable, termed "intergranular strain (IGS)", to determine the direction of the previous loading. This modification also, often referred as the "intergranular strain concept", has been implemented in finite element software. Currently, the hypoplastic model can be used for not only cohesionless materials but also clays and organic soils.

The hypoplastic model includes eight parameters that allow for a realistic description of granular material behaviour over a wide range of stresses and densities. It would be valuable to mention that three characteristic void ratios e_d, e_c and e_i denote the maximum densification, the critical state and the loosest state, respectively. These limiting void ratios reach the limit values e_{d0}, e_{c0} and e_{i0} at vanishing mean skeleton pressure and approach zero for very high mean skeleton pressure. Due to the difficulties in testing, these limit void ratios are assumed to be e_{min}, e_{max} and approximately 1.2 e_{max}. The intergranular strain concept (small strain behaviour) requires five additional parameters.

The required soil parameters defined in each model are introduced in the next section and more details about the background of different soil models have been explained in PLAXIS (2017), Dung (2009), Elmi Anaraki (2008) and Mašın (2010).

3 Material Properties

In this study, in order to assess the behaviour of piles during the static load testing, loose and dense Baskarp sand properties have been used in numerical modelling. As explained by Elmi Anaraki (2008), Baskarp sand is a uniform sand and the grain size has been classified as angular to sub-angular with D_{50} of approximately 140 μm. This sand has a total unit weight of 20 kN/m^3. Hypoplastic model parameters, including intergranular strain parameters and hardening soil model properties for this kind of soil are shown in Table 1. As mentioned by Elmi Anaraki (2008) in order to assess the density dependency of the soil behaviour, the initial void ratio varied between loose and dense conditions. Hence, in numerical modelling, incorporating Hypoplastic soil model, the initial void ratio has been selected 0.83 (D_r = 0.26) and 0.65 (D_r = 0.73) for loose and dense conditions, respectively.

Table 1. Baskarp sand properties (after Dung'2009)

Parameters	Hypoplastic model with intergranular strain concept	Hardening Soil model (dense)	Hardening Soil model (loose)
ϕ_c (°)	30	–	–
h_s (MPa)	4000	–	–
n	0.42	–	–
e_{d0}	0.548	–	–
e_{c0}	0.929	–	–
e_{i0}	1.08	–	–
α	0.12	–	–
β	0.96	–	–
m_T	2	–	–
m_R	5	–	–
R_{max}	0.0001	–	–
β_r	1	–	–
χ	2	–	–
E_{50}^{ref} (kPa)	–	40500	31000
E_{oed}^{ref} (kPa)	–	50000	33000
E_{ur}^{ref} (kPa)	–	121500	93000
ϕ (°)	–	37	31.3
ψ (°)	–	9	2
m	–	0.5	0.5
v_{ur}	–	0.2	0.2
p^{ref} (kPa)	–	100	100

It can be noted that all hypoplastic parameters were derived from the laboratory tests conducted by Elmi Anaraki (2008) while the material set of the hardening soil model was derived by Dung (2009) using the soil tests facility in PLAXIS 2D, in such a way that using simulating the drained triaxial and oedometer tests, the soil responses from the hardening soil model are similar to those obtained from the hypoplastic model. Using this method, by correlating the soil responses from the Hardening Soil and Mohr-Coulomb models in drained triaxial test (Fig. 3), Mohr-Coulomb model properties were derived and shown in Table 2.

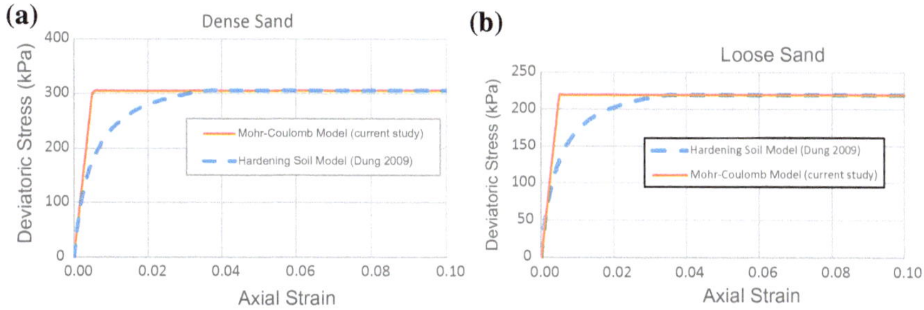

Fig. 3. Simulated drained triaxial test sand applying Mohr-Coulomb and Hardening Soil models for (a) dense sand (b) loose sand

Table 2. Mohr-Coulomb model properties for Baskarp sand

Parameters	Dense	Loose
E_i (kPa)	60000	45000
E_{50} (kPa)	33000	24750
υ	0.35	0.25
ϕ (°)	37	31.3
ψ (°)	9	2

In numerical modelling, the linear elastic model was assigned to the concrete pile with Poisson's ratio of 0.2, the elastic modulus and the unit weight of 30 GPa and 25 kN/m^3, respectively. For interface parameters of MC and HS models, as explained by Aghayarzadeh et al. (2017), the strength reduction factor R_{inter} was regarded as the main interface parameter, which is taken into account the strength decrease of the interface element in the corresponding soil layer. However, the interface parameters of HP model defined in PLAXIS include the interface oedometer modulus, E_{oed}^{ref}, the interface strength parameters such as cohesion, the friction angle and the dilation angle. Hence, in this model the interface shear strength is directly given in strength parameters instead of using a factor relating the interface shear strength to the soil shear strength as defined on Mohr-Coulomb and Hardening Soil models. In addition, in HP model, two parameters are included in interface parameters to enable stress-dependency of the interface stiffness according to a power law formulation (Eq. 3):

$$E_{oed}\left(\sigma_n^{'}\right) = E_{oed}^{ref}\left(\frac{\sigma_n^{'}}{UD - P^{ref}}\right)^{UD-Power} \tag{3}$$

where UD-Power is the rate of the stress dependency of the interface stiffness, UD-P^{ref} is the reference stress level (usually 100 kN/m^2) and $\sigma_n^{'}$ is the effective normal stress in the interface stress point. Interface parameters used in numerical modelling for different soil models are introduced in Table 3.

Table 3. Interface parameters of different soil models defined in numerical modelling

Parameters	Mohr-Coulomb Model	Hardening Soil Model	Hypoplastic Model (Loose Sand)	Hypoplastic Model (Dense Sand)
R_{int}	1	1	–	–
E_{oed}^{ref} (kPa)	–	–	33000	50000
$c_{ref}^{'}(kPa)$	–	–	0.1	0.1
$\phi'(°)$	–	–	31.3	37
ψ (°)	–	–	2	9
UD-Power	–	–	0	0
UD-P^{ref} (kPa)	–	–	100	100

4 Numerical Modelling

In this paper, all numerical simulations were performed using a two-dimensional finite element program in which the geometry was defined in an axisymmetric mesh. In the analysis, the length and the dimeter of concrete pile were assumed 10 m and 0.4 m, respectively. The finite element model of concrete pile is shown in Fig. 4. The soil cluster was modelled by 15-noded triangular elements and for boundary conditions. Horizontal fixity were applied to the left and right sides of the geometry, whereas the bottom nodes are fixed both horizontally and vertically. The ground water table was assumed at the ground surface and dense and loose sands were assigned to the soil with three different constitutive soil models, introduced above. At the first step, the mesh dependency of each constitutive soil model was evaluated by changing the size of mesh from very fine to very coarse in two cases with or without the interface element and the results are shown in Fig. 5. It should be noted that the load-displacement curves are drawn by applying prescribed displacements of 2, 4, 8, 16, 35 and 50 mm at the top of the pile. As can be seen in Fig. 5, it is clear that in the case of using interface element between soil and pile, all the models show less sensitivity to the mesh size compared to the cases in which interface element is removed. In addition, in case of using interface element, both Mohr-Coulomb and Hardening Soil models showed less sensitivity to element size compared with Hypoplastic model. Furthermore, for both MC and HS models in the case of using interface (R_{int} = 1), for high displacements few stairs occurred in load-displacement curve although the global shape of the load-displacement curves seem to stay realistic while HP model with the same conditions showed a smoother curve. Finally, since in modelling with the Mohr-Coulomb model, the tangent modulus (E_i) was used; hence, the occurred displacement compared with two other soil models was overestimated.

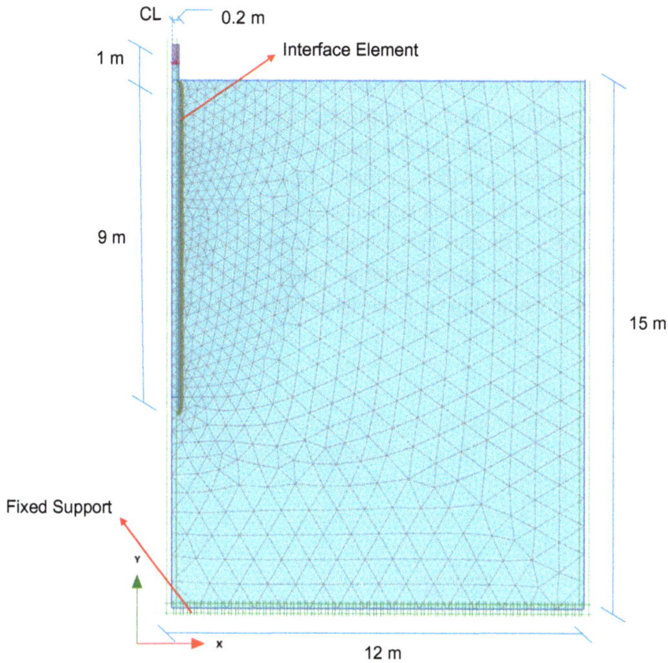

Fig. 4. Finite element model of the pile and the adjacent ground with the corresponding generated mesh

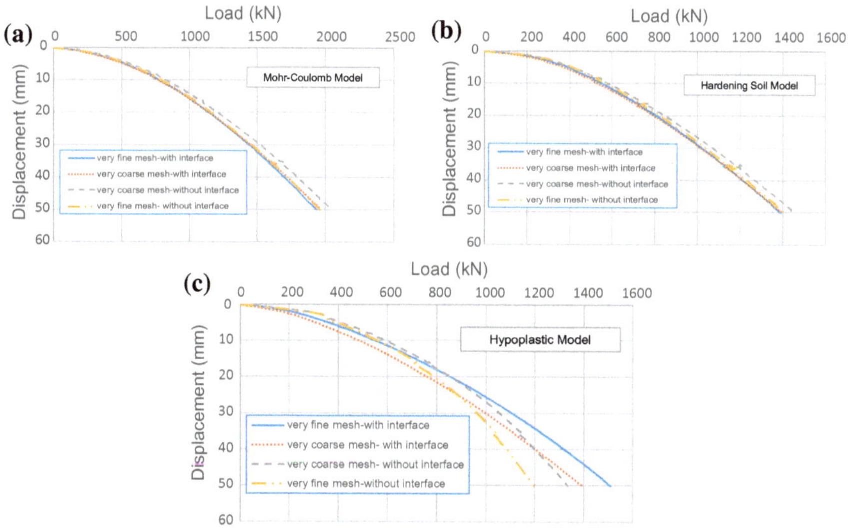

Fig. 5. Mesh size dependency of finite element models with (a) MC (b) HS and (c) HP constitutive models

After conducting mesh size sensitivity analysis, in the next step the load-settlement curves of concrete bored pile were obtained in both saturated dense and loose states for sand in four cases using different constitutive soil models including Mohr-Coulomb, Hardening Soil, Hypoplastic and Hypoplastic with Intergranular Strain (IGS) using a fine mesh. Load-Displacement curves using different soil models are presented in Fig. 6. It should be mentioned that in this section for modelling with Mohr-coulomb model the secant modulus (E_{50}) has been used. As can be seen in Fig. 6, there is a close correlation between MC, HS and HP models in both dense and loose sand conditions. However, by activating the Intergranular Strain (IGS) concept of the Hypoplastic model the load-displacement curve is considerably overestimated.

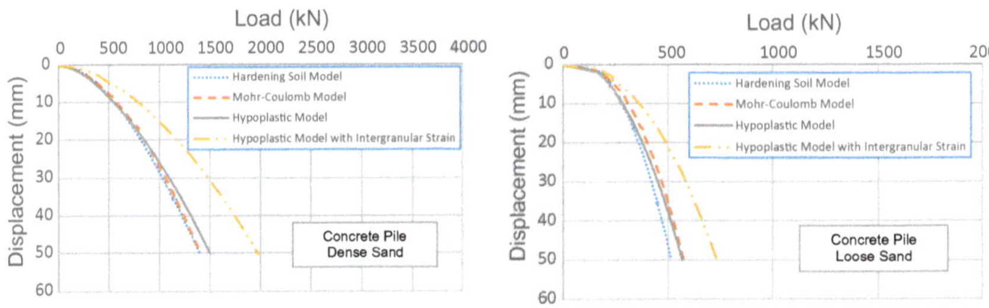

Fig. 6. Comparison of obtained Load-Displacement curves using different soil models in dense and loose sand soils

There are various methods of determining the ultimate capacity of piles from load-displacement plots, especially in compression. A common method takes Q_{static} as the load required to displace the pile 10% of its diameter. Using this method and comparing the bearing capacity of piles in displacement of 40 mm (i.e. 10% of the pile diameter) similar values can be obtained employing different soil models, except when IGS is activated in Hypoplastic model.

In order to evaluate the piles behaviour in more detail, the mobilized shaft friction (q_s), the portion of toe (Q_b) and shaft capacity (Q_s) were studied separately. As can be seen in Fig. 7a, firstly, strength reduction factor (R_{int}) was assumed to be equal to 1 and it is clear that in this case by increasing the settlement of the pile from 50 mm to 100 mm and 150 mm the mobilized shaft and toe capacity and the mobilized skin friction (q_s) along the shaft increases. Generally, in the current version of software it is possible to get the relationship between the applied displacement and the corresponding force of selected points. Since the selected points in the modelling process have been assumed at the pile head, hence it is not attainable to get the relationship between base resistance and tip displacement directly. However, it is possible to obtain the base load-displacement curve manually. For achieving this aim, as mentioned earlier, different displacement in different phases (2, 4, 8, 16, 35, 50, 100 and 150 mm) were applied over the pile head and the base resistance was calculated in the following procedure: if a pile is simulated as a plate, resistance at the base will be the axial force at the tip of the pile; and if pile is simulated as a volume pile (used in this study), resistance at the base will be calculated by integrating the effective vertical stress at the base of the pile with the tip area of the pile. Then, the relationship between the base resistance and tip displacement can be achieved by plotting the base resistance, which was calculated above, and the corresponding tip displacement. The shaft resistance (Q_s) can be calculated by subtracting the base resistance (Q_b) from the total resistance (Q):

$$Q_s = Q - Q_b \quad (4)$$

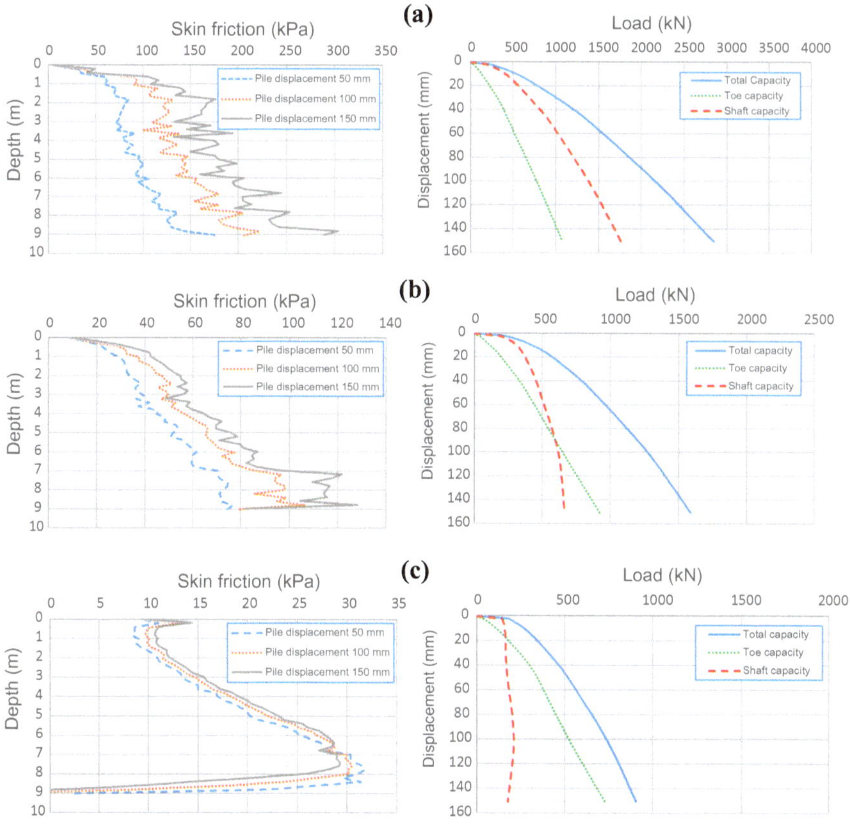

Fig. 7. Comparison of mobilized skin friction along the pile length and toe and shaft capacity of concrete pile in dense sand with Hardening Soil model and different interface reduction factors: (a) R_{int} = 1 (b) R_{int} = 0.9 (c) R_{int} = 0.8

In addition, for drawing the skin friction mobilized along the pile length (q_s), it is necessary to draw the shear stress along the pile at the pile surface. The results can be observed in Fig. 7. By decreasing the interface reduction factor (R_{int}) to 0.9 and 0.8, (Fig. 7b, c) it can be seen that the portion of the base capacity from the total capacity increases, while shaft capacity decreases. In strength reduction factor 0.8, the skin fiction is mobilized in low pile displacements (less than 50 mm) and by increasing the pile displacement to 100 and 150 mm the mobilized shaft friction does not change and shaft capacity (Q_s) remains constant. Similar to this behaviour is observed in loose dense sand in which with strength reduction factor 0.9 the shaft friction was mobilized and when increasing the pile head displacement no change was observed. However, the pile in loose sand showed less toe and shaft capacities in comparison to dense sand (Fig. 8). It should be regarded that disturbances, observed near the ground surface in skin friction graphs in low interface reduction factors is attributed to the free length movement of the pile (1 m above the ground surface).

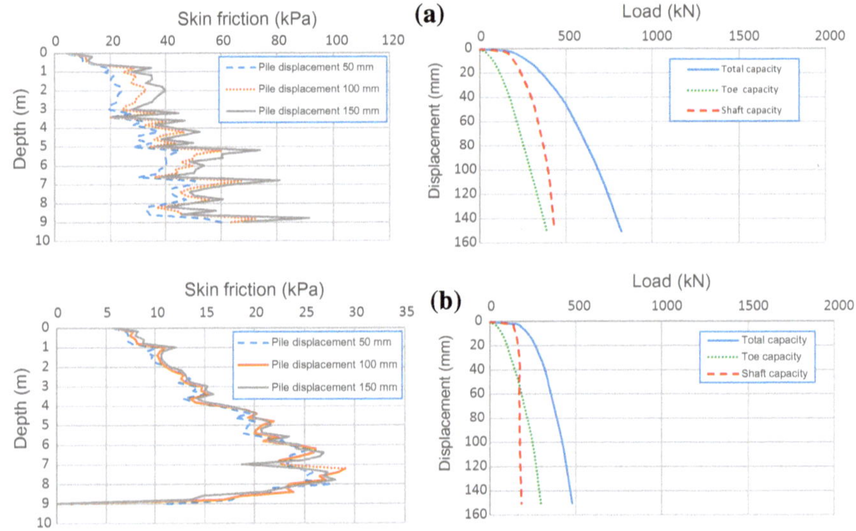

Fig. 8. Comparison of mobilized skin friction along the pile length and toe and shaft capacities of concrete pile in loose sand with Hardening Soil model and different interface reduction factors (a) $R_{int} = 1$ (b) $R_{int} = 0.9$

In order to evaluate the effect of different parameters of Hypoplastic model on the load-displacement curve obtained during static load testing of concrete bored pile in dense sand, each of the parameters of the Hypoplastic model, introduced earlier, was increased 30%, while remaining parameters were kept unchanged. The results of the parametric study have are in Fig. 9. Referring to Fig. 9, it is crystal clear that among the evaluated parameters in this study, the granular hardness (h_s) shows the least and the critical friction angle (φ_c) and the exponent relating to the sensitivity of granular skeleton to change of pressure (n) show the most influences on the load-displacement curve, respectively.

Finally, in order to assess the Intergranular Strain (IGS) concept in modelling of static pile load testing, the pile was simulated in both dense and loose concrete using Hypoplastic (HP) model. The normalized length of the intergranular strain tensors are depicted in Fig. 10 for dense sand and in Fig. 11 for loose sand. The normalized length of the intergranular strain tensor varies between 0, indicating the soil being inside the

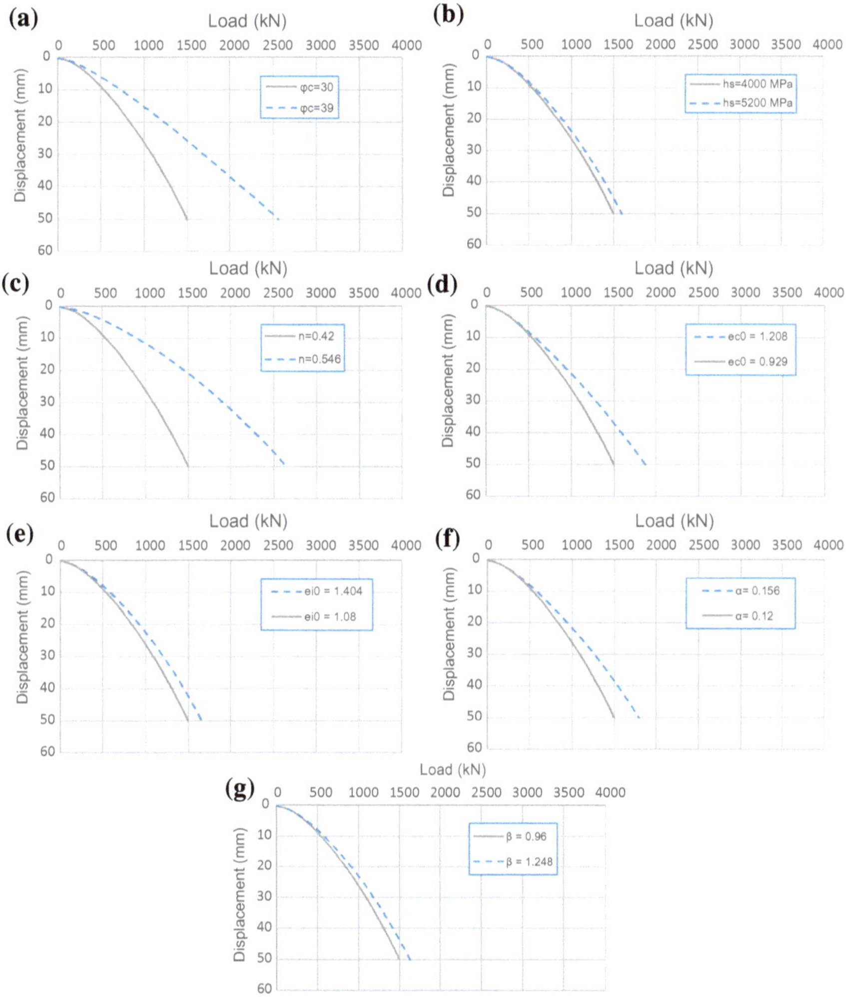

Fig. 9. Influence of different parameters of Hypoplastic model on load-displacement curve of concrete bored pile in saturated dense sand with interface reduction factor 1: (a) φ_c (b) h_s (c) n (d) e_{c0} (e) e_{i0} (f) α (g) β

elastic range, and 1, corresponding to the state swept-out of the small-strain memory. In the case of the normalized length of the intergranular strain being equal to 1 the soil behaviour is governed by the basic hypoplastic model. In other words, the normalized length of the intergranular strain tensor demonstrates how the small-strain stiffness is activated in different parts of the modelled geometry. Figures 10 and 11 show that the soil is in Hypoplastic state in the vicinity of the pile and further from the pile, the values of the normalized length of the intergranular strain tensor are very low and at these places the soil remains elastic. This finding correlates very well with the large strain behaviour of soil around the pile during the static load testing.

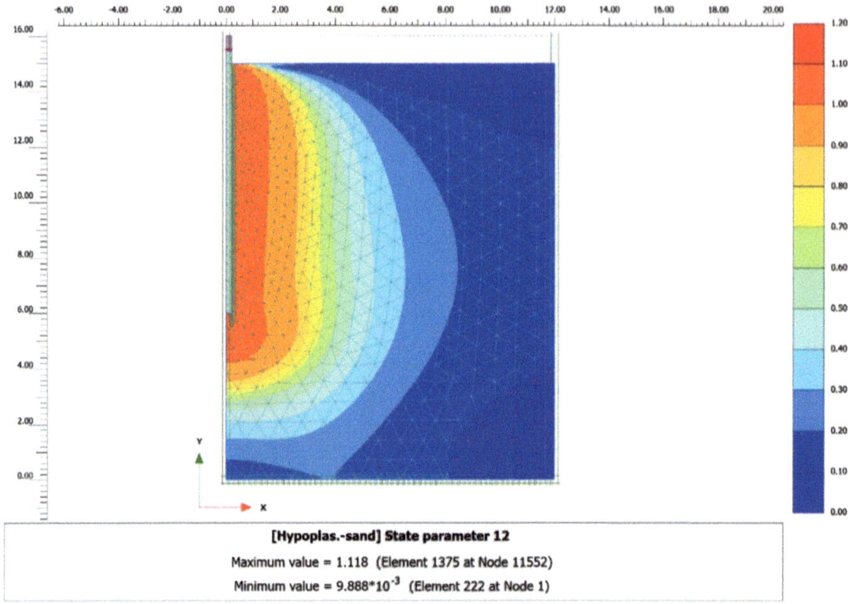

Fig. 10. Static load test of a concrete bored pile in saturated dense sand- normalized length of the intergranular strain tensor

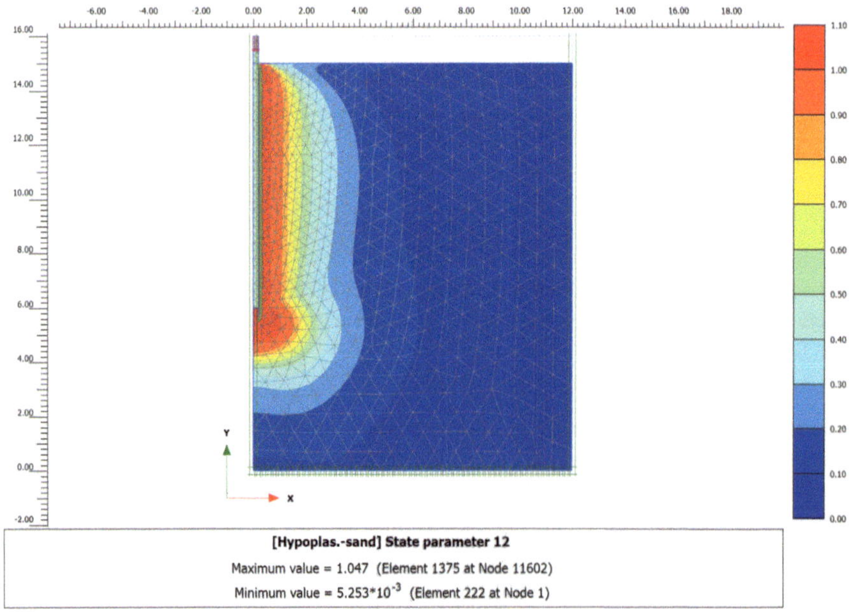

Fig. 11. Static load test of a concrete bored pile in saturated loose sand- normalized length of the intergranular strain tensor

5 Conclusions

This study is the part of an ongoing comprehensive investigation related to evaluating the solid and tubular pile behaviours under static and dynamic load testing. In this study, behaviour of concrete solid piles using different soil models including Mohr-Coulomb, Hardening Soil and Hypoplastic soil models were evaluated. Reasonable and similar predictions were observed applying these three soil models except hypoplastic with incorporation of intergranular strain concept. Since large strain behaviour is observed during static load testing in soil around the pile, the intergranular strain concept defined in hypoplastic model moderately overpredicts the settlement. Generally in dynamic load testing, soil around a pile experiences small strain behaviour; hence, applying this concept will be necessary to attain a better prediction. The details and elaborations are described in a follow up paper. Furthermore, among the parameters of hypoplastic model it has been found that the granular hardness (h_s) shows the least whereas the critical friction angle (φ_c) and the exponent relates to sensitivity of granular skeleton to change of pressure (n) show the most influences on the load-displacement curve obtained during the simulation of static load testing.

References

Aghayarzadeh, M., Khabbaz, H., Fatahi, B., Terzaghi, S.: Continuum numerical modelling of dynamic load test for steel pipe piles. International Conference on Advancement of Pile Technology and Pile Case Histories, Indonesia, 10 pages (2017)

Dung, P.H.: Modelling of Installation Effect of Driven Piles by Hypoplasticity (2009)

Elmi Anaraki, K.: Hypoplasticity Investigated. Master of Science, Delft University of Technology (2008)

Grand, B.A.: Types of piles: their characteristics and general use. Highw. Res. Rec. (333) (1970)

Gudehus, G.: A comprehensive constitutive equation for granular materials. Soils Found. **36**(1), 1–12 (1996)

Kolymbas, D.: A generalized hypoelastic constitutive law. In: Proceedings of XI International Conference on Soil Mechanics and Foundation Engineering, 2626 (1985)

Mašın, D.: PLAXIS Implementation of HYPOPLASTICITY (2010)

Niemunis, A., Herle, I.: Hypoplastic model for cohesionless soils with elastic strain range. Mech. Cohesive-Frictional Mater. **2**(4), 279–299 (1997)

PLAXIS, B.: Plaxis 2D Version. 2017. Reference Manual. Delft, The Netherlands (2017)

Schanz, T., Desrues, J., Vermeer, P.: Comparison of sand data on different plane strain devices. Proc. IS-Nagoya **97**, 289–294 (1997)

von Wolffersdorff, P.A.: A hypoplastic relation for granular materials with a predefined limit state surface. Mech. Cohesive-Frictional Mater. **1**(3), 251–271 (1996)

Wehnert, M., Vermeer, P.: Numerical analyses of load tests on bored piles. Numer. Methods Geomech.–NUMOG IX, 505–511 (2004)

Yu, F., Yang, J.: Base capacity of open-ended steel pipe piles in sand. J. Geotech. Geoenviron. Eng. **138**(9), 1116–1128 (2012)

Experimental Investigations on Uplift Capacities of Single and Group of Granular Anchor Piles

Mohit Kumar, Hans Raj Vashishtha, and Vishwas A. Sawant(✉)

Department of Civil Engineering, IIT Roorkee,
Roorkee 247667, Uttarakhand, India
{mhtal.pce2015, sawntfce}@iitr.ac.in,
vhansraj16@gmail.com

Abstract. Among a variety of ground improvement techniques, granular piles or stone columns find use in a variety of ground problems. Although Granular piles are economical and versatile in application, they are unable to provide safeguard the foundation against uplift or pull-out loads. In such situations, a modification over granular pile known as granular anchor piles is used. In the present study, a series of pull-out tests has been conducted on granular anchor piles by varying its diameter and length to diameter ratio. Pull-out tests have also been conducted on group of two granular anchor pile by varying the spacing between them. It was observed that, with increase in diameter and length to diameter ratio of granular anchor piles, pull-out capacity increases. The pull-out resistance of the group of two pile was found to be greater than that for a single pile having same diameter and length to diameter ratio. But it was lesser than twice the capacity of single pile. The value further increased when the spacing between piles was increased from three times diameter to four times diameter.

1 Introduction

Different site conditions and types of project require geotechnical engineers to design foundations on ground which may not be suitable for the specific project. Such ground conditions include loose sand, soft clay, expansive clay, etc. These soils may be problematic for construction by virtue of low bearing capacity, large settlement and low rate of consolidation. In case of saturated loose sand, liquefaction is also a threat. Other problem which may be encountered is that the superstructure may impose pull-out load on the foundation. Granular piles can be installed in both loose sand and soft clay and help in ground improvement by increasing the bearing capacity and reducing the settlement by virtue of strengthening and stiffening of soil (Phanikumar and Rao 1996). They increase the rate of consolidation by virtue of improved drainage in clays. Despite all these advantages, granular piles are not suited for sites where expansive soil is present or when the foundation is subjected to uplift or pull-out loads. An improvement over granular piles, known as Granular Anchor Piles (GAP), is suitable solution to such problematic conditions. They are even more versatile than granular soils and much more economical than concrete piles.

© Springer International Publishing AG, part of Springer Nature 2019
S. Shu et al. (eds.), *New Developments in Materials for Infrastructure Sustainability and the Contemporary Issues in Geo-environmental Engineering*, Sustainable Civil Infrastructures, https://doi.org/10.1007/978-3-319-95774-6_8

2 Gap as Modified Granular Pile

Granular piles, despite their wide array of applications, have certain limitations. These limitations have been overcome using a modification over the ordinary granular piles. A GAP, in addition to a granular pile comprises of a mild steel base plate and a central mild steel anchor rod embedded in the footing. This GAP-foundation system provides all the advantages provided by granular piles. In addition to this, GAPs can resist uplift forces produced by swelling soils and pull-out forces exerted by structures like transmission towers and high rise buildings.

GAPs comprise of following components as shown in Fig. 1:

- Mild steel base plate
- Mild steel central anchor rod
- Granular pile installed over the base plate around the central anchor rod.

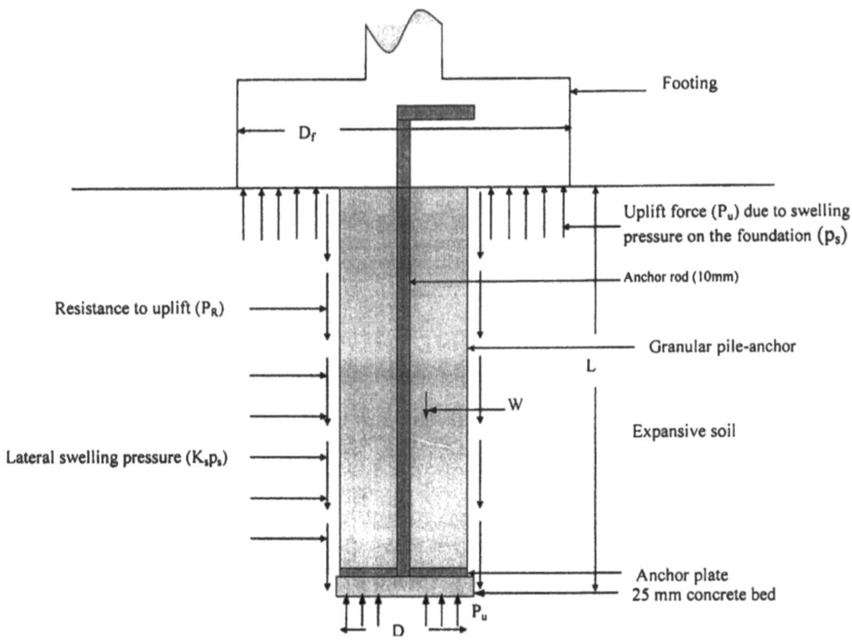

Fig. 1. Components of GAP and forces acting upon it (after Rao et al. 2007)

The central anchor rod of GAP is embedded into the foundation so the GAP and foundation act as a system. Description of all the forces acting on the GAP-foundation system is shown in Fig. 1. In case of swelling soil, uplift forces are applied on foundation. At the same time, the swelling soil applies additional lateral swelling pressure on the GAP surface. This increases the uplift capacity of GAP. Uplift is also resisted by the weight of the GAP.

Few experimental studies have been performed in last two decades. For a constant surface area, Phanikumar and Rao (1996) observed increase in diameter to be more effective in reducing heave. Phanikumar and Rao (2000) reported increase in pull-out resistance with increase in the interface friction angle from experimental investigations. In the experimental study reported by Rao et al. (2007), reinforcing expansive clay beds with GAPs reduced the amount of final heave and also increased the rate of heave of expansive clay beds. Krishna and Murthy (2013) reported the ultimate pull-out resistance of GAPs is comparatively more than double the value of concrete piles. Available literature indicated very few experimental studies on GAP in clayey soils. Response of group of GAP has not been fully explored yet. Present study is aimed to perform experimental study on a single GAP embedded in clayey soil with different length and diameter. Study is extended to consider group of two GAPs with varying spacing between piles.

3 Experimental Program

A series of pull-out tests on GAPs installed in clay were performed in laboratory. Variation of pull-out load with length is to diameter ratio (*L*/*D*) for different diameter of GAPs was analyzed. Pull-out tests were also performed on a group of two GAPs installed in clay for 50 mm diameter GAPs with varying spacing between piles as 3*D* and 4*D*.

3.1 Test Set-up and Materials

1. *Test tank*
   - Cylindrical tank of diameter 570 mm and depth 870 mm was used for pull-out test of single GAP.
   - Square tank of dimensions 1 m × 1 m × 1 m for pull-out test of group of 2 GAPs.
2. *Clay deposit*
   - Clay of medium compressibility (CI) as per Indian Standard Classification (ISC) system of classification was used. Other properties of clay have been listed in Table 1. All the tests were conducted in accordance with the respective Indian Standard (IS) codes.
3. *GAP material*
   - A mixture of coarse aggregates and sand with coefficient of uniformity (C_u) as 2.857 and coefficient of curvature (C_c) as 1.27 was used as granular material of GAP.
   - Central tendon rod of diameter 12.7 mm and length 1000 mm was used for both single and group of GAP.
   - Mild steel anchor plate of thickness 6 mm and diameters 50 mm, 75 mm and 100 mm were used.

Table 1. Different properties of soil used for laboratory testing

Property	IS Code	Value
Specific Gravity	IS:2720 (Part 3)-1980	2.473
Liquid Limit (%)	IS:2720 (Part 5)-1985	45.75
Plastic Limit (%)	IS:2720 (Part 5)-1985	19.75
Plasticity Index (%)	IS:2720 (Part 5)-1985	26
Inherent Moisture Content (%)	IS:2720 (Part 2)-1973	3.86
Maximum Dry Density (MDD) (kN/m^3)	IS:2720 (Part 7)-1974	17.6
Optimum Moisture Content (OMC) (%)	IS:2720 (Part 7)-1974	16.1
UCS (kPa)	IS:2720 (Part 10)-1991	120
Soil Classification as per ISC		CI

3.2 Preparation of Clay Bed

Clay was oven dried and powdered and thereafter its different properties as mentioned in Table 1 were determined through different laboratory tests. The clay was compacted to maximum dry density by mixing water equal to difference between optimum moisture content and inherent moisture content. Compaction of clay was done in four

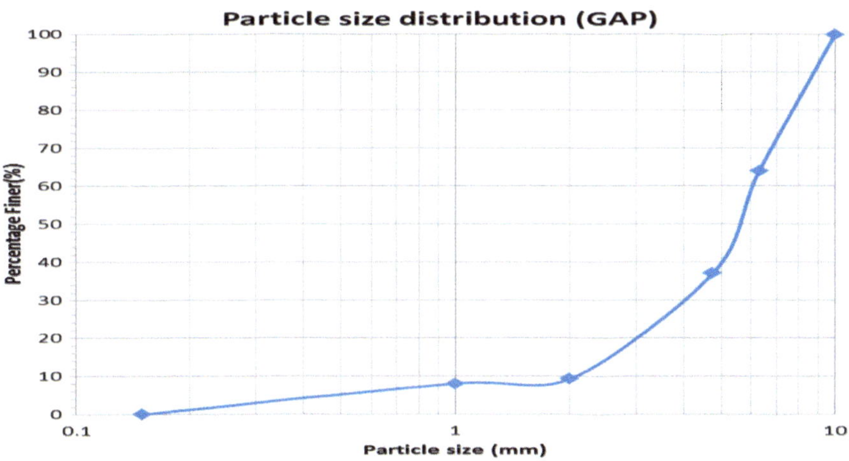

Fig. 2. Particle size distribution curve for granular material

layers (Fig. 2).

3.3 Installation of GAP

After installing clay bed, hole of diameter equal to diameter of GAP was augured. Thereafter, base anchor was installed in the bore hole. Granular pile was installed over the base in five layers. Each layer was tamped thoroughly using a metal rod.

3.4 Pull-Out Test

Pull-out test set-up consisted of a set of two pulleys mounted over a frame. These pulleys supported a cable attached to GAP at one end and loading pad at the other as shown in Fig. 3. Loads were applied on loading pad in small increments and soil-GAP

(a) Single GAP

(b) Group of 2 GAP

Fig. 3. Pull-out test set-up

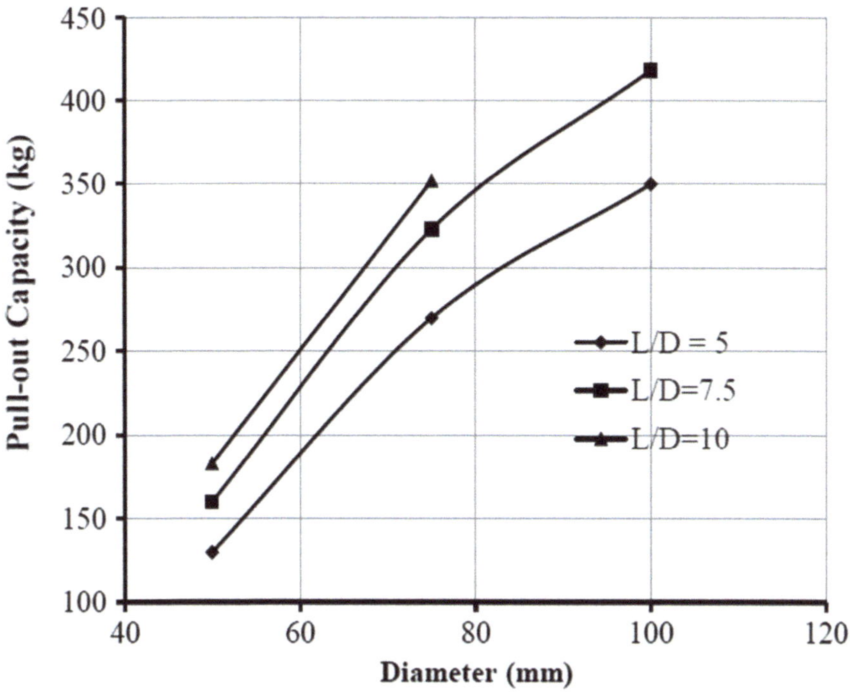

Fig. 4. Variation of pull-out capacity with L/D ratio for given diameter

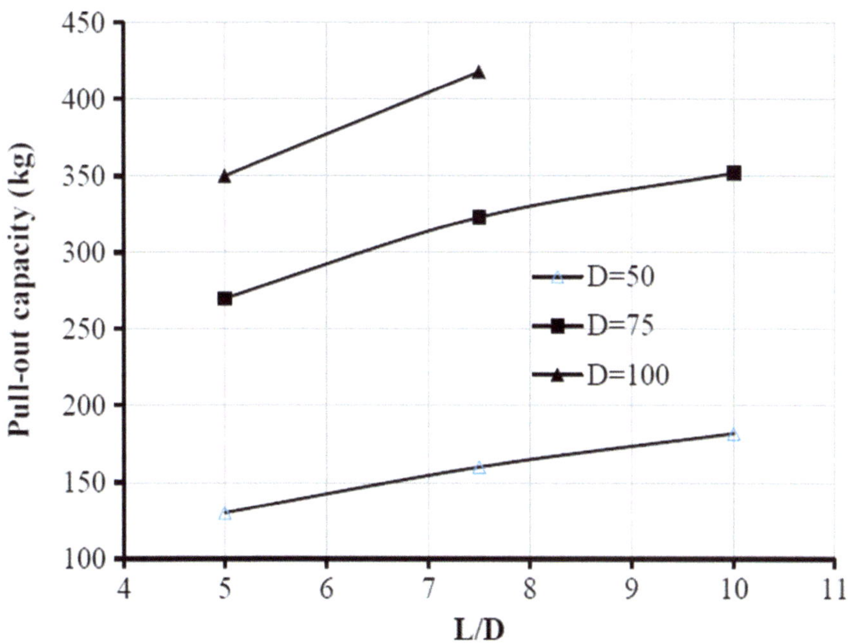

Fig. 5. Variation of pull-out capacity with diameter for given L/D ratio

system was allowed to deform for about 20 min for each loading. Test was continued till failure was realized. Failure was assumed to take place when the displacement became equal to one tenth the diameter of GAP. Tests were performed for GAPs of diameter 50 mm, 75 mm and 100 mm for *L*/*D* ratio of 5, 7.5 and 10. Two pull-out tests were conducted on a group of 2 GAPs of 50 mm diameter and *L*/*D* ratio 5 with spacing between the GAPs as 3*D* and 4*D* (Figs. 4 and 5).

4 Results and Discussions

The experimental program revealed the variation of GAP displacement with pull-out load as well as variation of pull-out capacity with *L*/*D* ratio and diameter of GAP. Table 2 shows the pull-out capacities of GAPs of different diameters and *L*/*D* ratio. It is evident from both table and graph that pull-out capacity increases with increase in *L*/*D* ratio but this effect is not significant after *L*/*D* ratio between 7.5 and 10. It can be observed from Table 2 that the increase in pull-out load at failure for 50 mm diameter GAP increases by a fair degree when *L*/*D* ratio increases from 5 to 7.5 (23.08% increase), but the change is not appreciable when *L*/*D* ratio increases from 7.5 to 10 (14.38%). A similar trend was observed for GAP of diameter 75 mm where the increase in pull-out load at failure increases by 19.63% when *L*/*D* ratio increases from 5 to 7.5. But the percentage increase falls to 9% when *L*/*D* ratio increases from 7.5 to 10. This behavior may be attributed to the fact that size of bulge in GAP is a function of diameter of GAP and not length. So, further increase in length does not increase the resistance due to bulging. It is also evident that increase in diameter also leads to increase in pull-out capacity. Table 2 indicates increase in pull-out load by 107.7% when diameter increases from 50 mm to 75 mm for *L*/*D* ratio of 5. For same *L*/*D* ratio, the percentage increase is 29.62% when diameter increases from 75 mm to 100 mm.

Table 2. Pull-out capacity of single GAPs

L/*D* →	5	7.5	10
Diameter (mm) ↓	Pull-out load at Failure (kg)		
50	130	160	183
75	270	323	352
100	350	418	–

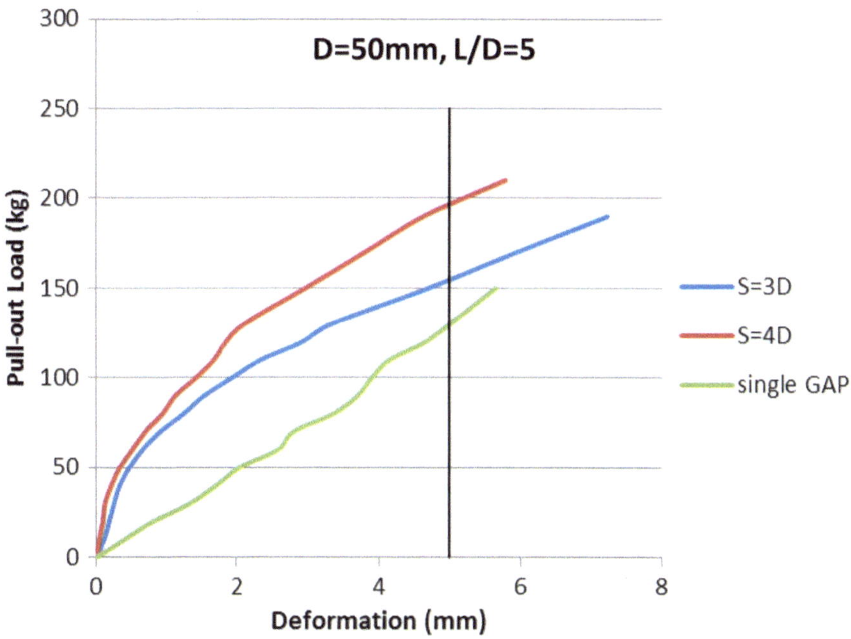

Fig. 6. Load vs Deformation curves for 50 mm diameter single GAP and group of 2 GAPs

Similarly, for *L*/*D* ratio of 7.5, the percentage increase is 101.87% when diameter increases from 50 mm to 75 mm while the increase is 29.41% for increase in diameter from 75 mm to 100 mm. This can happen because of increase in bulge diameter and length with increase in diameter of GAP. The load-displacement response of group of two piles indicate greater stiffness as compared to that of single GAP (Fig. 6). The failure load of group of 2 GAPs for spacing of 3*D* and 4*D* were found to be 155 kg and 198 kg respectively as compared to 130 kg for single pile of same dimension as shown in Fig. 4.

5 Conclusions

Following conclusions could be drawn from the experimental study conducted on GAPs in clay:

1. Pull-out capacity of GAPs increases with *L*/*D* ratio as the frictional resistance increases with increase in surface area.
2. The rate of increase in pull-out capacity of GAP with *L*/*D* ratio decreases between *L*/*D* ratio of 7.5 to 10.
3. Increase in diameter of GAP increases the pull-out capacity of GAP as increase in surface area increases frictional resistance.
4. Stiffness of soil increases for a group of 2 GAPs compared to single GAP of same dimensions.

5. Pull-out capacity of group of 2 GAPs is increased as compared to single GAP.
6. Pull-out capacity of group of 2 GAPs at spacing of 4*D* is more than GAPs at spacing of 3*D*.

Acknowledgments. All the authors want to express their deep sense of gratitude towards CSIR-CBRI for sponsoring the research project. Authors also want to acknowledge the help of Dr. Pradeep Kumar, Scientist E CSIR-CBRI for extending kind help in providing some experimental accessories.

References

IS:2720 (Part 2)-1973: Methods of Test for Soils-Determination of Water Content. Bureau of Indian Standards, New Delhi

IS:2720 (Part 3)-1980: Methods of Test for Soils-Determination of Specific Gravity. Bureau of Indian Standards, New Delhi

IS:2720 (Part 5)-1985: Methods of Test for Soils-Determination of Liquid and Plastic Limit. Bureau of Indian Standards, New Delhi

IS:2720 (Part 7)-1974: Methods of Test for Soils-Determination of Water Content-Dry Density Relation Using Light Compaction. Bureau of Indian Standards, New Delhi

IS:2720 (Part 10)-1991: Methods of Test for Soils-Determination of Unconfined Compressive Strength. Bureau of Indian Standards, New Delhi

Krishna, P.H., Murty, V.R.: Pull-out capacity of granular anchor piles in expansive soils. IOSR J. Mech. Civ. Eng. (IOSR-JMCE) **5**(1), 24–31 (2013)

Kumar, M.: An Experimental Study on Granular Anchor Piles, M. Tech Dissertation submitted to Indian Institute of Technology Roorkee (2017)

Phanikumar, B.R., Srirama Rao, A.: Granular pile anchors in expansive soils. In: Proceedings of a National Seminar on Partially Saturated Soils and Expansive Soils, Kakinada, India, pp. 15–22 (1996)

Phanikumar, B.R., Ramachandra Rao, N.: Increasing pull-out capacity of granular pile anchors in reactive soils using base geosynthetics. Can. Geotech. J. **37**(4), 870–881 (2000)

Rao, A.S., Phanikumar, B.R., Babu, R.D., Suresh, K.S.: Pull-out behaviour of granular pile-anchors in expansive clay beds in-situ. ASCE J. Geotech. Geoenviron. Eng. **133**(5), 531–538 (2007)

Generalized Solutions for Lateral Bearing Behavior of Large Diameter Monopile Foundation for Offshore Wind Turbine Considering Double Additional Moment Effects

Ming-Xing Zhu[1(✉)], Hong-Qian Lu[1], Guo-Liang Dai[2], Wei-Ming Gong[2], and Lei Wang[1]

[1] China Energy Engineering Group Jiangsu Electric Power Design Institute (JSPDI) Co., LTD, Nanjing, Jiangsu, China
phd_mxingzhu19856l@vip.163.com, {luhongqian, wanglei}@jspdi.com.cn

[2] School of Civil Engineering, Southeast University, Nanjing, Jiangsu, China
seudaigl@126.com, wmgong@seu.edu.cn

Abstract. To examine the influence of double additional moment effects on lateral response of large diameter monopile foundation, this work firstly establishes analytical constitutive relation of additional moment versus pile slope (M_s-θ) induced by vertical shaft resistance and constitutive model of additional moment to slope (M_g-θ) subjected to the gravity of soil core inside steel pipe pile, respectively. Furthermore, the analytical model and differential expressions of pile element are developed with consideration of the presented double additional moment effects. By means of transfer matrix method and Laplace transformation, this work deduces the analytical solutions of transfer matrix coefficients for pile in elastic and plastic state and obtains the responses of larger diameter monopile foundation. The proposed methodology is validated via comparing its predictions with the literatures results. Success of proposed methodology is illustrated, which implies that the proposed method can be used to evaluate the pile response under actions of vertical shaft resistance and axial force along pile shaft effectively. Finally, parametric study of additional moments is performed and the results reveal that, for the influence of vertical shaft resistance, decreasing ratio of pile responses will increase with increasing limit skin friction τ_u, pile diameter d or decreasing equivalent critical displacement $s_{u,eq}$, in which d and $s_{u,eq}$ have more notable influence. In other word, the influence of vertical shaft resistance cannot be ignored for monopile foundation with large diameter and for stiff soil mass around pile. Although factors, including stiffness, gravity, length of soil core, diameter and embedded length of pile, as well as the stiffness of soil mass around pile, have impact on the lateral bearing capacity of pipe pile when considering soil core effect, the influence of additional moment M_g due to soil core is too small (within 0.1%) to be neglected.

Keywords: Vertical shaft resistance · Additional moment effect
Monopile foundation · Lateral bearing capacity · Transfer matrix solution

© Springer International Publishing AG, part of Springer Nature 2019
S. Shu et al. (eds.), *New Developments in Materials for Infrastructure Sustainability and the Contemporary Issues in Geo-environmental Engineering*, Sustainable Civil Infrastructures, https://doi.org/10.1007/978-3-319-95774-6_9

1 Introduction

Over the past 10 years, as a kind of efficient clean energy, more than 200 offshore wind farms have been installed in European and China coastlines. Among all the foundation types to support upper structure, the most commonly used is monopile foundation whose utilization ratio exceeds 70% (Kallehave et al. 2015) due to simple design procedure and convenient construction. Normally, the approach to evaluate the horizontal bearing capacity of monopile under lateral force and overturning moment is the semiempirical method known as the *p-y* curve method on the basis of Winkler hypothesis (Zhang 2009). However, the additional moment effect on lateral bearing characteristic of monopile will become more remarkable as increasing pile diameter, which will lead to questionable results from directly adopting *p-y* curve method (Byrne et al. 2015). Dodds and Martin (2007), Lam and Martin (1986) all argued that the force couple termed as additional moment, arising from vertical side resistance along pile shaft, will have a more pronounced effect to diminish pile deflection as increasing pile diameter. By means of variational approach, Alikhanlou (1981) and Tseng (1984) analyzed the influence of vertical shaft resistance on lateral response of caisson in clay and sandy layers, respectively. Accordingly, the comparison reveled that the calculated results are closer to measured values with consideration of additional moment effect. Based on numerical simulation, Karapiperis and Gerolymos (2014) performed influence analysis of additional moment and correspondingly established additional moment constitutive model. Moreover, Gerolymos and Gazetas (2006a), Assimaki and Gazetas (2009) proposed elastic initial stiffness of additional moment versus slope and concluded that additional moment effect on horizontal bearing capacity of caisson cannot be ignored with increasing diameter which was also confirmed by Gerolymos and Gazetas (2006b). However, studies as discussed above were all postulated soil with elastic state and additional moment had no relation to vertical side friction.

Additionally, McVay and Niraula (2004) thought ultimate side resistance of pile shaft, except for pile diameter, also has an obvious significance to additional moment. As such, McVay and Niraula (2004) conduced centrifugal model tests for socketed piles and developed expression of additional moment-slope on the basis of artificial soft-rock τ-s curve. Besides that, Ashour and Helal (2013) described the contribution of vertical skin friction to large-diameter shafts in stiff clay and obtained analytical solution for resisting moment by assuming that the distribution of vertical side friction on the shaft cross-section could be a cosine function. But, these proposed models can only be used for pile in special soil layer.

Another additional moment effect, which is induced by the weight of soil core in steel pipe pile, will gradually become remarkable as increasing pile deflection and pile diameter. However, there are few literatures available to study this problem.

To analyze the influence of double additional moments on lateral behavior of monopile foundation, this work, firstly, presents numerical expressions of additional moment considering τ-s curve with any type. Then, normalized analysis with form of M_s/M_{su}-θ/θ_{ref} is first performed to establish constitutive models of additional moment versus slope on the basis of derived initial stiffness $k_{\theta,ref}$, ultimate additional moment M_{su} and reference slope θ_{ref} in this research. Moreover, the constitutive model of additional moment-slope (i.e., $M_g - \theta$) subjected to the influence of soil core is also analytically derived. To take into account the double additional moment effects, this paper obtains the semi-analytical solutions for laterally loaded monopile by means of transfer matrix approach (Nogami and Paulson 1985; Zhang et al. 2013; Zhu et al., 2017). Agreement between calculated results and test data is quite good, which implies that the influence of additional moment M_s induced by vertical skin resistance should not be neglected, especially for pile with large diameter or in stiffer soil layer. While, the additional moment M_g effect due to soil core could be neglected in the design procedure.

2 Numerical Solution for Additional Moment Due to T-S Curve

As shown in Fig. 1, Ashour and Helal (2013) assumed that the distribution of vertical shaft resistance only exist in the passive side of pile shaft, which is also validated by in situ test (Briaud et al. 1984) and numerical simulation (Li 2012). However, it is different to obtain analytical solution for additional moment due to uncertainty of τ-s curve model and nonlinear distribution of skin friction around pile. Thus, this work adopts dispersion sum technology to deduce numerical solution for additional moment depending on McVay's research (2004). The profile view of pile shaft at arbitrary depth are shown in Fig. 2, where arc AC and BC are discretized into n_β slices. As a result, each arc length l_i after dividing can be expressed as:

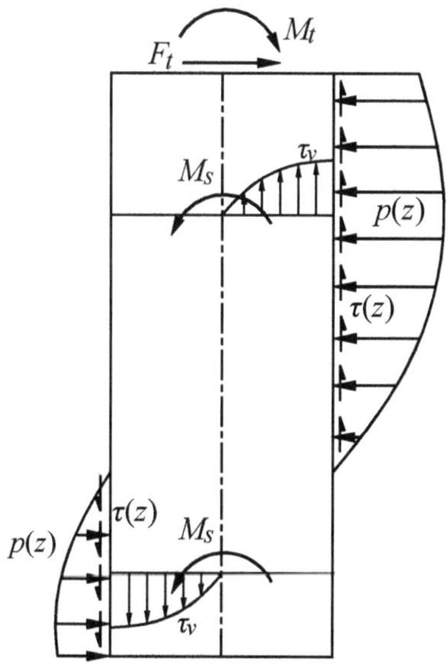

Fig. 1. Sketch of distribution of vertical shaft resistance

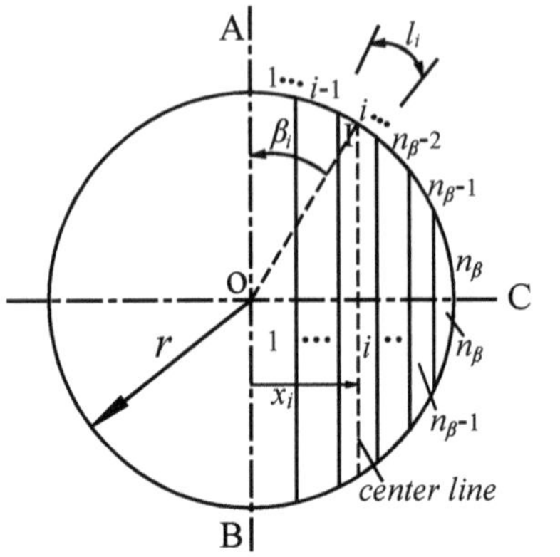

Fig. 2. Discretization of pile cross-section

$$l_i = \frac{\pi r}{2n_\beta} \tag{1}$$

where r refers to pile radius.

Accordingly, the angle between line OA and OI (i.e., the center point of ith arc slice to origin) is defined as β_i, which is given by:

$$\beta_i = \frac{(2i-1)\pi}{4n_\beta} \tag{2}$$

If pile section has a slope value θ, the average axial displacement between pile and soil will become as $s_i = x_i \times \theta$, where x_i ($= r\sin\beta_i$) represents the distance from center of cross-section to the center of ith arc slice. The value of vertical skin resistance τ_i for ith arc slice can be obtained easily for adopting any τ-s curve model. Thus, the total moment could be found by summing the moments acting on all arc slices:

$$M_s = \sum_{i=1}^{n_\beta} 2\tau_i l_i x_i = \frac{\pi r^2}{n_\beta} \sum_{i=1}^{n_\beta} \left[\tau_i \sin\left(\frac{(2i-1)\pi}{4n_\beta} \right) \right] \tag{3}$$

in which, M_s, named as additional moment, represents the total moment per unit length for arbitrary pile cross-section (McVay and Niraula 2004; Assimaki and Gazetas 2009).

3 Additional Moment-Slope Due to Skin Resistance

A wide variety of τ-s curve models have been proposed during past decades and different τ-s curve will perform unique behavior. As a result, it is unpractical to obtain a universal analytical constitutive model of additional moment which can be suitable for any form of τ-s curve. As such, this work will concentrate on establishment procedure of the normalized analysis with form of M_s/M_{su}-θ/θ_{ref} and will derive constitutive model on the basis of τ-s curve with linear elastoplastic type and hyperbolic type.

3.1 Pure Linear Elastic τ-S Model and Pure Plastic τ-S Model

For τ-s curve with pure linear elastic characteristic, as shown in Fig. 3a, the analytical solution for additional moment M_s is expressed as:

$$M_s = 2\int_0^{\pi/2} k_\tau \theta r^3 \sin^2 \psi d\psi = \frac{\pi d^3 k_\tau}{16}\theta = k_{\theta,ref}\theta \tag{4}$$

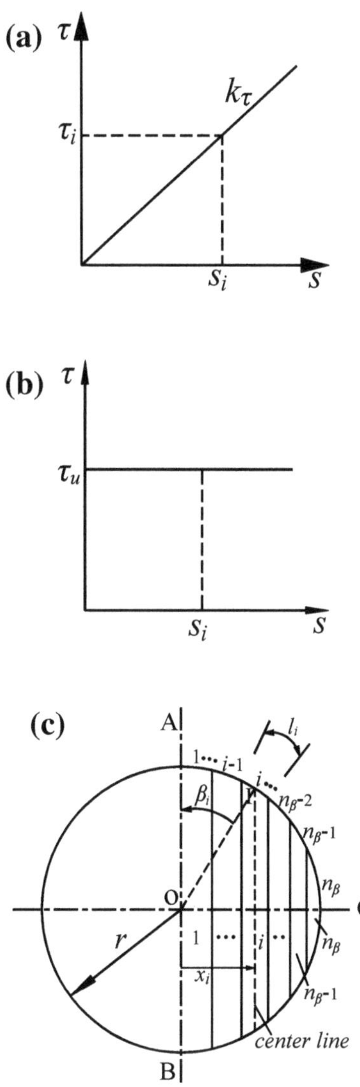

Fig. 3. Derivation of special τ-s curve models: (a) pure linear elastic model; (b) pure plastic model and **c** integral approach

in which, k_τ is the stiffness of linear elastic τ-s curve; d is pile diameter; θ refers to pile slope; $k_{\theta,\text{ref}}$ is defined as reference initial stiffness of additional moment induced by linear elastic τ-s curve, which equals to $\pi d^3 k_\tau/16$; ψ is the integration variable, as depicted in Fig. 3c.

Similarly, if τ-s curve has pure plastic behavior as shown in Fig. 3b, the expression of additional moment is described as:

$$M_{su} = 2\int_0^{\pi/2} \tau_u r^2 \sin^2\psi d\psi = \frac{\tau_u d^2}{2} \tag{5}$$

where M_{su} = ultimate additional moment; τ_{u} = ultimate shaft resistance.

By setting Eqs. (4) and (5) equal to each other, and special slope value can be obtained:

$$\theta_{ref} = \frac{M_{su}}{k_{\theta,ref}} = \frac{8}{\pi d} \times \frac{\tau_u}{k_\tau} = \frac{8 s_{u,eq}}{\pi d} \tag{6}$$

where θ_{ref} is defined as reference critical slope, which will be used as variable of normalized curve; $s_{\text{u,eq}}$ is defined as equivalent critical displacement of τ-s curve of shaft resistance, which equals to the ratio of ultimate shaft resistance τ_{u} to initial stiffness k_τ.

3.2 Linear Elastic-Plastic τ-s Model

Based on the definition of equivalent critical displacement in Eq. (6), the critical displacement s_{u} of linear elastoplastic τ-s curve (as shown in Fig. 4) is identical to equivalent critical displacement $s_{\text{u,eq}}$. As a result, the normalized curves with form of M_s/M_{su}-$\theta/\theta_{\text{ref}}$ is illustrated in Fig. 5 under the influence factors (listed in Table 1) including pile diameter d, equivalent critical displacement $s_{\text{u,eq}}$ and ultimate shaft resistance τ_{u}.

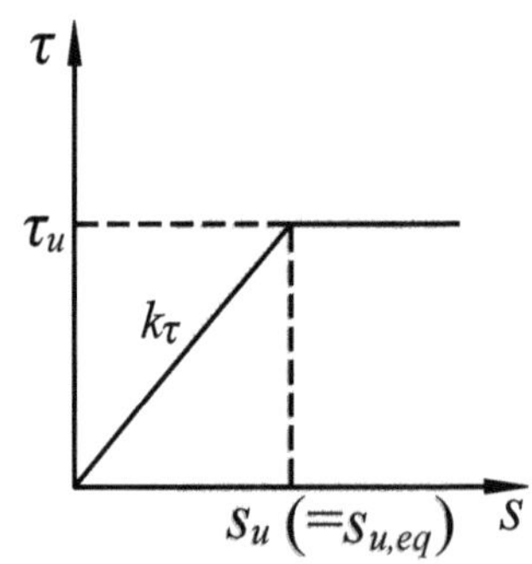

Fig. 4. Linear elastoplastic model of τ-s curve

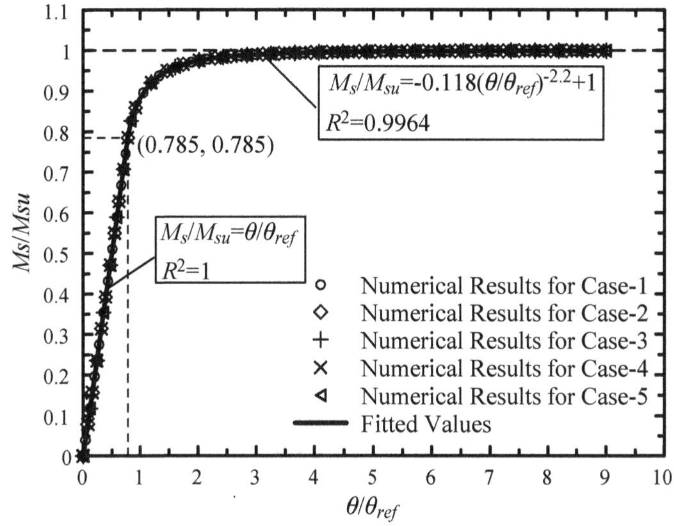

Fig. 5. Normalized curves of additional moment by using linear elastoplastic τ-s model

Table 1. Calculation cases of M_s derivation of constitutive model

No.	Pile diameter d/m	Ultimate skin resistance τ_u/kPa	Equivalent critical displacement $s_{u,eq}$/mm
Case-1	0.5	40	8
Case-2	1	40	8
Case-3	3	40	8
Case-4	1	40	16
Case-5	1	80	8

As shown in Fig. 5, all the calculated data points from numerical solution [i.e., Eq. (3)] have the same tendency. Accordingly, the analytical constitutive model of additional moment-slope could be expressed as:

$$\frac{M_s}{M_{su}} = \begin{cases} \frac{\theta}{\theta_{ref}} & \left(0 \leq \frac{\theta}{\theta_{ref}} \leq 0.785\right) \\ -0.118\left(\frac{\theta}{\theta_{ref}}\right)^{-2.2} + 1 & \left(\frac{\theta}{\theta_{ref}} > 0.785\right) \end{cases} \tag{7}$$

where ultimate additional moment M_{su} and reference critical slope θ_{ref} should be obtained by Eqs. (5) and (6), respectively. The critical value 0.785 could be obtained by $(2s_u/d)/\theta_{ref}$.

To verify the correctness of proposed additional moment expression in Eq. (7), four validation cases (as listed in Table 2) are performed considering influence parameters d, τ_u and $s_{u,eq}$. From the comparison of additional moment-slope (M_s-θ) as depicted in

Fig. 6, the analytical result from Eq. (7) is identical to that of numerical approach from Eq. (3) for each case, which implies that the proposed constitutive model for linear elastoplastic τ-s curve could be adopted to calculate additional moment M_s efficiently.

Table 2. Validation cases of additional moment M_s

No.	Pile diameter d/m	Ultimate skin resistance τ_u/kPa	Equivalent critical displacement $s_{u,eq}$/mm
Case-1	1.47	31.6	5.9
Case-2	2.53	31.6	5.9
Case-3	2.53	31.6	13.8
Case-4	2.53	23.7	5.9

Additionally, by comparing Case-1 with Case-2 in Fig. 6, it is could be found that if pile diameter d increase n-fold, the additional moment M_s in linear elastic state [i.e., the first formula of Eq. (7)] will increase n^3 times and ultimate additional moment M_{su} will rise n^2 times, which indicates that diameter d has a significant influence on additional moment. Moreover, M_s will enhance at the same ratio to the increased range of ultimate shaft resistance τ_u due to ultimate additional moment M_{su} only be impacted by τ_u as shown in curves of Case-2 and Case-4. Based on the comparisons of Case-2 and Case-3, it is evident that the smaller equivalent critical displacement $s_{u,eq}$ be set, the bigger value of M_s will be reached in the elastic stage of M_s-θ curve. However, the influence of $s_{u,eq}$ will fade away as increasing pile slope θ.

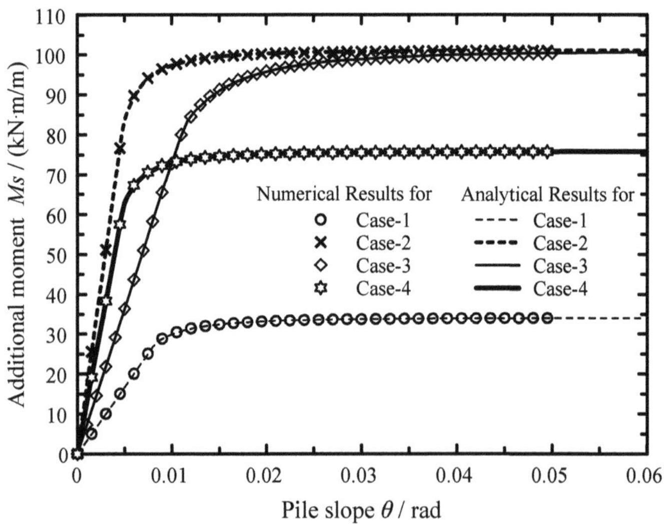

Fig. 6. Validation cases of M_s-θ curves for linear elastic-plastic model

3.3 Hyperbolic τ-s Model

Figure 7 is a typical shaft resistance τ-s model with hyperbolic type, and corresponding expression could be described as:

$$\tau = \frac{s}{\frac{1}{k_{ini}} + \frac{s}{\tau_u}} \tag{8}$$

where k_{ini} and τ_{u} are the initial stiffness and limit value of hyperbolic model.

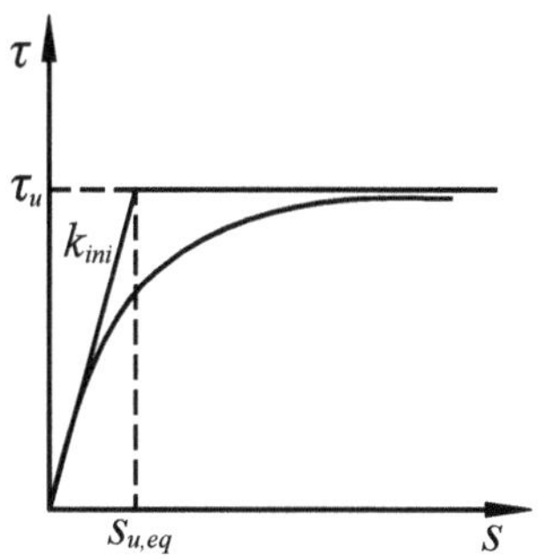

Fig. 7. Hyperbolic τ-s curve model

Assuming that the reference critical slope θ_{ref} equals to $\tau_{\text{u}}/k_{\text{ini}}$ based on the definition in Eq. (6), the comparisons of M_s/M_{su}-$\theta/\theta_{\text{ref}}$ curves by means of numerical solution [i.e., Equation (3)] are illustrated in Fig. 8. The numerical results show that all the cases have the same varying tendency under different influence factors listed in Table 1. As a result, the analytical model of additional moment-slope can be described as:

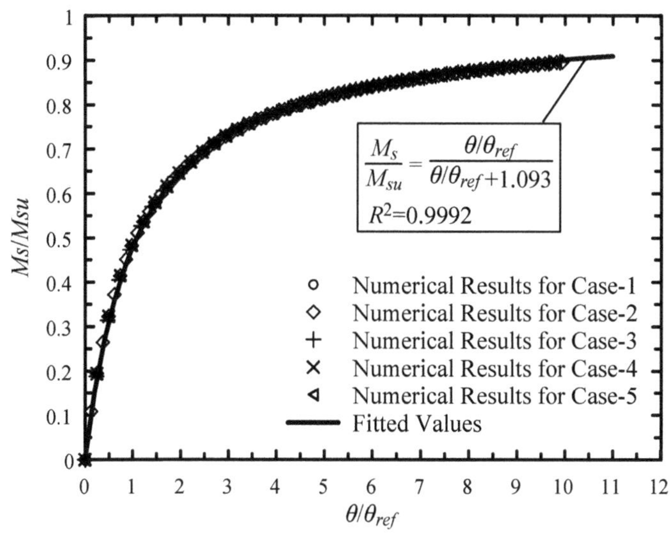

Fig. 8. Normalized curves of additional moment by using hyperbolic τ-s curve model

$$\frac{M_s}{M_{su}} = \frac{\theta/\theta_{ref}}{\theta/\theta_{ref} + 1.093} \tag{9}$$

Based on the validation cases listed in Table 2, the results of M_s-θ curve from Eq. (3) and Eq. (9) prove the correctness of proposed constitutive model for hyperbolic τ-s model, which is shown in Fig. 9. Meanwhile, the influence rule of diameter d, ultimate shaft resistance τ_u and equivalent critical displacement $s_{u,eq}$ on additional moment M_s is identical to that of comparisons are illustrated in Fig. 6. In other words, M_s will increase as increasing parameters d and τ_u or with decreasing $s_{u,eq}$.

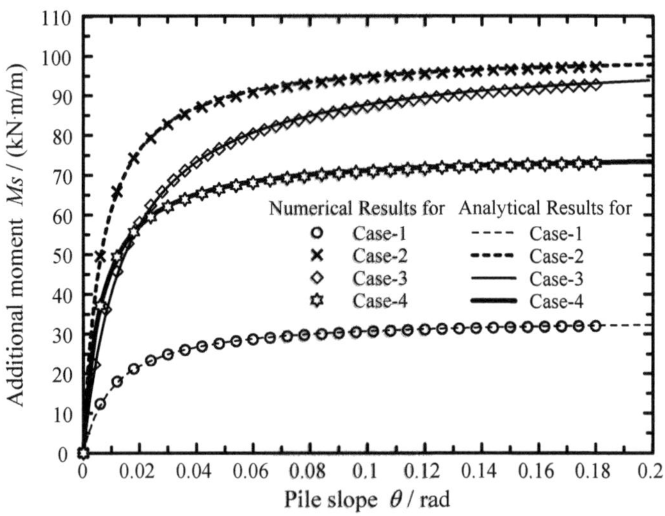

Fig. 9. Validation cases of M_s-θ curves for hyperbolic τ-s model

4 Additional Moment-Slope Induced by Soil Core

As shown in Fig. 10, the additional moment effect induced by the weight of soil core in steel pipe pile will gradually become remarkable as increasing pile deflection and pile diameter. In addition, the bending stiffness of steel pile commonly used should be the complex stiffness which is consist of steel and soil. However, there are few literatures to study this problem.

Based on the conclusion that there is no friction between soil core and inner wall of pipe pile under horizontal force (Li 2012), this work postulates that the additional moment is only resulted from the weight of soil core. According to the element model (as shown in Fig. 11), the equilibrium equation of bending moment can be expressed as:

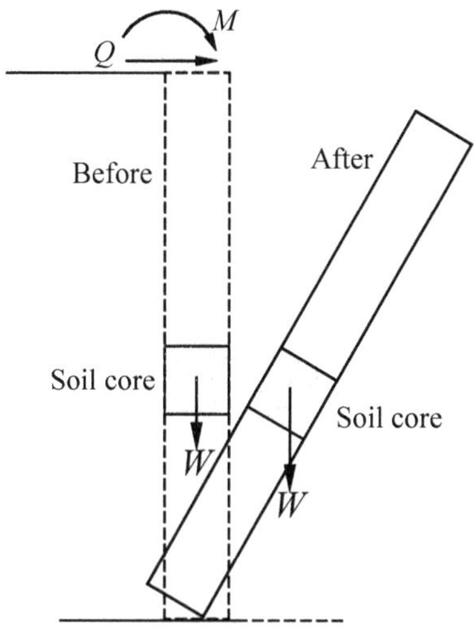

Fig. 10. Sketch of weight effect of soil core inside pipe pile

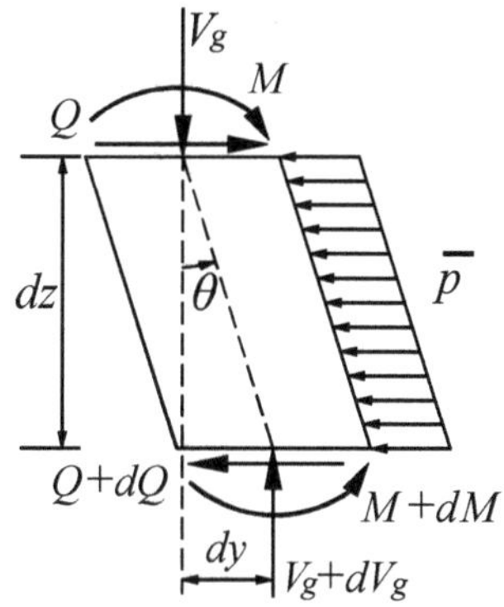

Fig. 11. Analysis of element of soil core

$$\frac{dM}{dz} = Q - V_g \frac{dy}{dz} \tag{10}$$

where y, M and Q represent the deflection, moment and shear force, respectively; V_g is the vertical force induced by weight of soil core, which will produce the additional moment with value of $V_g dy$. By setting $k_g = V_g$ and $\theta = dy/dz$, the additional moment can be described as:

$$M_g = k_g \theta \tag{11}$$

in which M_g = additional moment due to soil core effect, which represents the total moment per unit length for arbitrary pile cross-section; k_g = stiffness of M_g-θ curve, which is calculated by:

$$k_g(z_i) = \gamma_{si} A_{si} z_i + \sum_{j=1}^{i-1} \gamma_{sj} A_{sj} H_{sj} \tag{12}$$

where γ_{si}, A_{si} and H_{si} are the unit weight, cross-sectional area and length of *i*th soil core; z_i is the depth calculated from the top of *i*th soil core.

5 Application of Additional Moment-Slope Models

As discussed above, additional moment M_s is not only affected by τ-s curve, but also influenced by pile diameter. However, there are few literature available to study additional moment M_s effect and corresponding application. Byrne et al. (2015) successfully performed new design methods for large-diameter monopile considering M_s effect. But the work of Byrne et al. (2015) did not show how to evaluate additional moment M_s and how to solve pile response. For the influence of M_g arising from soil core on lateral response of pile, only numerical study (Li et al. 2017) could be found.

Assuming that a pile is discretized into many slices and corresponding length, diameter, bending stiffness for *i*th pile slice are h_i, d_i, and EI_i, respectively. Attention should be attached is that EI_i should equals to EI_{pi}+ EI_{si} if considering the soil core influence, EI_{pi} and EI_{si} are the bending stiffness of pipe pile and soil core. During the derivation procedure, positive directions of M_s and M_g are clockwise and anticlockwise, and other assumptions are the same with Zhu's work (2017).

5.1 Transfer Matrix Coefficient for Pile-Soil Interaction in Elastic State

In order to derive the transfer matrix coefficient by transfer matrix method (Nogami and Paulson 1985; Zhang et al. 2013; Zhu et al. 2017), the force status of element as shown in Fig. 12 could be written as the following differential equation set:

$$\begin{cases} \dfrac{dy_i}{dz_i} = \theta_i \\ \dfrac{d\theta_i}{dz_i} = \dfrac{M_i}{EI_i} \\ \dfrac{dM_i}{dz_i} = Q_i + M_{s,ave(i)} - M_{g,ave(i)} \\ \dfrac{dQ_i}{dz_i} = -k_{ave(i)} y_i \end{cases} \tag{13}$$

in which y_i, z_i, θ_i, M_i and Q_i are the pile deflection, depth, slope, bending moment and shear force of the bottom of *i*th pile slice; As shown in Fig. 12, $k_{ave(i)}$, $M_{s,ave(i)}$ and

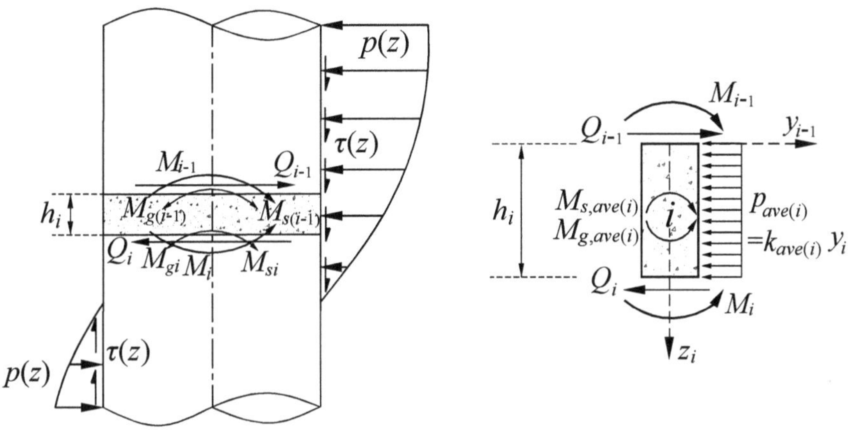

Fig. 12. Analytical model of discretization of pile shaft

$M_{\mathrm{g,ave}(i)}$ are the average subgrade reaction moduli, additional moment induced by shaft resistance and soil core for ith pile slice, which could be described as:

$$\begin{cases} k_{ave(i)} = \left(k_{(i-1)} + k_{(i)}\right)/2 \\ M_{s,ave(i)} = \left(M_{s(i-1)} + M_{s(i)}\right)/2 \\ M_{g,ave(i)} = \left(M_{g(i-1)} + M_{g(i)}\right)/2 \end{cases} \tag{14}$$

where the subscripts $i - 1$ and i for k, M_{s} and M_{g} refer to the top and bottom of ith pile slice.

As mentioned above, additional moments induced by shaft resistance and soil core (i.e., M_s and M_g) are associated with slope θ, which makes it difficult to obtain analytical transfer matrix coefficient. To simplify the deriving process, this work assumes that $M_{\mathrm{s,ave}(i)}$ and $M_{\mathrm{g,ave}(i)}$ are calculated by the slope value of previous iteration results. As such, for the first time to iterate, $M_{\mathrm{s,ave}(i)}$ and $M_{\mathrm{g,ave}(i)}$ should be set to be zero. Thus, $M_{\mathrm{s,ave}(i)}$ and $M_{\mathrm{g,ave}(i)}$ could be regarded as constants and corresponding transfer matrix coefficient on the basis of transfer matrix method and Laplace transformation (Zhu et al. 2017) can be expressed as:

$$\mathbf{U}_i = \begin{bmatrix} \varphi_1 & \dfrac{\varphi_2}{2\beta_i} & \dfrac{2\varphi_3\beta_i^2}{k_{ave(i)}} & \dfrac{-\beta_i\varphi_4}{k_{ave(i)}} & \chi_1 \\ \beta_i\varphi_4 & \varphi_1 & \dfrac{2\varphi_2\beta_i^3}{k_{ave(i)}} & \dfrac{2\varphi_3\beta_i^2}{k_{ave(i)}} & \chi_2 \\ \dfrac{-k_{ave(i)}\varphi_3}{2\beta_i^2} & \dfrac{k_{ave(i)}\varphi_4}{4\beta_i^3} & \varphi_1 & \dfrac{\varphi_2}{2\beta_i} & \chi_3 \\ \dfrac{-k_{ave(i)}\varphi_2}{2\beta_i} & \dfrac{-k_{ave(i)}\varphi_3}{2\beta_i^2} & \beta_i\varphi_4 & \varphi_1 & \chi_4 \\ 0 & 0 & 0 & 0 & 1 \end{bmatrix} \tag{15}$$

in which $\beta_i = [k_{\mathrm{ave}(i)}/(4EI_i)]^{0.25}$; χ_1, χ_2, χ_3, and χ_4 can be given by:

$$\begin{cases} \chi_1 = -\beta_i \varphi_4 \left(M_{s,ave(i)} - M_{g,ave(i)}\right) / k_{ave(i)} \\ \chi_2 = 2\beta_i^2 \varphi_3 \left(M_{s,ave(i)} - M_{g,ave(i)}\right) / k_{ave(i)} \\ \chi_3 = \varphi_2 \left(M_{s,ave(i)} - M_{g,ave(i)}\right) / (2\beta_i) \\ \chi_4 = (\varphi_1 - 1)\left(M_{s,ave(i)} - M_{g,ave(i)}\right) \end{cases} \tag{16}$$

In the above two equations, the parameters φ_1, φ_2, φ_3 and φ_4 can be expressed as:

$$\begin{cases} \varphi_1 = \cos(\beta_i h_i)\cosh(\beta_i h_i) \\ \varphi_2 = \cos(\beta_i h_i)\sinh(\beta_i h_i) + \sin(\beta_i h_i)\cosh(\beta_i h_i) \\ \varphi_3 = \sin(\beta_i h_i)\sinh(\beta_i h_i) \\ \varphi_4 = \cos(\beta_i h_i)\sinh(\beta_i h_i) - \sin(\beta_i h_i)\cosh(\beta_i h_i) \end{cases} \tag{17}$$

5.2 Transfer Matrix Coefficient for Pile-Soil Interaction in Plastic State

As shown in Fig. 12, if the pile-soil interaction reaches to the plastic state (i.e., $p_{\mathrm{ave}(i)} = p_{\mathrm{u,ave}(i)}$), the differential equation set of Eq. (13) could be re-written as:

$$\begin{cases} \dfrac{dy_i}{dz_i} = \theta_i \\ \dfrac{d\theta_i}{dz_i} = \dfrac{M_i}{EI_i} \\ \dfrac{dM_i}{dz_i} = Q_i + M_{s,ave(i)} - M_{g,ave(i)} \\ \dfrac{dQ_i}{dz_i} = -p_{u,ave(i)} \end{cases} \tag{18}$$

where $p_{\mathrm{u,ave}(i)}$ is the average value of ultimate soil resistance of top and bottom of corresponding ith slice, which is described as

$$p_{u,ave(i)} = \left(p_{u(i-1)} + p_{u(i)}\right)/2 \tag{19}$$

By the same solving approach mentioned above, the transfer matrix coefficient for pile-soil interaction in plastic state can be described as:

$$\mathbf{U}_i = \begin{bmatrix} 1 & h_i & \dfrac{h_i^2}{2EI_i} & \dfrac{h_i^3}{6EI_i} & \dfrac{4h_i^3\left(M_{s,ave(i)} - M_{g,ave(i)}\right) - p_{u,ave(i)} h_i^4}{24EI_i} \\ 0 & 1 & \dfrac{h_i}{EI_i} & \dfrac{h_i^2}{2EI_i} & \dfrac{3h_i^2\left(M_{s,ave(i)} - M_{g,ave(i)}\right) - p_{u,ave(i)} h_i^3}{6EI_i} \\ 0 & 0 & 1 & h_i & \dfrac{2h_i\left(M_{s,ave(i)} - M_{g,ave(i)}\right) - p_{u,ave(i)} h_i^2}{2} \\ 0 & 0 & 0 & 1 & -p_{u,ave(i)} h_i \\ 0 & 0 & 0 & 0 & 1 \end{bmatrix} \tag{20}$$

5.3 Solution for Laterally Loaded Pile Considering Double Additional Moment Effect

Combined Eqs. (15) and (20) with transfer matrix principle (Zhang et al. 2013; Zhu et al. 2017), the transfer function for the whole pile can be obtained by:

$$\boldsymbol{S}_n = \boldsymbol{U}_n\boldsymbol{U}_{n-1}\ldots\boldsymbol{U}_i\boldsymbol{U}_{i-1}\ldots\boldsymbol{U}_1\boldsymbol{S}_0 \tag{21}$$

in which $\boldsymbol{S}_n = [y_n, \theta_n, M_n, Q_n, 1]^{\mathrm{T}}$ and $\boldsymbol{S}_0 = [y_0, \theta_0, M_0, Q_0, 1]^{\mathrm{T}}$ represent the pile deflection, slope, bending moment and shear force of pile end and pile head, respectively. $\boldsymbol{U}_i$ is the elastic transfer matrix coefficient calculated by Eq. (15) or the plastic transfer matrix coefficient given by Eq. (20), which is depended on the state of soil-pile interaction.

To solve Eq. (21), it is necessary to know the boundary conditions for pile head and pile tip which are listed below:

$$\begin{cases} M_0 = M_t \quad and \quad Q_0 = F_t\,(head\text{−}free) \\ \theta_0 = 0 \quad and \quad Q_0 = F_t\,(head\text{−}fixed) \end{cases} \tag{22}$$

and

$$\begin{cases} M_n = 0 \quad and \quad Q_n = 0\,(tip\text{−}free) \\ y_n = 0 \quad and \quad \theta_n = 0\,(tip\text{−}fixed) \end{cases} \tag{23}$$

where M_{t} and F_{t} are the overturning moment and lateral force acting on pile head.

Substituting one of the pile head boundary conditions and one of the pile base boundary conditions into Eq. (21), the response of $\boldsymbol{S}_0$ could be obtained. Thus, the responses of pile at any depth can be expressed as:

$$\boldsymbol{S}_i = \boldsymbol{U}_i\boldsymbol{U}_{i-1}\ldots\boldsymbol{U}_1\boldsymbol{S}_0 \tag{24}$$

in which $\boldsymbol{S}_i = [y_i, \theta_i, M_i, Q_i, 1]^{\mathrm{T}}$ represents the pile responses of pile deflection, slope, bending moment and shear force of *i*th pile slice, respectively.

6 Case Studies

6.1 Validation of Additional Moment Induced by Shaft Resistance

Bhushan et al. (1979) reported a in situ test of laterally loaded drilled shaft in stiff clay soil with pile length of 4.995 m, pile diameter of 1.22 m and bending stiffness of 2.25×10^6 kN m^2. The pile head is subjected to a horizontal force F_{t} and moment $0.23F_{\mathrm{t}}$, respectively. The boundary conditions of pile head and bottom are free. The *p-y* curve referring to the pile-soil interaction is recommended by API clay model (1991). For this *p-y* curve, undrained shear strength C_{u} = 227 kPa, unit weight γ = 18.8 kN/m^3, special strain value ε_{50} = 0.0072. In the appendix of API (1991), the ultimate skin friction τ_{u} is suggested to be $0.5C_{\mathrm{u}}$ if the value of undrained shear strength C_{u} exceeds 72 kPa, which leads to τ_{u} has a value of 113.5 kPa. For the critical displacement

s_u, Mokwa (1999) argued that s_u should range from 2.54 to 7.62 mm. While, Juirnarongrit and Ashford (2006) thought that such critical displacement s_u is independent of pile length, diameter, and the value of s_u is recommended from 5 to 8 mm. In this case study, s_u is adopted by 8 mm from the perspective of safety. To evaluate the influence of vertical shaft resistance, τ-s models with linear elastoplastic type and hyperbolic type are taken into account, as shown in Fig. 13a, and corresponding M_s-θ curves are illustrated in Fig. 13b. As a result, by setting $M_g = 0$, the curves of pile head force versus pile deflection are shown in Fig. 14 based on the solutions mentioned above.

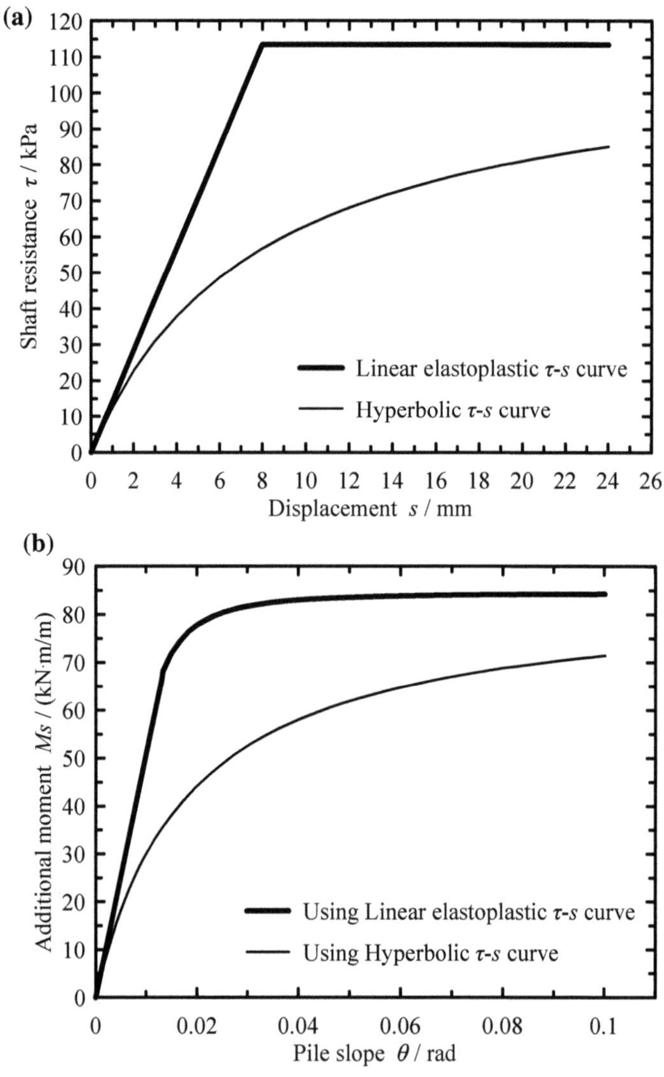

Fig. 13. Sketch of cases for different shaft curves: (a) cases of τ-s curves and (b) cases of corresponding M_s-θ curves

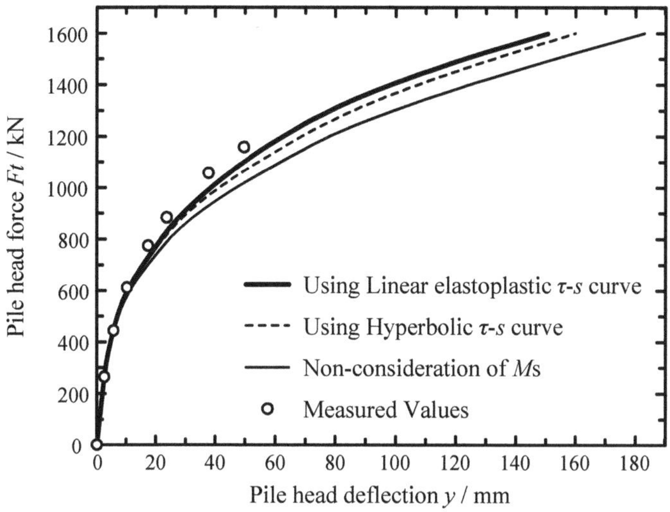

Fig. 14. Comparisons of curves of pile head force F_t vs deflection y under different cases

From the comparisons, it is evident that pile deflections provided by presented approach considering additional moment M_s effect are more closer to the measured values, which not only implies that the influence of vertical shaft resistance cannot be ignored, but also ensures the validity of the solving solution. Moreover, the difference of pile deflections affected by M_s of elastoplastic τ-s model and hyperbolic model is very small, despite the two M_s-θ curves (as shown in Fig. 13b) have obviously different behaviors.

6.2 Validation of Additional Moment Induced by Soil Core

In view of no analytical studies on the additional moment M_g induced by soil core, this work will be verified by the case of vertically and laterally loaded pile which is similar to the influence of soil core. A series tests of model pile with length 0.8 m, diameter 0.016 m, bending stiffness 0.189 kN m^2 was performed by Wu and Zhao (2001). Zhu et al. (2017) also studied this case with the following p-y model:

$$p = \begin{cases} n_{hs}(z_0 + z)^m y & (y < y_u) \\ n_{hs}(z_0 + z)^m y_u & (y \geq y_u) \end{cases} \tag{25}$$

where n_{hs} (=5.84 × 10^3 kN/m^3) is the stiffness coefficient of p-y curve; y_u = 3d/80 is the critical displacement; z_0 (= 0.3 m) is the equivalent depth.

Zhu et al. (2017) assumed that the axial force along pile shaft could be regarded as a constant, which equals to the vertical force V_t applied at pile head. Depending on Eq. (11), V_t can be treated as k_g, and corresponding comparisons of bending moment-pile curves are shown in Fig. 15. The predicted moments from proposed approach in this search are not only identical to those of Zhu's method, but also differ little from model tests, which ensures the correctness of deduction.

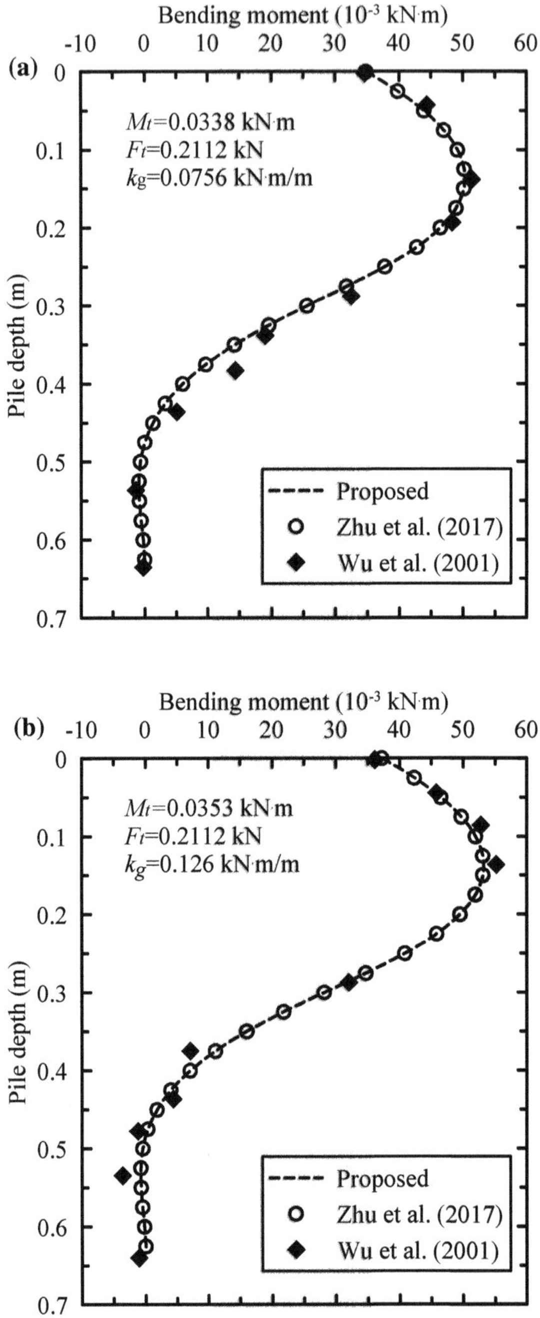

Fig. 15. Comparisons of moment between calculated and measured results for (a) the first pile and (b) the second pile

7 Parametric Analysis

To study the additional moment M_s and M_g effects on the bearing behavior of laterally loaded pile, a 30 m-long, 5 m-diameter and 4.88×10^8 kN m^2-bending stiffness steel pipe pile is assumed to be embedded into sandy soil with inner friction angle $\varphi = 35°$ and effective unit weight $\gamma = 10.3$ kN/m^3. The API sand p-y model (1991) is utilized to simulate the pile-soil interaction. For this sand *p-y* curve, initial stiffness of soil spring $k = 2.2 \times 10^4$ kN/m^3, C_1, C_2 and C_3 are 2.97, 3.42 and 53.79, respectively. The forces applied at pile head are lateral force F_t and moment $M_t = 20F_t$.

7.1 Influence of Parameters on Additional Moment M_s

Influence of Skin Resistance τ-s Curves on Lateral Behavior

To analyze the influence of equivalent critical displacement $s_{u,eq}$ and ultimate shaft resistance τ_u of linear elastoplastic τ-s model on the lateral behavior of large-diameter, three study cases are listed in Table 3, and corresponding M_s-θ curves are illustrated in Fig. 16.

Table 3. Cases of parametric analysis for considering different τ-s models

No.	Ultimate skin resistance τ_u/kPa	Equivalent critical displacement $s_{u,eq}$/m
Case-1	67	0.05
Case-2	96	0.05
Case-3	96	0.01

Notes The value of $s_{u,eq}$ with 0.05 m is calculated by 0.01d according to API (1991)

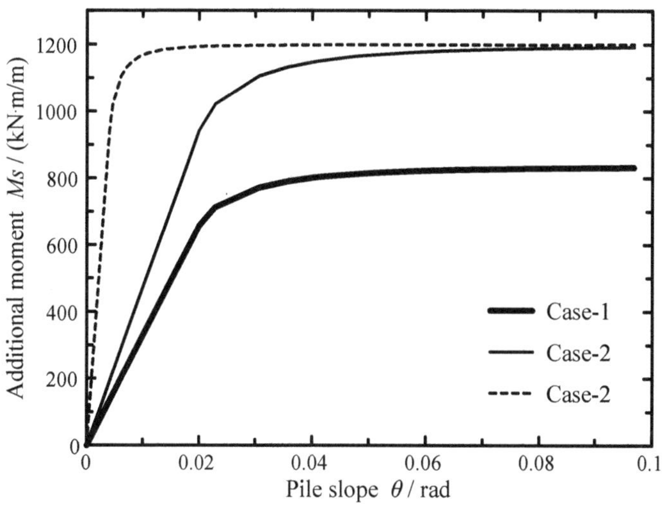

Fig. 16. M_s-θ curves of different τ-s models

By comparing Case-1 and Case-2, it is quite clear that the extent of reduction of pile deflection in mud-line (Fig. 17a) and maximum bending moment (Fig. 17b) is proportional to the increasing ratio of ultimate shaft resistance τ_u, which means that the larger τ_u will be set, the smaller pile responses will be obtained. Notes that the F_{tmax} (=37145 kN) represents the ultimate bearing capacity of laterally loaded pile without considering additional moment effect.

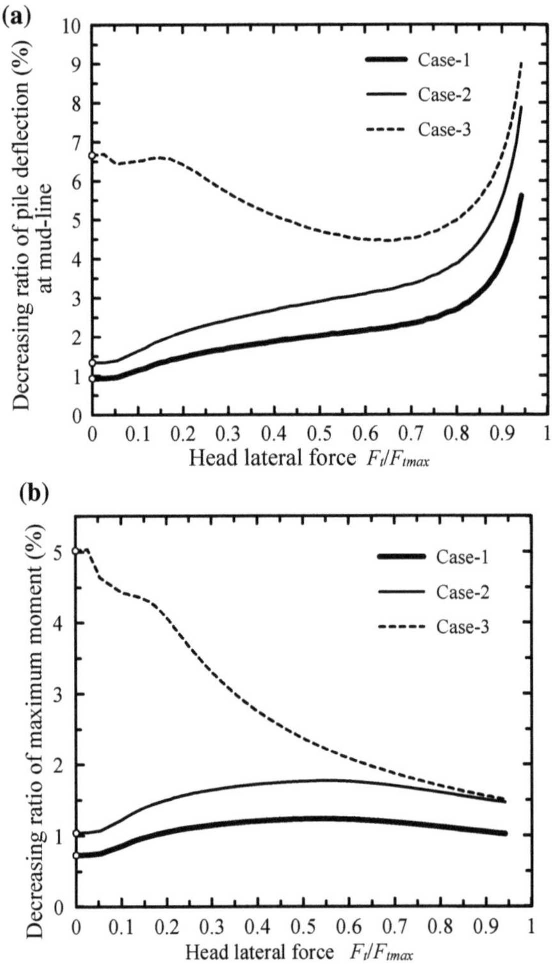

Fig. 17. Influence of τ-s models on: (a) pile deflection and (b) maximum moment

However, the decreasing levels of deflection and maximum moment for Case-3 are obviously different from Case-2, which have a higher value of decline level. Attention should be drawn to the fact that the difference between Case-2 and Case-3 will gradually disappear due to the same pile diameter and ultimate shaft resistance τ_u.

As discussed above, it could be concluded that pile responses will decrease as increasing ultimate shaft resistance τ_u or reducing equivalent critical displacement $s_{u,eq}$ which has more notable influence.

Influence of Pile Diameter on Lateral Behavior

By assuming pile diameter with 0.5, 3, 5 and 8 m, this section will perform the influence of diameter on the lateral behavior of pile foundation. Ultimate shaft resistance τ_u and equivalent critical displacement $s_{u,eq}$ are 96 kPa and 0.025 m, respectively.

From the comparisons of Fig. 18, it could be found that the additional moment M_s effect is too small to be ignored for small-diameter pile; however, the reduction degree of pile deflection in mud-line (Fig. 18a) and maximum bending moment (Fig. 18b) will dramatically increase with increasing pile diameter. Thus, additional moment M_s effect induced by vertical shaft resistance should be taken into account in the design procedure of large-diameter monopile, which is also confirmed by Byrne et al. (2015).

7.2 Influence of Parameters on Additional Moment M_g

This section is to analyze the factors, including stiffness E_s, gravity γ, length L_s of soil core, pile diameter d, pile embedded length L_b and the stiffness of soil mass (controlled by inner friction angle φ) around pile, impact on the lateral bearing capacity of pipe pile. By ignoring the influence of additional moment M_s, the soil moduli E_s of soil core inside pipe pile is 20 MPa, and other parameters are the same with those mentioned above.

The results of reduction level of pile deflection (i.e., compared with non-consideration of M_g) at mud-line for six cases (as shown in Fig. 19) reveal that the influence of additional moment M_g subjected to soil core on laterally loaded monopile is too small to be neglected (within 0.1%), despite factors of six cases can slightly affect on lateral behavior of pile. As a result, for laterally loaded pile, the design approach normally used without consideration of additional moment M_g effect is reasonable. Notes that: (1) The 'displacement control line' attached in Fig. 19 is the displacement design criterion of offshore wind turbine, which is calculated by $L_b/500$ (Sun 2016); (2) The decreasing ratio of pile deflection considering M_g effect, as marked in the vertical axis, is the result that by comparing with the non-consideration of M_g condition.

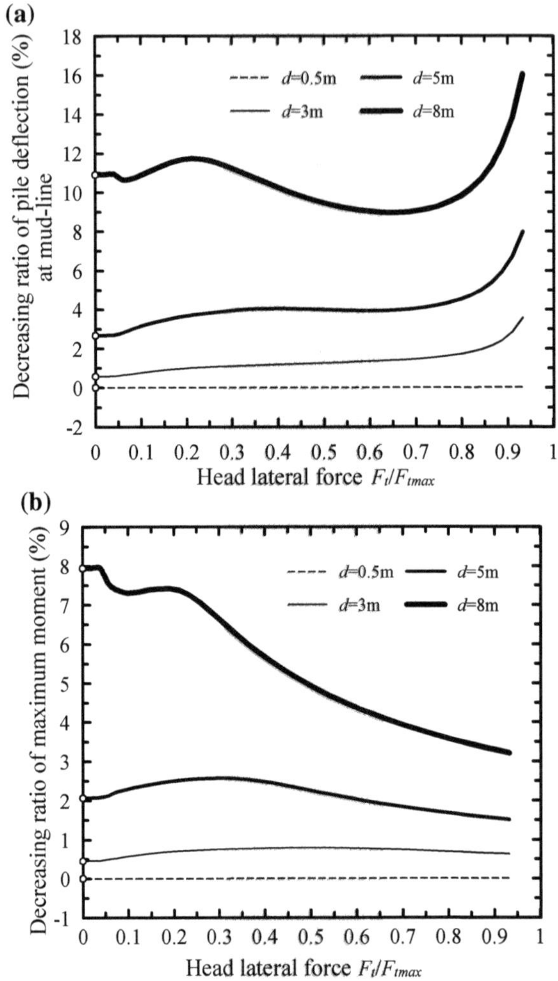

Fig. 18. Influence of pile diameters on: (a) pile deflection and (b) maximum moment

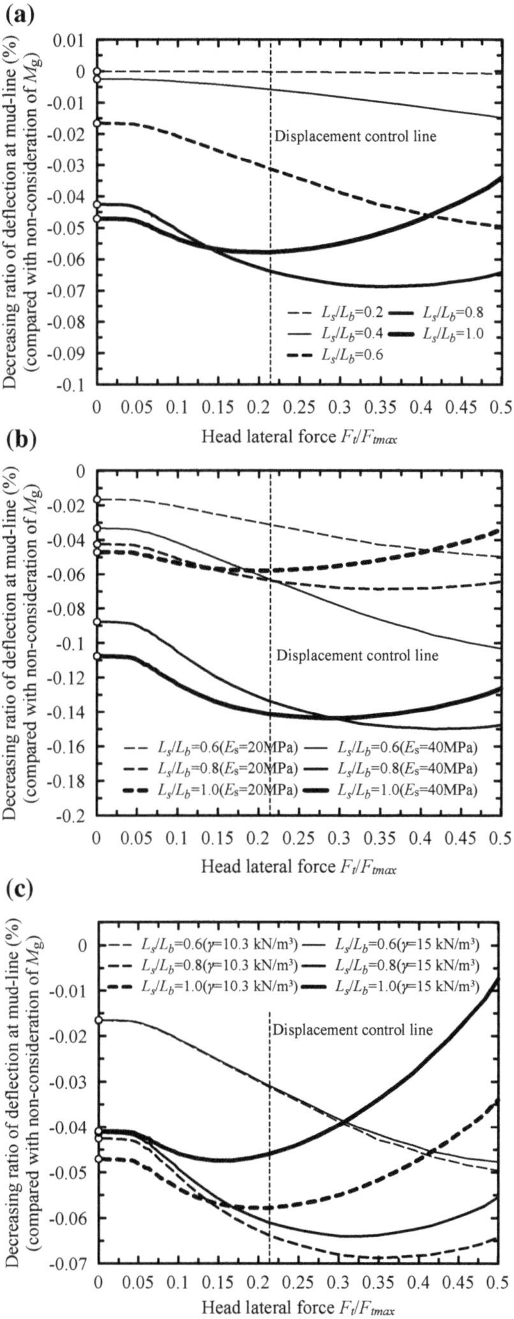

Fig. 19. Additional moment *M*g effect on pile deflection considering influence factors: (a) soil core length inside pipe pile; (b) elastic modulus of soil core; (c) soil core gravity; (d) stiffness of soil mass around pile; (e) pile diameter and (f) pile embedded length

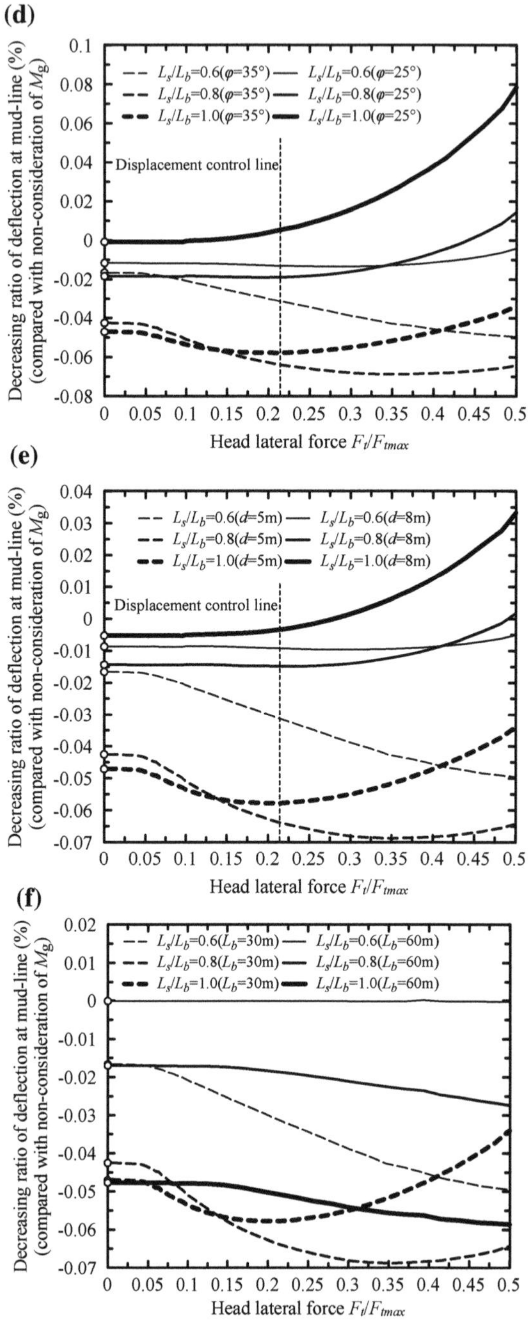

Fig. 19. *(continued)*

8 Conclusions

To investigate the influence of vertical shaft resistance and weight of soil core inside pipe pile on the lateral behavior of monopile foundation, this work, firstly, introduced the concept of additional moment and derived the analytical constitutive models of additional moment M_s induced by skin resistance and additional moment M_g subjected to soil core, as well as the numerical solution for M_s. Furthermore, the generalized solutions for laterally loaded monopile considering double additional moment effects were obtained by transfer matrix approach. Validation of the proposed research is illustrated by comparing its results with those of model tests and numerical method. These comparisons ensure the correctness and validity of research in this paper, and corresponding finding were as below:

(1) The normalized method, with form of M_s/M_{su}-θ/θ_{ref}, for deriving constitutive model of additional moment-slope induced by shaft resistance is efficient and recommendable.
(2) Additional moment M_s and reduction level of pile responses will increase as increasing pile diameter d and ultimate skin friction τ_u or with decreasing equivalent critical displacement $s_{u,eq}$. While, the influence of diameter d is more remarkable.
(3) Although factors, which is consisted of stiffness, gravity, length of soil core, diameter and embedded length of pile, as well as the stiffness of soil mass around pile, have slight impact on the lateral bearing capacity of pipe pile, such influence of additional moment M_g induced by soil core is too small to be ignored.

Acknowledgements. The research work presented herein is supported by China Postdoctoral Science Foundation Funded Project (Grant No: 2017M611955), Jiangsu Planned Projects for Postdoctoral Research Funds (Grant No: 1701028B), Jiangsu Power Design Institute Technology Project (Grand No: 32-JK-2016-001) and National Basic Research Program of China (973 Program, Grant No.2013CB036304), for which we are grateful.

References

Alikhanlou, F.: A Discrete Model for the Analysis of Short Pier Foundations in Clays. Texas Tech University, Lubbock (1981)

American Petroleum Institute (API).: Recommended practice for planning, designing and constructing fixed offshore platforms. API Recommended Practice 2A (RP 2A), 19th edn. Washington, DC (1991)

Ashour, M., Helal, A.: Contribution of vertical skin friction to the lateral resistance of large-diameter shafts. J. Bridge Eng. **19**(2), 289–302 (2013)

Assimaki, D., Gazetas, G.: A simplified model for lateral response of large diameter caisson foundations-linear elastic formulation. Soil Dyn. Earthq. Eng. **29**(2), 268–291 (2009)

Bhushan, K., Fong, P.T., Haley, S.C.: Lateral load tests on drilled piers in stiff clays. J. Geotech. Eng. Div. **105**(8), 969–985 (1979)

Briaud, J.L., Smith, T., Mayer, B.: Laterally loaded piles and the pressuremeter: comparison of existing methods. In: Laterally Loaded Deep Foundations: Analysis and Performance. ASTM, West Conshohoken, PA, pp. 97–111 (1984)
Byrne, B.W., McAdam, R., Burd, H.J., et al. New design methods for large diameter piles under lateral loading for offshore wind applications. In: 3rd International Symposium on Frontiers in Offshore Geotechnics (ISFOG 2015), Oslo, Norway, June. pp. 705–710 (2015)
Dodds, A.M., Martin, G.R. Modeling pile behavior in large pile groups under lateral loading. Rep. FHWA Contract Number DTFH61-98-C-00094. Multidisciplinary Center for Earthquake Engineering Research (2007)
Gerolymos, N., Gazetas, G.: Static and dynamic response of massive caisson foundations with soil and interface nonlinearities-validation and results. Soil Dyn. Earthq. Eng. **26**(5), 377–394 (2006a)
Gerolymos, N., Gazetas, G.: Winkler model for lateral response of rigid caisson foundations in linear soil. Soil Dyn. Earthq. Eng. **26**(5), 347–361 (2006b)
Juirnarongrit, T., Ashford, S.A.: Soil-pile response to blast-induced lateral spreading. II: analysis and assessment of the p–y method. J. Geotech. Geoenviron. Eng. **132**(2), 163–172 (2006)
Kallehave, D., Byrne, B.W., LeBlanc Thilsted, C., Mikkelsen, K.K.: Optimization of monopiles for offshore wind turbines. Phil. Trans. R. Soc. A **373**(2035), 20140100 (2015)
Karapiperis, K., Gerolymos, N.: foundations. Comput. Geotech. **56**, 100–120 (2014)
Lam, I.P., Martin, G.R. Seismic design of highway bridge foundations. Report No. FHWA/RD-86/102, design procedures and guidelines, vol. 2. U.S. Department of Transportation, Federal Highway Administration, Springfield, Virginia, U.S.A (1986)
Li, C.: A Study on Bearing Capacity Performance of Large-Diameter Monopile Foundation under Lateral Loads. Dalian University of Technology, Dalian (2012). (in Chinese)
Li, X., Dai, G., Gong, W., et al. A numerical study into the impact of soil coring on lateral capacity of piles with large diameter. In: The 27th International Ocean and Polar Engineering Conference, International Society of Offshore and Polar Engineers (2017)
Mcvay, M.C., Niraula, L. Development of P-Y curves for large diameter piles/drilled shafts in limestone for FBPIER. Civil and Coastal Engineering Dept, University of Florida (2004)
Mokwa, R.L.: Investigation of the resistance of pile caps to lateral loading. Virginia Polytechnic Institute and State University, Virginia (1999)
Nogami, T., Paulson, S.K.: Transfer matrix approach for nonlinear pile group response analysis. Int. J. Numer. Anal. Meth. Geomech. **9**(4), 299–316 (1985)
Sun, Y.X.: Experimental and numerical studies on a laterally loaded monopile foundation of offshore wind turbine. Zhejiang University, Zhejiang (2016). (in Chinese)
Tseng, T.J.: Laterally loaded rigid piers in sand and sandy soils. Texas Tech University, Lubbock (1984)
Wu, M., Zhao, M.H.: Study on pile-soil interaction under large deflection and its model test. Chin. J. Geotech. Eng. **23**(4), 436–440 (2001). (in Chinese)
Zhang, L.: Nonlinear analysis of laterally loaded rigid piles in cohesionless soil. Comput. Geotech. **36**(5), 718–724 (2009)
Zhang, L., Zhao, M., Zou, X.: Behavior of laterally loaded piles in multilayered soils. Int. J. Geomech. **15**(2), 06014017 (2013)
Zhu, M.X., Zhang, Y.B., Gong, W.M., et al.: Generalized solutions for axially and laterally loaded piles in multilayered soil deposits with transfer matrix method. Int. J. Geomech. **17**(4), 04016104 (2017)

Full-Scale Lateral Load Tests to Determine Load-Displacement Characteristics of Driven Piles in Soft Clay

Gong Chaosittichai(✉) and Pongpipat Anantanasakul

Department of Civil and Environmental Engineering, Mahidol University, Nakhon Pathom 73170, Thailand
gong.cho@student.mahidol.ac.th,
pongpipat.ana@mahidol.ac.th

Abstract. This paper presents the details and preliminary results of full-scale lateral load tests on driven piles in soft Bangkok clay. Steel pipe piles with a diameter of 0.25 m and a depth of 12.5 m were instrumented with strain gauges at various depths. Load tests were performed to large pile displacements. Based on the framework of elastic beam on foundation, the strain gauge outputs were employed to determine pile bending moments, deflections and soil reactions. These calculated pile deflections were validated by means of the results of inclinometer measurements during the course of testing. The experimental pile responses were compared to the numerical predictions of the computer program LPILE. It was found that the numerical model predicted unrealistically stiff load-deflection behavior. The reason was likely the use of Matlock $p - y$ model for soft clay which significantly overestimated the experimental $p - y$ relationships of Bangkok clay particularly near the ground surface.

1 Introduction

Pile supported structures such as high-rise buildings, bridges and piers are commonly subject to lateral loads created by wind and wave actions, berthing forces, movement of vehicles, etc. The response of a laterally loaded pile is influenced by the combined effects of pile flexural stiffness and stress-strain and strength characteristics of the foundation soil.

Broms (1965) introduced an analytical method to determine a safe embedded length of short rigid piles or long piles with plastic hinges. This method considers passive and active earth pressures acting on different parts of the pile as it rotates under lateral loading. Hetenyi (1946) presented an approach to analyze the load-deflection behavior of laterally loaded piles based on their analogy to an elastic beam on foundation. Analytical solutions of the differential governing equations have been introduced for some simple boundary conditions. In this method, the response of foundation soil to the deflection of pile is represented by a non-linear $p - y$ relationship. Computer codes based on the finite difference method, such as LPILE (Reese et al. 2004), have been developed to solve the pile governing equations for more realistic boundary conditions.

© Springer International Publishing AG, part of Springer Nature 2019
S. Shu et al. (eds.), *New Developments in Materials for Infrastructure Sustainability and the Contemporary Issues in Geo-environmental Engineering*, Sustainable Civil Infrastructures, https://doi.org/10.1007/978-3-319-95774-6_10

Along with this process, several $p-y$ models have been proposed for different foundation materials (Matlock 1970; Welch and Reese 1972; Cox et al. 1974 and Reese et al. 1975 among others). In recent years, the finite element method has increasingly been employed to analyze the behavior of laterally loaded piles (Trochanis et al. 1991; Brown et al. 1989; Georgiadis and Georgiadis 2012). The pile and foundation soil are modeled as three-dimensional deformable bodies. Their mechanical responses and the behavior at the interface between the pile and soil are represented by proper constitutive laws.

Full-scale load testing serves as a means of paramount importance not only to better understand the actual pile behavior under lateral loads but also to validate the results of the aforementioned analytical and numerical methods. In spite of this, a very limited number of lateral load tests on piles or drilled shafts in Bangkok clay have been performed. The majority of these are tests on large-diameter bored piles. It is impossible to test such shafts to large displacements or deflections because excessive lateral forces would be required. As such, full details of the load-deflection behavior of deep foundations in this particular urban soil are not available. The present study was undertaken to fill such a gap. Three field load tests on instrumented steel piles of moderate size were performed. The piles were loaded to head displacements up to 125 mm. The load-deflection responses of test piles were investigated, and the applicability of Matlock $p-y$ model for soft clay to predict the behavior of laterally loaded piles in Bangkok clay was primarily assessed.

2 Site Information and Properties of Foundation Soils

Full-scale lateral load tests were performed to study the load-deflection responses of driven piles in Bangkok clay. The site for these field load tests was located in Mahidol University, Nakhon Pathom, Thailand. The university is located in the western part of the Greater Bangkok area with an average elevation of 2 m above mean sea level (MSL).

A dedicated subsurface exploration program was conducted to characterize the ground conditions and to obtain the mechanical properties of foundation soil. Two boreholes 20 m deep were drilled by wash boring. One borehole was located at the center of test site, while the other was located 50 m away. The subsurface information obtained from the two boreholes was used to cross-establish a profile of the ground conditions. Undisturbed samples were obtained every 1.5 m using Shelby tubes from a depth of 1.5 m to 13.5 m. At greater depths, the soil was relatively stiff, and samples were collected by split-barrel samplers. Standard penetration tests were performed every 1.5 m from a depth of 13.5 m toward the ends of boreholes. The groundwater level was observed to be at 1 m below the ground surface. Laboratory index tests, consolidation tests, unconfined compression tests (UC), unconsolidated-undrained triaxial tests (UU) and consolidated-undrained triaxial (CU) tests were performed on the retrieved soil samples. The results of subsoil investigation and laboratory characterization are presented in form of a representative log of boring in Fig. 1.

A layer of soft-to-medium clay (CH) was encountered from the ground surface to a depth of 12.5 m. A seam of silty clay (CL) 1.5 m in thickness was found at a depth of

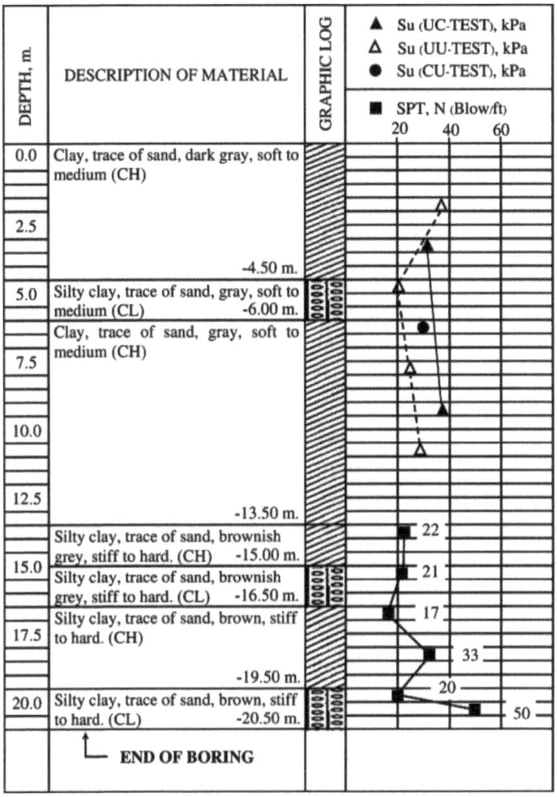

Fig. 1. Profile of foundation soils at test site and relevant strength parameters.

4.5 m. The top clay layer was underlain by a sequence of stiff clays of high and low plasticity. These stiff clay layers extended beyond the end of boring. A trace of sand was observed in all these cohesive soils. The undrained shear strengths of the top soft-to-medium clay varied from 40 kPa near the ground surface to about 30 kPa at the bottom of the layer. The top clay layer was slightly overconsolidated. The overconsolidation ratios (OCR) decreased from about 3 at a depth of 2.5 m to 2 at 7 m. Past research (Horpibulsuk et al. 2007) has indicated that the sequence of stiff clays had been created by the interplay of alluvial, fluvial and deltaic activities during the late Pleistocene. The uppermost clay layer was, however, formed during the Holocene transgression where the sea level increased and reached its maximum height of about 4 m above the present mean sea level about 6,000 years ago. The rising sea interacted with the existing fluvial environments, and this resulted in the progradation of tidal flats and created delta sediments. The level of Holocene sea gradually decreased and reached the present mean sea level around 1,500 years ago. This process exposed the uppermost layer of marine clay. This particular deposit of cohesive soil is usually referred to as soft Bangkok clay. The thickness of this clay layer varies between 10–20 m in the Greater Bangkok area.

3 Pile Instrumentation and Test Set-up

The test piles were 12.5 m long, 0.25 m in inner diameter and 9 mm thick. Each pile was fabricated by welding two steel pipes with a length of 6 m together. Conical protective shoes 0.5 m long were fitted to the bottom ends to prevent the piles from being damaged during driving. The pipe material conformed to the specifications of ASTM A53 standard.

The instrumentation scheme of the test piles is shown in Fig. 2. To investigate the load-deflection characteristics, strain gauges were installed, in pair, on the opposite sides of piles. They were integrated into the data acquisition system using a quarter-bridge type of circuitry. Four strain gauges were, therefore, used to sense the flexural behavior at one pile depth. This arrangement allowed for measurement of the deflections of pile under both compression and tension and ensured the availability of output signals even when some strain gauges might be damaged. Strain gauges were installed every 0.5 m from the ground surface to a depth of 4 m. At greater depths where small pile deflections were expected, strain gauges were located further apart, i.e., every 1 m for depths of 4 to 7 m and every 2 m thereafter. Two steel C channels were attached onto the opposite sides of each pile, thus covering the strain gauges and serving as protective cases during pile driving. A separate calibration bending test was performed on a 3 m - long section of test pile to investigate the bending moment-curvature response. The yield bending moment was observed to be 210 kN-m.

A load cell was employed to monitor lateral loads applied to the pile head at 0.25 m above the ground surface. The head displacements were tracked by three LVDTs attached to the pile at different elevations. The output signals from the strain gauges, load cell and LVDTs were processed and stored in a dedicated data acquisition computer. An inclinometer was also utilized to monitor pile lateral displacements. Inclinometer casings 12 m long were attached to the insides of test piles using welded U-bolts. High-strength epoxy was used to fix the casings and bolts together, thus preventing slippage of the casings during pile installation and testing. Inclinometer readings were recorded manually every 0.5 m for the entire pile depth.

A preliminary analysis of piles subject to lateral loads was performed using the computer program LPILE. The numerical results suggested that the test pile would start to yield in bending at an applied head load of 13 ton. Concrete piles of various cross-sections and lengths were analyzed. The use of four square concrete piles 0.3 m wide and 13 m long each with a safe lateral load of 3.25 ton as the reaction elements was found to be optimum for the load tests.

The test and reaction piles were driven using a 5.2 ton drop hammer. All piles were installed to a depth of 12 m. As shown in Fig. 3a, seven reaction piles were employed. They were installed in two perpendicular rows, each consisting of four piles. The center-to-center spacing among the concrete piles was 0.7 m, thus rendering a width of 2.4 m for the pile group. The test piles and corresponding reaction piles were located 2.2 m apart to accommodate a horizontal loading assemblage. This assemblage consisted of a hydraulic actuator, load cell and transfer beams. Its arrangement is shown in Fig. 3b. A pin support was used at the connection between the test pile and the loading assemblage, and no moment was transferred among these structural elements. After

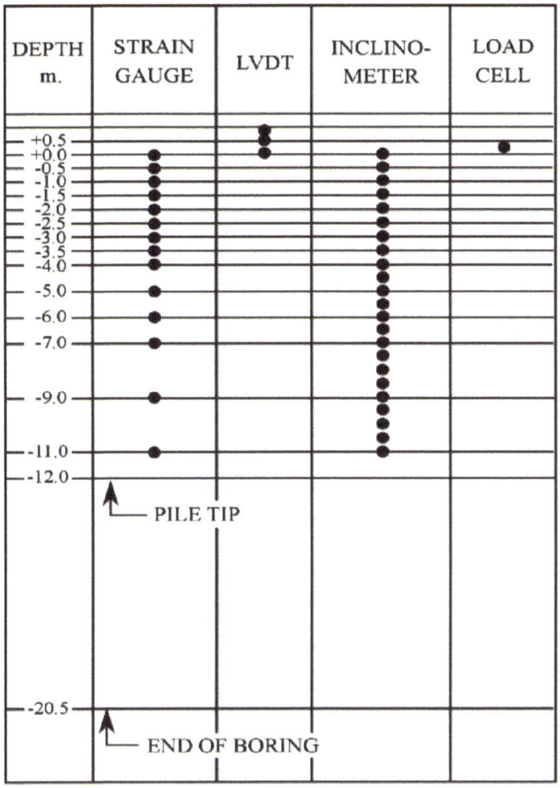

Fig. 2. Instrumentation scheme of test piles.

installation, the piles were left for 30 days before the first load test was performed. This set period was to allow the deep foundations to gain more realistic load carrying capacities as the excess pore pressures in the foundation soils, which were created during pile driving, dissipated with time.

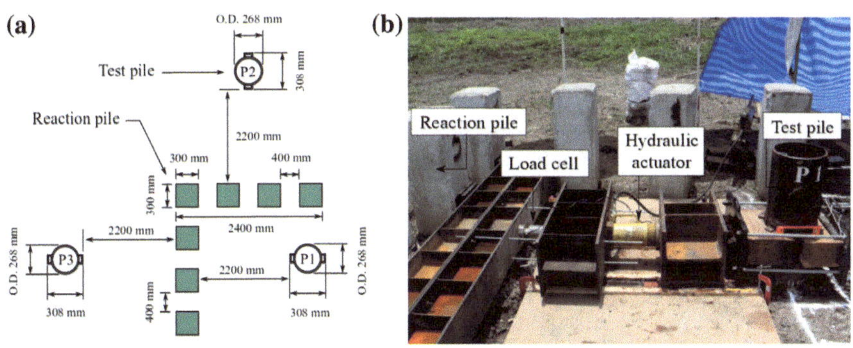

Fig. 3. Arrangement of test and reaction piles (a) and load providing assemblage (b).

4 Testing Program

Horizontal loads were applied in steps of 1 ton until maximum loads of about 10 to 11 ton were attained. In each step, the hydraulic actuator was extended at rates close to 1 mm/min until the target load was achieved. This rate was identical to that used to determine the undrained shear strengths of foundation soils from laboratory UC and UU triaxial tests. This protocol for load application complied with the specifications of ASTM D3966-07 standard. It should be noted that the load tests were discontinued at the aforementioned maximum loads because the observed bending moments in the test piles were close to the yield value (210 kN-m), and the hydraulic actuator approached the end of its stroke of 150 mm. The lateral load was held constant for 15 min for each step. This was carried out to allow the interaction between pile and foundation soils to reach equilibrium. Observations made during the time of testing showed no significant changes of pile displacements after these hold periods. Inclinometer readings were obtained and used to cross-check the pile displacements determined from strain gauge results. The lateral load tests were performed every three weeks. The duration of three weeks allowed sufficient time for preparation of each test. The testing conditions were carefully kept as identical as possible, so the repeatability of experimental results could be investigated.

5 Results of Load Tests

5.1 Analysis of Strain Gauge Results

The strain gauge results were analyzed within the framework of elastic beam on foundation. According to this framework, the pile under consideration is assumed to be elastic with a constant bending stiffness (EI). The deflections of pile are symmetric about the longitudinal neutral plane, and the resulting strains are small. The curvature ($\emptyset$) is expressed as $\emptyset = 2\varepsilon/B$ where ε is the longitudinal strain on the surface of pile and B is the pile diameter. The bending moment is computed as $M = EI\emptyset$. The shear force (V) and perpendicular external load or soil reaction (p) in this application can be determined by differentiating the bending moment once and twice, i.e., $V = dM/dz$ and $p = dV/dz = d^2M/dz^2$. The pile slope (S) and deflection (y) are computed from integration of the curvature, i.e., $S = \int \emptyset dz = (1/EI) \int Mdz$ and $y = \int Sdz = (1/EI)$ $\int\int (Mdz)dz$. In these formulations, z denotes the depth or pile coordinate.

The strain gauge readings were used to compute the curvatures and bending moments at various pile depths. The pile slopes and deflections were determined by approximating the analytical integration using a trapezoidal rule:

$$\int fdz \approx \sum_{i=1}^{n} \left(\frac{f_i + f_{i-1}}{2} \right) (z_i - z_{i-1}) \tag{1}$$

In Eq. (1), n is the number of elevations of strain gauges. The shear forces and soil reactions were calculated using a two-point finite difference formula:

$$\frac{df}{dz} \approx \frac{f_{i+1} - f_{i-1}}{z_{i+1} - z_{i-1}} \quad (2)$$

5.2 Head Displacements

The relationships between lateral load (P) versus head displacement at ground surface (δ) are shown in Fig. 4. The curves suggest stiff load-displacement responses of piles at small displacements. At large displacements, the curves continue to gradually increase, however, at smaller rates. The tests were discontinued at lateral loads of 10 to 11 ton. Maximum head displacements of 115–125 mm were observed.

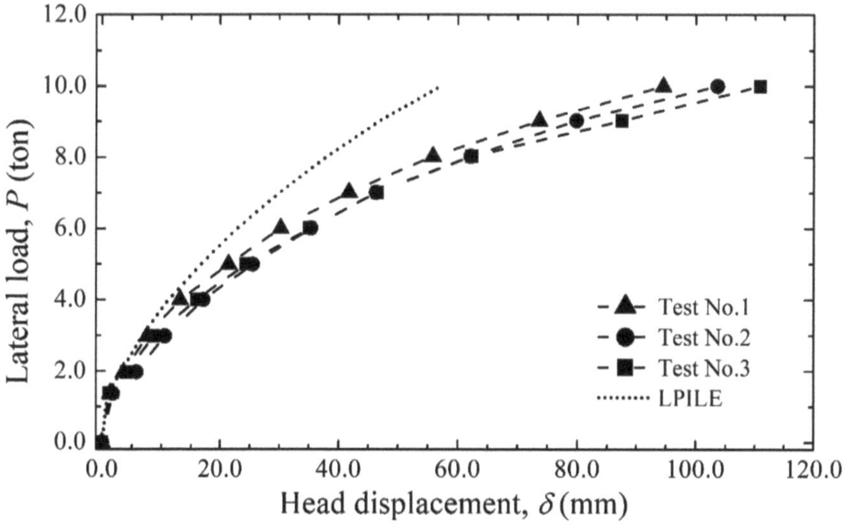

Fig. 4. Lateral load-head displacement relations at ground surface obtained from LVDT results. Numerical predictions of computer program LPILE are also plotted for comparison.

As shown in Fig. 4, the head displacements at ground surface measured by LVDTs obtained from the three tests vary within 10% of each other. This experimental scatter is fairly small, and it suggests that the adopted testing methodology can provide repeatable test results. The load-displacement responses of all tests are essentially identical. It follows that an increase in set time did not result in any significant change in the load-displacement characteristics of test piles. This finding supports the testing assumption that the excess pore pressures in the foundation soils created during pile driving were largely and sufficiently dissipated within the primary set period of 30 days.

The experimental $P - \delta$ curves are compared to the numerical predictions of the computer program LPILE. In this software, the pile is discretized into elements and each is attached to a Winkler spring in the perpendicular direction. The spring responses are characterized by non-linear $p - y$ relations. The accuracy of this type of numerical analysis depends on the ability of the employed $p - y$ model to predict the actual pile-soil behavior. The Matlock $p - y$ model for soft clay (Matlock 1970) was used in the LPILE analysis. This particular $p - y$ model is widely used within the local geotechnical engineering community to analyze the load-deflection behavior of piles and drilled shafts in Bangkok clay. It can be observed that the program significantly over-predicts the experimental curves. In other words, the head displacements are under-predicted. The difference between numerical and experimental results increases with the magnitude of P.

5.3 Pile Deflection Behavior

The pile bending moments, deflections and soil reactions of load steps of 4, 7 and 11 ton are plotted versus depth in Fig. 5. It can be observed that the maximum bending moments increase and take place at slightly greater depths with increasing P. In these tests, the locations of maximum bending moments are within a depth of 3.5 m or 13B from the ground surface. This depth is greater than the upper limit of 10B usually reported in the literature (Duncan et al. 1994). The corresponding bending moments of the maximum load of 11 ton is about 220 kN-m, and this is slightly larger than the yield bending moment of 210 kN-m. This observation indicates that the pile responses were essentially elastic throughout the course of field load tests.

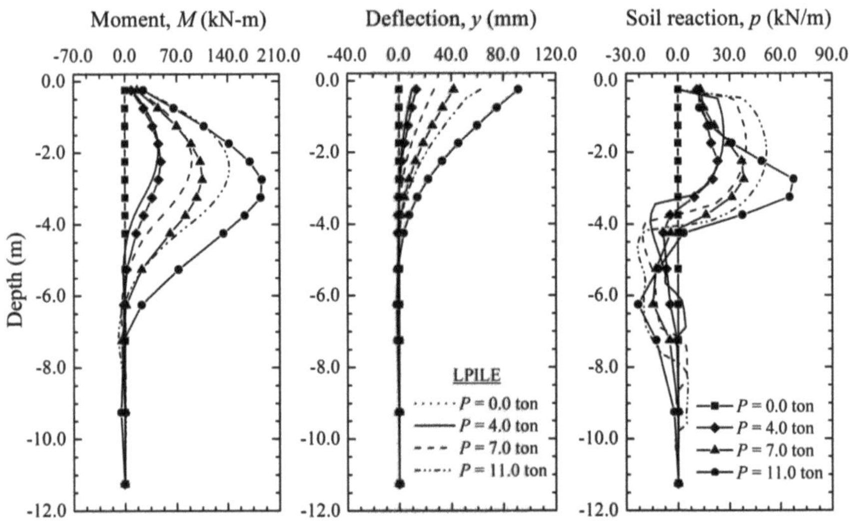

Fig. 5. Bending moments, pile displacements and soil reactions for different lateral loads. Numerical predictions of computer program LPILE are plotted for comparison.

Near the ground surface, the test piles deflect in the direction of lateral loads. The trend, however, reverses and the piles deflect in the opposite direction at greater depths. The magnitude of pile deflection, not surprisingly, increases with P. The variation of soil reaction with depth generally follows the trend of bending moment. The locations of maximum soil reactions, however, are at slightly greater depths. The numerical predictions of LPILE are also plotted for comparison. It can be observed that the model significantly under-predicts the bending moment responses particularly at depths below 6B. The pile deflections are also underestimated. The deviations between experimental results and model predictions generally increase with the magnitude of *P*. The magnitudes of maximum soil reactions for $P \leq 7$ ton can be matched by the numerical results. At greater *P*, however, the soil reactions become significantly larger particularly at pile depths of 10–14B. This, therefore, results in under-prediction of *p* on these parts of piles (Fig. 5c). The predicted locations of maximum soil reactions are about 6B higher than the field measurement data.

The experimental $p - y$ relationships determined at pile depths of 1 and 2 m are show in Fig. 6. It can be observed that the $p - y$ curve for $z = 1$ rapidly increases for *y* less than 5 mm. Beyond that, the curve marginally increases to a maximum value and remains relatively constant through large deflections. At a depth of 2 m, however, the curve becomes stiffer and continuously increases. Predictions of the Matlock $p - y$ model are also plotted for comparison. It can be observed that the model predicts unrealistically stiff $p - y$ responses. This over-prediction of $p - y$ relationships is more pronounced when *z* is small, i.e., near the ground surface. This is likely because the Matlock $p - y$ model assumes an unrealistic failure mechanism of the soil surrounding the top portion of pile. This results in a prediction of excessive ultimate resistance of soil and, therefore, the stiff $p - y$ relationships. Such stiff $p - y$ responses result in unrealistically small pile deflections and, thus, the under-prediction of maximum bending moments as already reported.

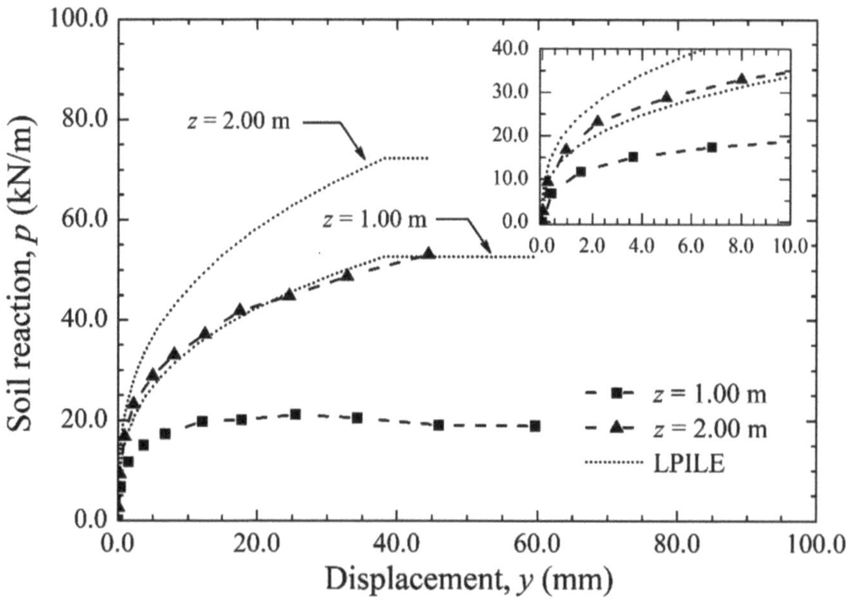

Fig. 6. Experimental and numerical $p - y$ relationships of pile depths of 1 m and 2 m.

5.4 Repeatability and Accuracy of Load-Deflection Characteristics

The variations of bending moment of the load step of 3 ton are shown in Fig. 7a. The bending moment curves of the three load tests plot fairly close to each other. Experimental scatter less than 10% is observed. The corresponding pile deflections and soil reactions are plotted in Fig. 7b and c. Similar magnitudes of experimental scatter can be observed. The pile deflections measured by inclinometer are also plotted in Fig. 7b. The deflection curves determined from the strain gauge readings and that of the inclinometer measurements are in good agreement. This suggests that the computed values of y and, thus, M and p are of good accuracies. These patterns of small variability and good accuracy can be observed for the results of other load steps. As such, the experimental results presented herein are accurate and of good repeatability.

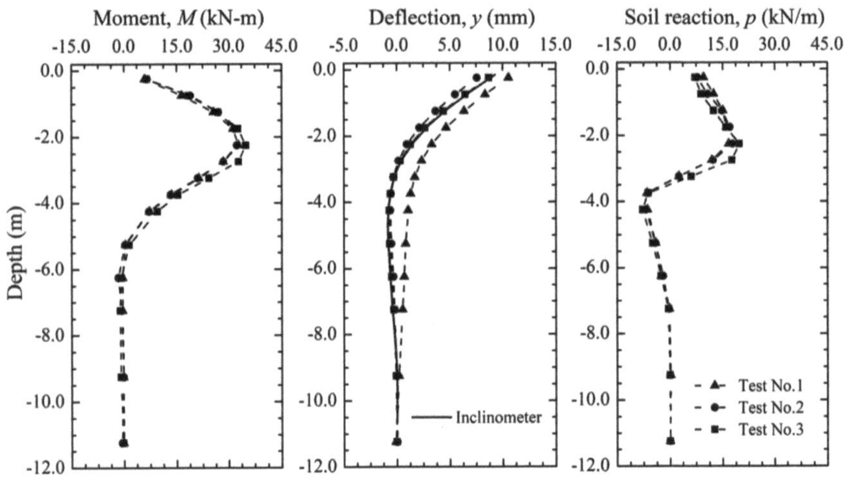

Fig. 7. Variability of load-deflection characteristics for $P = 3$ ton.

6 Conclusions

Three identical lateral load tests on driven piles in Bangkok clay were performed. The variability of head displacements, bending moments, pile deflections and soil reactions among the load tests was fairly small. Experimental scatter less than 10% was observed. The experimental results were of good accuracies. The responses of test piles were compared to the numerical predictions of the computer program LPILE. It was observed that the numerical model significantly under-predicted the pile deflections, head displacements and maximum bending moments. The model also predicted unrealistically stiff $p-y$ responses particularly near the ground surface. The Matlock $p-y$ model likely assumed an unrealistic failure mechanism of the soil surrounding the top portion of pile. This resulted in a prediction of excessive ultimate soil resistance and the stiff $p-y$ relationships. Such stiff $p-y$ responses rendered unrealistically small pile deflections and, thus, under-prediction of maximum bending moments.

Acknowledgments. The authors would like to acknowledge the financial support from the Thailand Research Fund (TRF) and SEAFCO Public Company Limited. We would like to also thank all members of the Geotechnical Engineering Laboratory, Mahidol University for their assistance during the course of field load tests.

References

ATSM, D 3966-07: Standard test methods for deep foundation under lateral load. American Society for Testing and Materials, West Conshohocken, Pennsylvania

Broms, B.B.: Design of laterally loaded piles. ASCE J. Soil Mech. Found. Div. **91**(3), 79–99 (1965)

Brown, D.A., Shie, C.F., Kumar, M.: P − Y curves for laterally loaded piles derived from three dimensional finite element model. In: Proceedings of 3rd International Symposium on Numerical Models in Geomechanics, Niagra Falls, pp. 683–690 (1989)

Cox, W.R., Reese, L.C., Grubbs, B.R.: Field testing of laterally loaded piles in sand. In: Proceeding of Offshore Technology Conference, Houston, Texas, Paper No. OTC 2079, pp. 459–472 (1974)

Duncan, J.M., Evans, L.T., Ooi, P.S.K.: Lateral load analysis of single piles and drilled shafts. ASCE J. Geotech. Eng. **120**(6), 1018–1033 (1994)

Georgiadis, K., Georgiadis, M.: Undrained lateral pile response in sloping ground. J. Geotech. Geoenviron. Eng. **136**(11), 1489–1500 (2012)

Hetenyi, M.: Beams on Elastic Foundation. University of Michigan Press, Ann Arbor, Michigan (1946)

Horpibulsuk, S., Shibuya, S., Fuenkajorn, K., Katkan, W.: Assessment of engineering properties of Bangkok clay. Can. Geotech. J. **44**(2), 173–187 (2007)

Matlock, H.: Correlations for design of laterally loaded piles in soft clay. In: Proceeding of the 2nd Annual Offshore Technology Conference, Houston, Texas, Paper No. OTC 1204, pp. 577–594 (1970)

Reese, L.C., Cox, W.R. Koch, F.D. Field testing and analysis of laterally loaded piles in stiff clay. In: Proceeding of the 7th Annual Offshore Technology Conference, Houston, Texas, Paper No. OTC 2132, pp. 672–690 (1975)

Reese, L.C., Wang, S.T., Isenhower, W.M.: Computer Program LPILE Plus Version 5 Technical Manual. Ensoft Inc., Austin, Texas (2004)

Trochanis, A.M., Bielak, J., Christiano, P.: Three-dimensional nonlinear study of piles. ASCE J. Geotech. Eng. **117**(3), 429–447 (1991)

Welch, R.C., Reese, L.C.: Laterally loaded behavior of drilled shafts. Research Report No. 3-5-65-89, Center for Highway Research, University of Texas at Austin (1972)

An Experimental Study on Strength Characteristics of Cohesionless Soil Under Small Gravity Fields

Guoqing Zhou[2], Feng Gao[1(✉)], and Pin-Qiang Mo[1,2]

[1] School of Mechanics and Civil Engineering, China University of Mining and Technology, Xuzhou 221116, Jiangsu, China
yantugaofeng@163.com

[2] State Key Laboratory for Geomechanics and Deep Underground Engineering, China University of Mining and Technology, Xuzhou 221116, Jiangsu, China

Abstract. Many countries around the world have accelerated the paces of space exploration in recent years. The small gravity field on the surface of the Lunar, Mars and other stars has shown its great influence on the mechanical properties of subsurface soils. A type of magnetic cohesionless soil, made from the cement and Fe_3O_4 magnetic powder, was prepared and adopted to carry out a series of triaxial compression tests under a confining stress condition within the range of 5-30 kPa. The tests were conduct under four different gravity fields (i.e. 0 g, 1/3 g, 2/3 g and 1 g; g indicates the earth gravity acceleration), by using the geotechnical magneto-gravity model testing equipment developed by China University of Mining and Technology. The experimental results show that the mechanical properties of cohesionless soil, including tangential modulus, failure strength, cohesion and friction angle, tend to increase with the gravity field during the shearing process, when a constant confining pressure is applied. In addition, relationships between the mechanical properties and gravity field were proposed. This research provides some insights into the mechanics under various gravity-induced stress gradients, and therefore contribute to the further development of surface excavation, base construction and resource exploitation on the Lunar, Mars and other planets.

1 Introduction

Mineral resources are the key to the global economy. There is an urgent need to consider developing potential lunar resources as the population continues to grow and resources are increasingly depleted (Ruess et al. 2006). The first unmanned lunar exploration mission of China is known as ‘Chang’e project’, operated by the China National Space Administration (CNSA). After completing its very first soft-landing and roving exploration on the Moon, the project aims to implement the next mission to collect and return the lunar sample (Gao 2008). Lunar base construction and mineral resources exploitation must involve a series of excavation and slope support and other geotechnical engineering problems. Strength characteristics as an important indicator of soil mechanical properties, its determination will play an important role in the China’s lunar exploration project and the subsequent infrastructure construction.

© Springer International Publishing AG, part of Springer Nature 2019
S. Shu et al. (eds.), *New Developments in Materials for Infrastructure Sustainability and the Contemporary Issues in Geo-environmental Engineering*, Sustainable Civil Infrastructures, https://doi.org/10.1007/978-3-319-95774-6_11

Mechanical properties of cohesionless soil under small gravity field have been studied by theoretical analysis, physical modelling and numerical simulation, with focuses on both strength and deformation characteristics. The bearing capacity and earth pressure of foundations under 1/6 g and 1 g levels were analyzed by Kobayashi (2005), through dimensional analysis and slip line methods with plasticity theory, showing that the gravity field has a significant influence on the development of soil destruction. Costes et al. (1971) carried out static cone penetration tests in a type of simulated lunar soil under different gravity fields, including 1/6 g, 1 g and 2 g by the dive flight method, which indicated the increase of cone tip resistance with the gravity field. The same method was also applied to investigate the tensile strength characteristics of JSC-1 simulated lunar soil by using square samples (Arslan et al. 2008). Additional research is also conducted to study the strength, macroscopic deformation and carrying capacity and other basic physical and mechanical properties of the TJ-1 and other simulated lunar soils (Jiang et al. 2013, 2016; Li et al. 2008). In terms of numerical simulation, the influence of gravity field, confining pressure and density on the strength of JSC-1A simulated lunar soil was studied by numerical triaxial compression test (Hasan and Alshibli 2010). The simulation and analysis of the interaction between sampler structure and lunar soil under 1/6 g condition were carried out by discrete element method (Zou et al. 2014). The simulation of static cone penetration tests by discrete element method showed that the damage zone is concentrated on the upper part of the cone tip, whereas the normalized cone tip resistance shows a growth trend when the gravity field decreases (Jiang et al. 2014). Tan et al. (2012) reported that the gravity field has a great significant effect on the macroscopic indicators of soil through DEM simulation of simulated lunar soil.

The existing research, however, is mainly based on ground mechanics or the use of discrete elements method. It tends to be difficult to reveal the effect of gravity field on the mechanical properties of soils systematically, reliably and accurately; due to the limitation on the simulation of small gravity field environment. Therefore the development of geotechnical magneto-gravity model testing equipment based at China University of Mining and Technology (CUMT), not only provides the verification for the numerical simulation results, but also acts as the key to reveal the strength and deformation characteristics of the cohesionless soil. In this paper, a type of magnetic cohesionless soil was employed to carry out a series of triaxial compression tests under low confining pressure levels between 5 to 30 kPa, together with the theory and method of magnetic gravity field physical model test (Zhou et al. 2012a, b). The experimental study on the influence rules of gravity field on the strength of cohesionless soil sample by analyzed the results of partial stress-axial strain and other data during the shear deformation process.

2 Testing Method

2.1 Testing Equipment

The experimental tests presented in this paper were conducted using the geotechnical magneto-gravity model testing equipment, developed by CUMT, as shown in Fig. 1. The testing system can realize the simulation of different gravity fields by controlling the magnetic field strength in the testing space and the magnetic properties of the test materials in a normal temperature, static and stable environment.

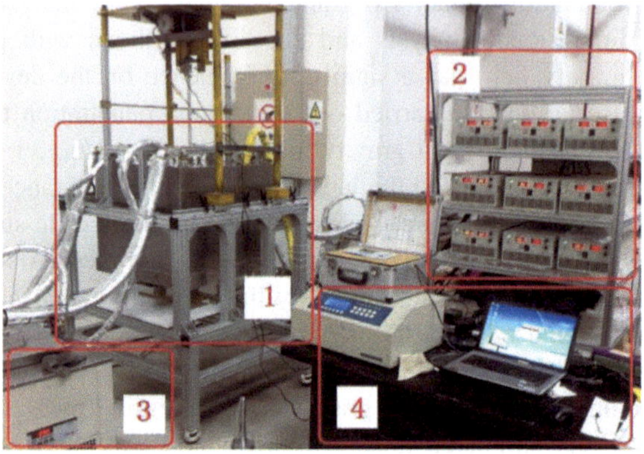

1: Testing container 2: Power supply system

3: Cooling system 4: Data acquisition system

Fig. 1. The small gravity field test system

The axial load is applied by the quasi-static loading method of the strain-controlled stepper motor. The TJL-1 dynamometer is installed between the pressure chamber and the axial loading system to realize the collection of axial load, while the axial displacement is measured by the high precision LVDT mounted on the loading bar. Confining pressure is provided by the TSZ-2 automatic triaxial instrument, in which the drainage volume is automatically collected by the sensor and data conversion device. During the test, the pressure chamber was fixed at the center of the small gravity field testing equipment, by controlling the magnetic field and then to achieve the simulation of different gravity fields.

2.2 Experiment Material

A large number of experimental tests on manufacturing the simulants for the magnetic model tests were conducted by the research group of CUMT (Li et al. 2012; Li 2012). On this basis, the selected magnetic cohesionless soil is composed of 62.5R type high strength aluminate cement and BMMF-1 series Fe_3O_4 magnetic powder together. Fe_3O_4 magnetic powder is used as a fine particle medium to optimize the particle size gradation of cement, and also as a magnetic medium to ensure the magnetization of cohesionless soil material (Fig. 2).

Physical properties of cohesionless soil materials is shown in Table 1, and the particle distribution curve of the sample is provided in Fig. 3. The different small gravity fields are simulated by changing the strength and morphology of the magnetic field in the testing space, including 1/6 g, 1/3 g, 2/3 g and 1 g.

Fig. 2. Magnetic sensitive cohesionless soil

Table 1. Physical properties of the cohesionless soil

Specific density	Maximum dry density ρ_d (g cm^{-3})	Minimum dry density ρ_d (g cm^{-3})	Maximum void ratio e_{max}	Minimum void ratio e_{min}
2.930	2.840	1.083	1.704	0.032

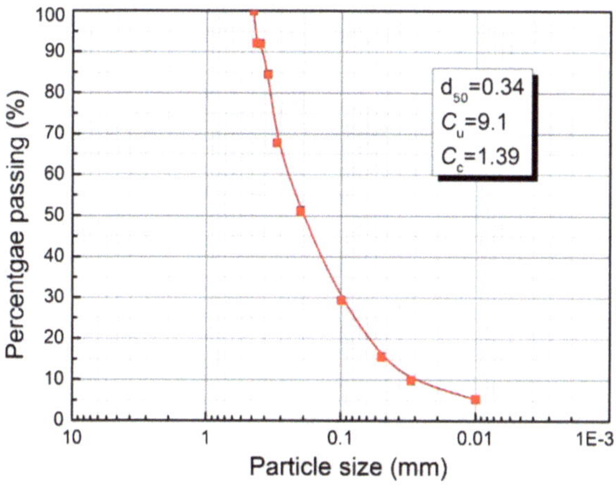

Fig. 3. Particle distribution curve of the cohesionless soil

2.3 Test Plan and Procedure

The maximum particle size of the cohesionless soil is 0.45 mm, while the sample size of the triaxial compression test is Φ 39.1 mm × 80 mm. The confining pressure of the sample has four levels, including 5 kPa, 10 kPa, 20 kPa and 30 kPa respectively. It should be noted that the test focuses on the study of the strength characteristics of the cohesionless soil under different gravity field, and the relevant test parameters are the effective stress index. In addition, the test using consolidated drainage shear mode (CD) due to the involved dry and cohesionless soil during the test process. The test procedure is listed briefly as follows:

(1) Sample preparation: in the base of the pressure chamber placed permeable stone, filter paper, latex film and open round mold, and then calculate the quality of the sample according to the dry density of the sample is 1.30 g/cm^3. The soil sample was divided into three equal parts, and the first soil sample is uniformly filled into the latex film and filled into the design height of the layer eventually, followed by the second and third layers. The top of the sample needs to be flatten, and then put the filter paper, permeable stone, loading cap, remove the open round mold and tighten the latex film after finishing the third layer (Fig. 4).

Fig. 4. Schematic of sample

(2) Test process: First, the isotropic consolidation of the sample is carried out, and then the magnetic field strength in the test space is adjusted to obtain a stable small gravity field. In the experiment, the axial loading rate was set to 0.05 mm/s, regardless of the effect of loading rate on the strength and deformation characteristics of the sample. Automatic collection of axial load, axial displacement and the drainage volume of pressure chamber and other data during the test.
(3) End of test: When the axial strain of the sample reaches 25%, the loading end test is stopped. In order to ensure the reliability of the test data, the triaxial compression test under the gravity field of each group is carried out at least three groups, and a total of 48 sets of triaxial compression test. After the individual discrete test data were excluded, the mean value of each group of test data was used for subsequent analysis.

3 Test Results and Analysis

3.1 Stress-Strain Characteristics Under Different Gravity Fields

In the triaxial compression tests, the deviator stress q is the difference of principal stress $\sigma_1 - \sigma_3 = F/A_a$, and the axial strain is the percentage of the axial displacement Δh versus the initial height of the sample $\varepsilon_a < (\Delta h/h_0) \times 100\%$. For tests with a certain confining pressure, the stress-stain curves of the cohesionless soil are shown in Fig. 5. It can be seen that the shear failure process of the sample could be divided into three stages, including the initial elastic stage, the elastic-plastic deformation stage and the strain hardening stage.

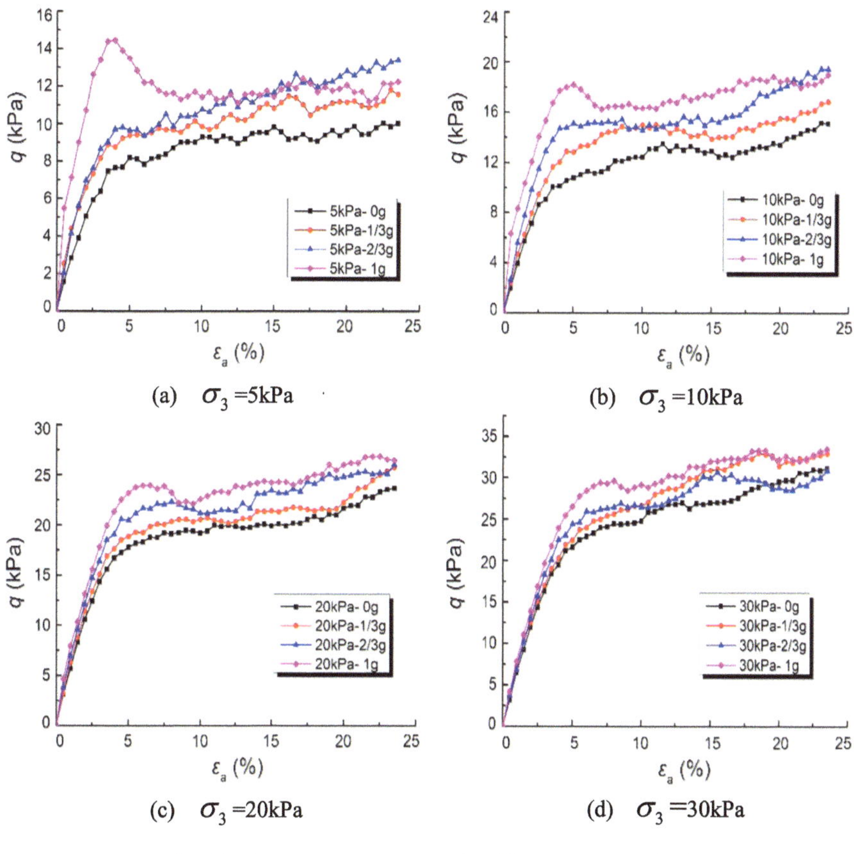

Fig. 5. Stress-strain curves under different gravity field

Initial elastic stage: the stress-strain curve is approximately linear when the axial strain $\varepsilon_a < 2.0\%$. Under the same confining pressure, the stress-strain curve of the sample becomes steep as the gravity field increase. At the same time, the axial strain corresponding to the sample into the elastic-plastic deformation stage shows an increasing trend with the increase of gravity field.

Elastic-plastic deformation stage: the stress-strain curve of the sample is gradually inclined from the diagonal line to the yield step, and the curve is gradually bent upward when the axial strain $\varepsilon_a > 2.0\%$. The simulated lunar soil sample also showed similar yield step in the shear deformation process with the confining pressure of 25–75 kPa (Zou et al. 2011). Under the same confining pressure, the stress-strain curve of the sample gradually appears peak value in the elastic-plastic deformation stage with the increase of the gravity field. When the gravity field is 1 g, the sample exhibits a strain softening characteristic, which indicates that the effective stress level of the sample particles increases with the increase of the gravity field.

Strain hardening stage: due to the restraint of the sample under the confining pressure, the micro-cracks are gradually compacted result in the strength of sample increase, and the sample exhibits strain hardening. With the increase of the confining pressure, the stress-strain curves of the sample tend to be closer under different gravity field, which indicates that the influence of the gravity field on the mechanical properties is gradually weakened.

3.2 Tangential Modulus

According to the stress-strain curve of cohesionless soil sample in the shear process, the influence of gravity field on the tangential modulus during the shear deformation process was analyzed. The tangential modulus is calculated as follows:

$$E = \frac{\Delta\sigma}{\Delta\varepsilon} = \frac{\Delta q}{\varepsilon_a} = \frac{\Delta(\sigma_1 - \sigma_3)}{\varepsilon_a} \quad (1)$$

where E is the tangential modulus and σ_3 is the confining pressure.

In the initial elastic deformation stage of the sample, the tangential modulus is calculated by the deviatoric stress q of the corresponding axial strain ε_a= 1.00%, and the value is shown in Table 2.

Table 2. Tangential modulus

E (kPa)	0 g	1/3 g	2/3 g	1 g
σ_3 = 5 kPa	283.73	439.38	414.46	713.62
σ_3 = 10 kPa	393.26	466.93	559.27	831.36
σ_3 = 20 kPa	567.54	628.56	694.53	792.40
σ_3 = 30 kPa	644.45	709.46	686.15	778.27

It can be seen that when the confining pressure is constant, the tangential modulus of the sample exhibits a linear relationship with the gravity field, and the experimental data is further analyzed to obtain the relationship between the tangential modulus, confining pressure, and gravity field similarity constant respectively:

$$E = (-12.24\sigma_3 + 488.14)N_g + (15.55\sigma_3 + 205.29) \quad (2)$$

$$N_g = g'/g \quad (3)$$

where N_g is the gravity field similarity constant, dimensionless; g' is the magnetic small gravity field, and unit is m/s^2; g is the earth gravity field, which equals to 9.8 m/s^2.

Figure 6 shows the measured and predicted values of the relationship between the tangential modulus and the gravity field. It can be seen that when the confining pressure is constant, the tangential modulus increases gradually with the increase of gravity field in the initial elastic deformation stage. When the gravity field increases from 0 g to 1 g, the tangential modulus of the sample increases significantly. The tangential modulus growth rate of the sample gradually decreases with the increase of confining pressure (see Fig. 7). Therefore, it is necessary to consider the influence of gravity field on the mechanical properties such as tangential modulus during the study of the mechanical properties of soil under small gravity field and low confining pressure.

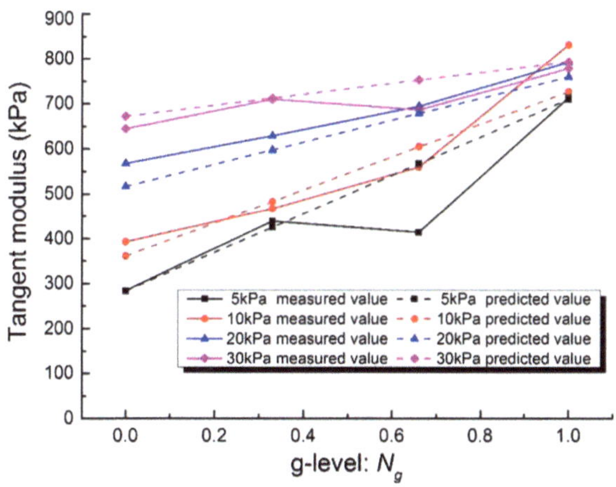

Fig. 6. Relationship between tangential modulus and gravity field

3.3 Failure Strength

From the stress-strain curves of the sample in the shear deformation process, it can be seen that there is no obvious peak value at the low confining pressure level, and the sample is characterized as strain hardening as a whole. When the confining pressure increases to 30 kPa, the sample shows a certain degree of strain softening characteristics, but the overall change is not significant. Therefore, according to the relevant provisions of "Geotechnical Test Method Standard" (GB/T50123-1999), the stress-strain curve has no obvious peak value, and the partial stress is used as the failure value at axial strain ε_a = 15%. The magnitude of deviatoric stress is taken as the failure strength of the sample (see Table 3).

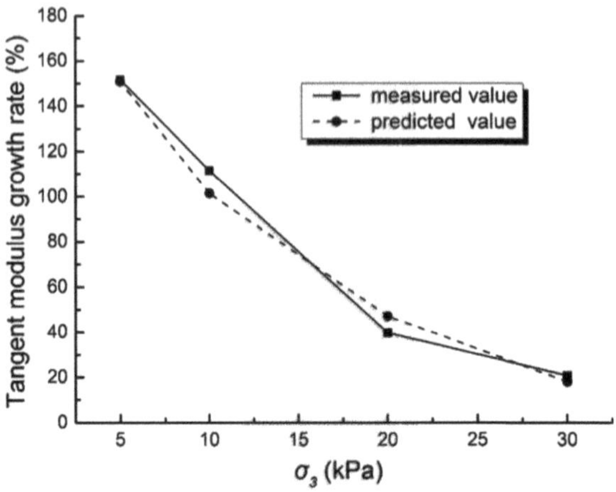

Fig. 7. Relationship between tangential modulus growth rate and confining

Table 3. Failure strength

σ_f (kPa)	0 g	1/3 g	2/3 g	1 g
σ_3 = 5 kPa	9.82	10.83	11.65	11.42
σ_3 = 10 kPa	12.93	13.88	15.41	17.28
σ_3 = 20 kPa	19.89	21.39	23.45	24.20
σ_3 = 30 kPa	26.99	30.89	29.93	31.99

It can be seen that when the gravity field is constant, the failure strength increases with the confining pressure during the shear deformation process. When the confining pressure is constant, the failure strength of the sample increases with the increase of the gravity field, and the experimental data are further analyzed to obtain the relationship between the failure strength, confining pressure and gravity field respectively:

$$\sigma_f = (0.08\sigma_3 + 2.44)N_g + (0.72\sigma_3 + 5.97) \tag{4}$$

Figure 8 shows the measured and predicted values of the relationship between the failure strength and the gravity field. It can be seen that the slope of the failure strength-gravity field curve shows a decreasing trend as a whole when the confining pressure increases from 5 kPa to 30 kPa. This indicates that the restraint effect of the confining pressure on the shear deformation process of the sample is enhanced, and the effective stress level of the sample particles is gradually weakened due to the gravity field.

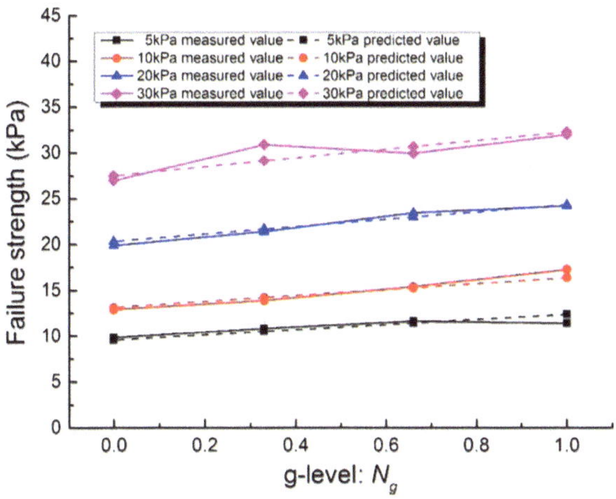

Fig. 8. Relationship between failure strength and gravity field

3.4 Cohesion and Friction Angle

According to the failure strength of each group of tests under different confining pressure, and the cohesion and friction angle of the macroscopic mechanical indexes of the sample under four gravity fields were obtained (Table 4).

Table 4. Relationships between cohesion, friction angle and gravity field

g-level: N_g	0	1/3	2/3	1
Cohesion (kPa)	2.38	2.29	3.08	3.07
Friction angle (°)	14.87	16.68	15.65	16.59

It can be seen that the gravity field has a significant effect on the cohesion and friction angle of the cohesionless soil sample. The cohesion of the sample increases with the increase of the gravity field, and the cohesion increases by 30% when the gravity field increases from 0 g to 1 g. Simultaneously, the relationship between the friction angle and gravity field of the sample also shows a similar tendency to the cohesion. The friction angle of the sample increases by 12% when the gravity field is increased from 0 g to 1 g. Figure 9 shows the linear correlation between the cohesion, friction angle and gravity field.

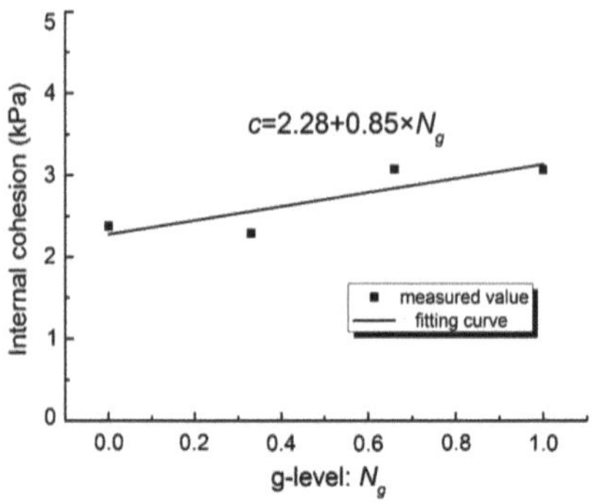

(a) Cohesion fitting curve

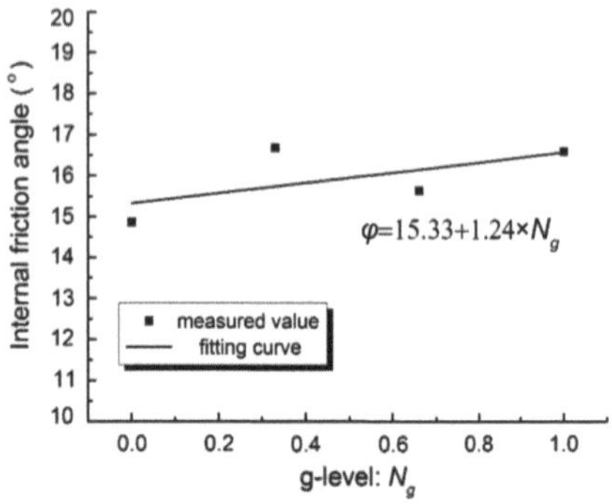

(b) Friction angle fitting curve

Fig. 9. Predicted relationships between cohesion, friction angle and gravity field

4 Conclusions

In this paper, the influence of gravity field on the strength characteristics of the cohesionless soil during the shear failure process is investigated by carrying out a series of triaxial compression tests under small gravity fields, which lead to the following conclusions.

At a low confining pressure level, the stress-strain curves of the cohesionless soil can be divided into three stages during the shear failure process: initial elastic stage, elastic-plastic deformation stage and strain hardening stage. In addition, the stress-strain curves congest with the increase of confining pressure under small gravity fields. This indicates that the gravity field has a large influence on the mechanical properties such as strength, deformation and bearing capacity of cohesionless soil at a lower stress level. When the confining pressure is constant, tangential modulus, failure strength, cohesion and friction angle of cohesionless soil are gradually increased with the increase of gravity field. On the other hand, the influence of confining pressure level on the tangential modulus is gradually reduced during the shear process with the increase of gravity field. The cohesion and friction angle of the sample appear to have linear correlations with the gravity field, which are therefore proposed in this paper.

Acknowledgments. The author would like to acknowledge financial support from 111 Project (B14021), 973 Project Title (2012CB026103), the National Natural Science Foundation of China (No. 51323004, 41672343).

References

Arslan, H., Sture, S., Batiste, S.: Experimental simulation of tensile behavior of lunar soil simulant JSC-1[J]. Mater. Sci. Eng. **478**(1–2), 201–207 (2008)

Costes, N.C., Cohron, G.T., Moss, D.C.: Cone penetration resistance test—an approach to evaluating in-place strength and packing characteristic of lunar soils. In: Proceedings of the Second Lunar Science Conference. Houston: [s.n.], pp. 1973–1987 (1971)

Gao, D.: Study on the Cislunar Transfer Trajectory of the Lunar Probe.National University of Defence Technology, 1–2 (2008)
Hasan, A., Alshibli, K.A.: Discrete element modeling of strength properties of Johnson Space Center (JSC1A) lunar regolith simulant. J. Aerosp. Eng. **23**(3), 157–165 (2010)
Nahashima, H., Shioji, Y., Kobayashi, T., et al.: Determining the angle of repose of sand under low-gravity conditions using discrete element method. J. Terrramech. **48**(1), 17–26 (2011)
Jiang, M., Dai, Y., Wang, X.: DEM analysis of cone penetration tests under low gravity conditions. Chin. J. Geotech. Eng. **36**(11), 2045–2053 (2014)
Jiang, M., Dai, Y., Zhang, H., et al.: Field experimental research on bearing properties of TJ-1 lunar soil simulant. Rock Soil Mech. **34**(6), 1529–1535 (2013)
Jiang, M., Xi, B., Sun, D., et al.: Experimental study of bearing behavior of TJ-1 lunar soil simulant using physical model. J. Tongji University (Natural Science) **44**(2), 167–172 (2016)
Kobayashi, T., Ochia, H., Yasufuku, N., et al.: Prediction of soil collapse by lunar surface operations in reduced gravity environment. In: Proceedings of the 15th International Conference of the ISTVS. Hayama (2005)
Li, J., Zou, M., Jia, Y., et al.: Lunar soil simulant for vehicle- terramechanics research in labtory. Rock Soil Mech. **29**(6), 1557–1561 (2008)
Li, R., Zhou, G., Chen, G., et al.: Development of a new geo- mechanical magnetic sensing soils similar material. Funct. Mater. **19**(43), 2620–2623 (2012)
Li, R.: Development of Similar Materials for Magnetic Sensitive Soil and Study on Its Magnetic Gravity Characteristics. China University of Mining and Technology (2012)
Ruess, F., Schaenzlin, J., Benaroya, H.: Structural design of a lunar habitat. J. Aerosp. Eng. **19** (13), 133–157 (2006)
Tan, S., Duan, L., Gao, H.: Effect analysis of gravity environment on simulated triaxial test results of lunar soil. In: National Conference on Computational Mechanics of Granular Materials, vol. 6 (2012)
Zhou, G., Chen, G, Li, R., et al. A Geotechnical Engineering Model Test Gravity Field Simulation Device and Method: China, CN201210294586.8[P]. 2012-12-26 (2012a)
Zhou, G., Li, R., Chen, G., et al.: A magnetic susceptibility similar material and its preparation method: China, CN201210294587.2[P]. 2012-02-06 (2012b)
Zou, M., Li, J., Liu, G., et al.: Experimental study of terra- mechanics characters of simulant lunar soil. Rock Soil Mech. **32**(4), 1057–1061 (2011)
Zou, M., Xu, S., Zhao, Z., et al.: Simulation analysis of the relationship between the lunar regolith and sampler actuator. J. Tsinghua University(Science Technology) **54**(7), 865–870 (2014)

Reasons for Mid-Span Failure of Pile Supported Bridges in Case of Subsurface Liquefaction

P. Mohanty[1(✉)] and S. Bhattacharya[2]

[1] University of Surrey, Guildford, UK
p.mohanty@surrey.ac.uk
[2] Geomechanics, University of Surrey, Guildford, UK

Abstract. The collapse of pile supported bridges in case of liquefied soil still continue to happen in the event of an earthquake. One of the recurrent observations in such failures is the mid span collapse of bridges with decks falling into the river while the piers close to the abutment and the abutment itself remain stable. Therefore, the present paper proposes a plausible failure mechanism for such mid-span collapse of pile supported bridges. It has been found that as the soil liquefies, the overall natural period of vibration of the piers increases as compared to that of before liquefaction. Subsequently, the lateral displacement demand on the pile head increases with the advent of liquefaction. Lastly, a quantitative analysis for the bridge failure has been carried out for the Showa Bridge (Niigata Earthquake 1964) and the Panshan Bridge (Haicheng Earthquake 1975). It has been observed that the displacement demand on the middle piers of these bridges increased by more than 100%, which led to its collapse, whereas the displacement demand on the other piers increased only by near around 50%. Therefore, the study suggests that the foundation of the middle piers of the bridges need to be additionally stiffened to restrict the increase in their natural period and to avoid its implications.

1 Introduction

Collapse of pile-supported river bridges in liquefiable soils are still observed after most major earthquakes, see for example Fig. 1a–d taken from earthquake observations between 1964 and 2011. It can be easily observed from the figures that the middle piers of the bridges collapsed without significantly affecting the other parts. The aim of the paper is to investigate the plausible failure mechanisms behind such collapse. Table 1 lists mid span failure for 10 other bridges in the aftermath of seismic liquefaction where similar observations were noted. These examples indicate that seismic loads and its effects considered for design are not fully understood and current guidelines for pile design under such conditions may be inadequate.

© Springer International Publishing AG, part of Springer Nature 2019
S. Shu et al. (eds.), *New Developments in Materials for Infrastructure Sustainability and the Contemporary Issues in Geo-environmental Engineering*, Sustainable Civil Infrastructures, https://doi.org/10.1007/978-3-319-95774-6_12

Fig. 1. Collapse of bridges due to liquefaction: (a) Zhuacun bridge (1976 Tangshan earthquake); (b) Rio-viscaya bridge (1991 Telire-limon earthquake); (c) San Felipito Railway Bridge (2010 El Mayor-Cucapah earthquake); (d) Rokko bridge (2011 Tohoku earthquake)

Aim and scope of the work:
Therefore, the aim of the present work is as follows:

(a) Present state-of-the-art of practice for pile design in liquefiable soils; (b) Postulate a plausible mechanism for such recurring mid-span failures; (c) Verifying the mechanism through analytical calculations; (d) Validate the calculation with two well-documented case studies of Showa Bridge collapse (1964 Niigata earthquake, Japan) and Panshan Bridge (1976 Tangshan earthquake, China)

2 Current Understanding of the Failure of Piles in Liquefied Soil

The current understanding of pile failure is based primarily on a bending mechanism where lateral loads due to inertia and slope movements (lateral spreading) induce bending in the pile. Permanent lateral deformation or lateral spreading is reported to be the main source of distress in piles (Boulanger et al. 1999; Wilson et al. 2000; Hamada 2000; Abdoun and Dobry 2002; Rollins et al. 2005; Janalizadeh and Zahmatkesh 2015). Japanese Code of Practice (JRA 2002) was the first national code to include checks on bending moments in piles due to lateral spreading of the ground. The code advises practising engineers to design piles against bending failure assuming that the non-liquefied crust exerts passive earth pressure on the pile and the liquefied soil offers 30% of the total overburden pressure.

Table 1. Abridged list of bridge failures due to seismic liquefaction

Earthquake (Eq.)	Bridge	Remarks
Tohoku Eq. (2011)	Rokko Bridge	This steel girder bridge collapsed by the effects of strong ground motion. (See Fig. 1d)
El Mayor-Cucapah earthquake (2010) (Mexico)	San Felipito Bridges	The railway road bridge bent translated towards the river, which exceeded the seating length of the bearings. (See Fig. 1c)
Wenchuan Eq. (2008)	Miaoziping Bridge	One of the five spans of Miaoziping Bridge had collapsed due to the earthquake
Wenchuan Eq. (2008)	Gaoyuan Bridge	Middle span fell off the piers due to liquefaction.
Costa Rica Eq. (1991)	Rio Viscaya Bridge	One internal supporting pier was missing. (See Fig. 1b)
Phillipines Eq. (1990)	Magsaysay Bridge	Piers settled and failed and the bridge fell into the river
Tangshan Eq. (1976)	Zhuacun Bridge	The girders of the middle spans collapsed. (See Fig. 1a)
Tangshan Eq. (1976)	Shahe Bridge	Bridge girder supports collapsed
Phillipines Eq. (1976)	Quirino Bridge	Midspan collapsed for the truss bridge
Haicheng Eq. (1975)	Panshan Bridge	One of the middle piers sank, causing collapse
Niigata Eq. (1964)	Showa Bridge	Few middle span collapsed, see Fig. 6

The engineers often use a simplified method normally referred to as 'Beam-on-Nonlinear-Winkler-Foundation', BNWF to account for the lateral force due to the flowing soil. It is done by adopting a generalized Winkler soil spring model with a load transfer function called p-y curve as shown in Fig. 2. Wilson et al. (2000) carried out centrifuge tests on single piles and pile group supported structures in liquefied sand and reported that there is a considerable uncertainty in any simplified representation of p-y characteristics of liquefied soil and this uncertainty must be considered in design. Conversely, there are three methods for considering the influence of liquefaction on p-y curves in sand. In the first case, the lateral resistance of liquefiable sand is assumed to be zero. This method may sometimes lead higher construction costs which may be very conservative. Another approach is to treat liquefiable sand as undrained soft clay and use the p-y curves for soft clay (Varun 2010). The third method for the simulation of pile response in liquefiable soils is the use of reduction factors in p-y curve, called p-multipliers (as shown by the factor 'β' in the Fig. 2.), which value usually ranges from 0.01 to 0.3 (Brandenberg et al. 2007). This method has been quite popular because of the simplified approach and easy computation. Later on different researchers like Dash et al. (2017) carried out experimental and numerical work to modify the ongoing API p-y curve to account for liquefied soil. The modified p-y curve for the liquefied soil given by Dash et al. (2017) is shown in Fig. 2 as well.

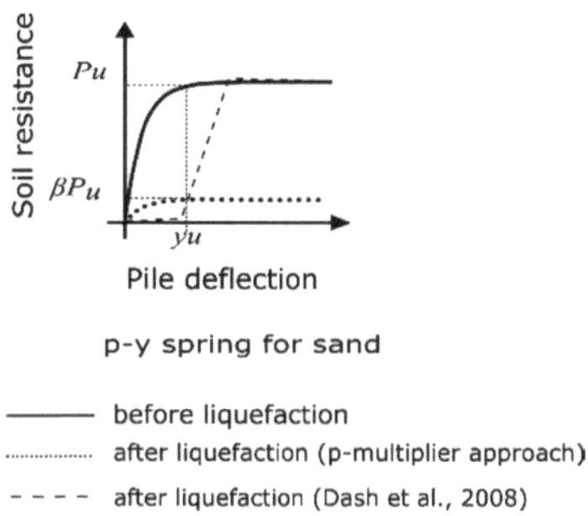

Fig. 2. p-y curves for nonliquefied and liquefied soil

The study carried out by (Shanker et al. 2007; Kimura and Tokimatsu 2009; Bhattacharya et al. 2014) found an alternative mechanism of failure of piles in case of liquefied soil. It has been found that the piles may fail because of buckling if a significant portion of the length of piles becomes unsupported in case of liquefied soil. In such a case, the piles behave as slender columns, in the absence of soil support. So if the axial load acting on the top of piles is near to the value of critical load, the pile may fail due to global buckling.

As the soil liquefies, if the degraded skin friction and end bearing resistance provided by the pile is less than the axial load acting, the pile may settle further to mobilize them. In such a case settlement failure happens. The structure may suffer from the differential settlement as different piles will undergo different degree of displacement as per the demand. The next section of the paper discusses about the failure of piers due to the change in their natural period and its subsequent implications.

3 Failure of Piles Due to the Effects Related to Change in Natural Period Due to Liquefaction

Recently, the study carried out by Mohanty et al. (2017) found that the piers of the pile supported bridges may fail because of the effects related to the elongation of natural period of piers due to the liquefaction. It has been observed that as the liquefaction sets in, the unsupported length of the pile increases as compared to that of before liquefaction. This has been schematically shown in Fig. 3a, b. This effect in turn increases the natural period of the piers.

Due to the natural riverbed profile, water depth increases as we move from abutments towards the center of the river channel. This would lead to relatively higher unsupported length of piles for central piers. Secondly, due to continuous scouring and in the absence of scour protection work, water depth may increase at the center of the river channel over time. As the unsupported length of a pile is a function of water depth and depth of liquefaction, it may easily be derived from the above considerations that the

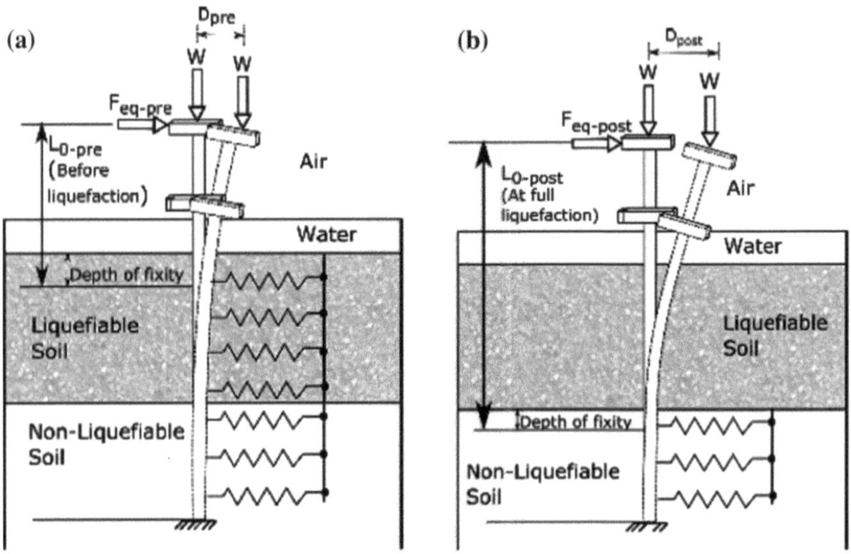

Fig. 3. Illustration of a single pile (a) Pre-liquefaction Stage; (b) Post-liquefaction stage

piles supporting central piers will have higher unsupported length as soil liquefies. Hence, as the soil liquefies the mid-span piers will have higher elongation in their natural periods. Subsequently, the displacement demand for these pier increases more as compared to the adjacent ones and it may dislodge the deck if adequate seating length is not provided.

The methodology adopted and the parameters required for calculating the natural period and its ensuing displacement demand of the bridge piers have been shown in Fig. 3 and are explained as follows.

Following the AASHTO code, equivalent static analysis (single-mode method) has been used for carrying out the analysis. Figure 3 shows mathematical idealisations of the problem which is inspired from the Showa and Rokko Bridge configuration (See Figs. 1d and 7) and the assumptions used in the analysis are listed below:

(a) Each pier is considered separately and the interactions due to the adjacent piers are neglected; (b) The pier and the pile beneath are treated to be continuous for simplicity of the analysis. As a result, same section property is used throughout for pile and pier as shown in Fig. 3; (c) Piles are considered to be flexible and as result there will be bending and no rigid body rotation; (d) The effect of group action on the individual pile is ignored.

(e) Pile is axially stable i.e. enough base capacity to resist bearing failure; (f) Piles are laterally unsupported in liquefiable zone and as a result there are no Winkler springs (p-y springs) in the liquefiable part (as shown in Fig. 3b).

The next section of the paper describes the methodology adopted to carry out the calculations to estimate the natural period of the piers before and after liquefaction and their subsequent displacement demand.

I. Calculate Depth of Liquefaction (H_{liq})

The depth of liquefaction can be determined from the ground profile obtained by carrying out field investigations or standard SPT tests and with the application of simplified methodology depicted in different codes of practice (Idriss and Boulanger 2008).

II. Calculate Unsupported Length of the Pile ($L_{0\text{-}Pre}, L_{0\text{-}Post}$)

Before liquefaction only the portion of the piers, which is above the ground, remains laterally unsupported. The depth of fixity must also be added to the length of this portion to obtain unsupported length of the pile before liquefaction ($L_{0\text{-}pre}$) (see Fig. 3a). At full liquefaction, the depth of liquefiable soil strata, calculated in the previous step and corresponding depth of fixity is considered in addition to find out the total unsupported length of the pile ($L_{0\text{-}post}$). It can be estimated by the method as established in (Bhattacharya and Goda 2013) and is schematically shown in Fig. 3b.

III. Calculate Natural Period of Vibration (T_{pre}, T_{post})

Based on (Indian Road Congress (IRC) 2014) the natural period of the pier can be obtained using Eq. (1) where the pier-pile is idealised as a single cantilever beam carrying the superstructure mass, resting on a foundation.

$$T = 2.0\sqrt{\frac{D}{1000F}} \tag{1}$$

where D = Appropriate dead load of the superstructure and live load in kN before liquefaction

F = Horizontal force in kN required to be applied at the centre of mass of superstructure for earthquake in transverse direction and the force to be applied at the top of the bearings for the earthquake in the longitudinal direction for one mm horizontal deflection at the top of the pier/abutment.

However, Eurocode 8: Part 2 (2011) recommends that the period of the pier can be evaluated by Flexible Deck Model as shown by Eq. (2).

$$T = 2\pi\sqrt{\frac{\sum M_i d_i^2}{g\sum M_i d_i}} \tag{2}$$

M_i = mass at the ith nodal point of the pier as when discretized

d_i= displacement in the direction under examination when the structure is acted upon by force gM_i acting at all nodal points in the horizontal direction considered before liquefaction.

Following the work of Lombardi and Bhattacharya (2014), the natural period for the pier is estimated based on idealization depicted in Fig. 4, where the stiffness K_{lat2} and K_{lat1} are being contributed by the upper liquefiable soil layer and the underlying non-liquefiable soil layer respectively. The damping of the non-liquefiable and the liquefiable soil is being represented by ζ_1 and ζ_2. For the pre-liquefaction stage, the stiffness offered by the both the layers contributes to the natural period of the pile-pier system. The Eq. (3) can be used to find out the natural period for such a case, where $K_{e\text{-}pre}$ is the stiffness of the equivalent pile–pier system before liquefaction and M_e is the equivalent mass lumped at the top of the pier as shown in Fig. 4a.

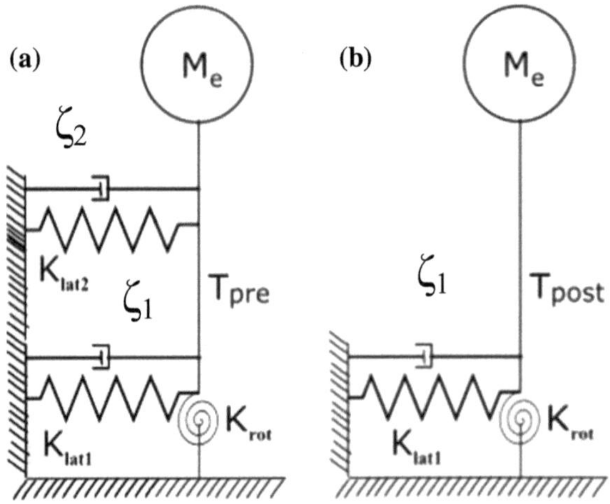

Fig. 4. Idealization of a bridge pile-pier system as a SDOF system (a) Pre-liquefaction Stage (b) Post-liquefaction stage

$$T_{pre} = 2\pi\sqrt{\frac{M_e}{K_{e-pre}}} \tag{3}$$

The complete pile-pier system can be idealized as a fixed cantilever with an unsupported length $L_{0\text{-}pre}$. So the stiffness before the liquefaction becomes

$$K_{e-pre} = \frac{3EI}{L_{0-pre}^3} \tag{4}$$

where EI is the flexural stiffness of the pile.

Conversely, the pile is laterally supported by only the underlying non-liquefiable soil layer after liquefaction. So, in post liquefaction condition, the contribution K_{lat2} of the liquefiable soil layer vanishes as shown in Fig. 4b.

Hence, for the post liquefaction case, the natural period can be estimated using equivalent post liquefaction stiffness $K_{e\text{-}post}$ and equivalent lumped mass M_e at the top as shown in Fig. 4b. So the natural period becomes

$$T_{post} = 2\pi\sqrt{\frac{M_e}{K_{e-post}}} \tag{5}$$

Where, the complete pile-pier system can be idealized as a cantilever with an unsupported length $L_{0\text{-}post}$. Similarly, the stiffness of the combined pile-pier system at full liquefaction:

$$K_{e-post} = \frac{3EI}{L_{0-post}^3} \tag{6}$$

The estimation of natural period by the Indian Road Congress ((IRC) 2014) tends to give a lower natural period as compared to the formulation given by Lombardi and Bhattacharya (2014). In addition, the formula proposed by the Eurocode 8: Part 2 (2011) requires the estimation of displacement at different nodal points of the structure when it is acted upon by inertial force at all nodal points in the horizontal direction, which is usually unknown. Hence, the formulation given by Lombardi and Bhattacharya (2014) has been used in the current analysis.

IV. Calculate Peak Acceleration (A_{pre}, A_{post})

The peak acceleration before the liquefaction A_{pre} is determined from T_{pre} using the design response spectra given in various regional codes; e.g. JRA (2002) and Indian Road Congress (IRC) (2014). As the design response spectra are determined by taking the maximum values of the response of the single degree of freedom system, the applicability of it for the current study is quite pertinent. Similarly, the peak acceleration at full liquefaction A_{post} is determined using T_{post} and the response spectrum provided in the codes of practice as mentioned already.

V. Calculate Peak Displacement (D_{pre}, D_{post})

The peak displacement D_{pre} is determined from A_{pre} and T_{pre} by using the following relationship.

$$D_{pre} = \left[\frac{T_{pre}}{2\pi}\right]^2 . A_{pre} \tag{7}$$

Similarly, the peak displacement at full liquefaction D_{post} is determined from A_{post} and T_{post} by using the following relationship.

$$D_{post} = \left[\frac{T_{post}}{2\pi}\right]^2 . A_{post} \tag{8}$$

Furthermore, Eqs. (7) and (8) can be combined to compare the enhanced spectral displacement at full liquefaction and is shown in Eq. (9). It may be noted that the change in spectral displacement is a function of two parameters: (a) T_{post}/T_{pre}; (b) A_{post}/A_{pre}. However, based on Eq. (9), it is quite clear that elongation of natural period has a greater influence on the lateral displacement relative to the reduction in spectral acceleration.

$$\frac{D_{post}}{D_{pre}} = \left[\frac{T_{post}}{T_{pre}}\right]^2 \frac{A_{post}}{A_{pre}} \tag{9}$$

Figure 5 shows a graphical representation of Eq. (9) for different values of A_{post}/A_{pre} and T_{post}/T_{pre}. It can be easily noticed that as T_{post}/T_{pre} will be higher for mid-span piers relative to the piers adjacent to the abutments, the correlation illustrated in Fig. 5 may explain higher lateral displacement demand.

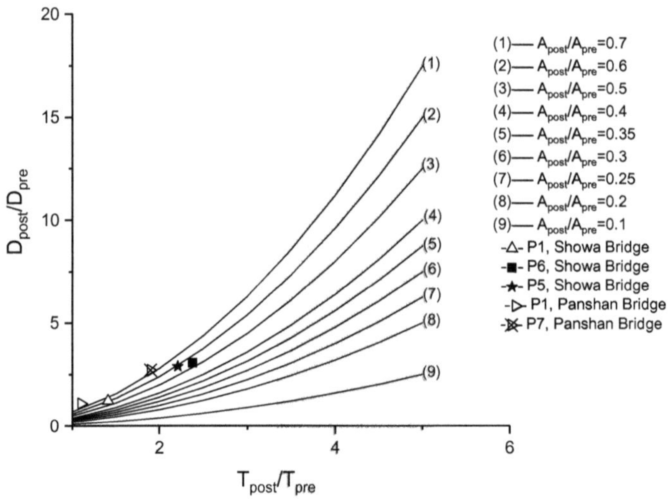

Fig. 5. Variation of peak displacement with increase in natural period

4 Case Studies of Bridge Failures in Liquefied Soil

The case studies of the dynamic failure of Showa Bridge during 1964 Niigata earthquake and the failure of Panshan Bridge during the 1975 Haicheng earthquake have been given as examples to corroborate the findings of the study. The details of the earthquake, bridge superstructure and substructure details along with the damages are also discussed.

4.1 Failure of Showa Bridge in Liquefiable Soil

This case study is very well documented (Bhattacharya et al. 2004, 2014; Bhattacharya and Kappos 2014). The Niigata earthquake occurred on 14th of June 1964 and registered a moment magnitude of 7.6. The Showa Bridge, spanning across the Shinano River, was located some 55 km from the epicentre and it collapsed as a result of the earthquake. The total length of the bridge was about 307 m. The bridge had 12 composite girders and its breadth was about 24 m. Main span length was about 28 m and side span length was about 15 m (Fukuoka 1966). The collapsed view of the Showa Bridge is shown in Fig. 6.

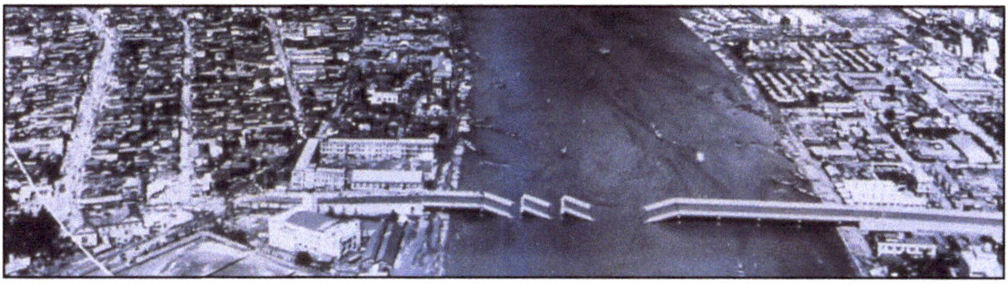

Fig. 6. Collapse of Showa Bridge

4.1.1 Liquefaction Profile

The details of the ground profile, SPT data and the water table can be obtained from the published work of (Iwasaki 2239). O'Rourke and Hamada (1992) estimated the ground liquefaction profile and it has been shown in Fig. 7. The soil at the site liquefied to a maximum depth of about 10 m below the riverbed, and this depth decreased towards the abutments (Bhattacharya et al. 2014) (Fig. 8).

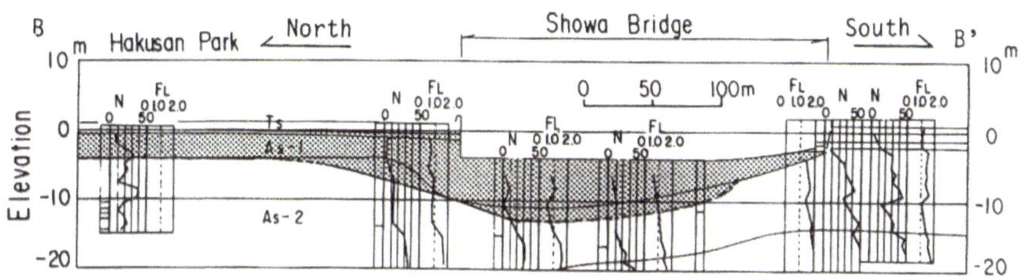

Fig. 7. Soil liquefaction profile (in grey) adapted from (O'Rourke and Hamada 1992)

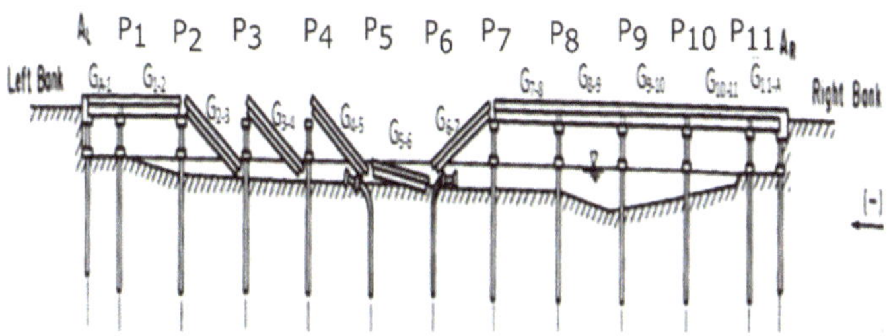

Fig. 8. Schematic diagram of the collapse of the bridge along with the deflections of the pile caps (Iwasaki 1986)

4.1.2 Foundation and Structural Details of the Bridge

The foundation of each supporting pier was a single row of 9 tubular steel piles connected laterally by a pile cap. Each pile was 25 m long with outer diameter of 0.609 m. The wall thickness of the pile varied from 9 to 16 mm. The material of the

Showa Bridge piles can be obtained from (Bhattacharya et al. 2014). The axial load carried by each pile was estimated as 740 kN. The design live loads are ignored as there was no significant traffic on the bridge during its failure. The bridge collapsed just one month after construction, which ensures less uncertainty regarding material strength as degradation of piles due to corrosion was not expected.

4.1.3 Estimation of Pertinent Dynamic Design Parameters of Showa Bridge

I. Unsupported length of the pile ($L_{0\text{-}pre}$, $L_{0\text{-}post}$):

The unsupported length of the pile in pre-liquefaction and post liquefaction stage is determined with the appropriate depth of fixity as per study given by Bhattacharya and Goda (2013). The unsupported length for the piles are estimated as per the data available from the available literatures (Bhattacharya et al. 2014) and is presented in Table 2.

Table 2. Analysis for the Showa Bridge (Niigata Earthquake 1964)

Pier No.	H_{air} (m)	H_{water} (m)	H_{liq} (m)	$L_{0\text{-}pre}$ (m)	T_{pre} (sec)	D_{pre} (m)	$L_{0\text{-}post}$ (m)	T_{post} (sec)	D_{post} (m)	% increase in D	Remarks*
P1	6	0	5	9	1.60	0.95	13.4	2.91	1.17	23.15	NC
P2	6	2.5	5	11.5	2.31	1.2	15.9	3.77	1.76	46.66	NC
P3	6	3	6.5	12	2.47	1.3	17.9	4.50	2.8	**115.3**	C
P4	6	3	8	12	2.47	1.3	19.4	5.08	3.3	**153.8**	C
P5	6	3	9	12	2.47	1.3	20.4	5.48	3.78	**190.7**	C
P6	6	3	10	12	2.47	1.3	21.4	5.88	4.0.	**207.6**	C
P7	6	4	4.5	13	2.78	1.6	16.9	4.13	2.5	56.25	NC
P8	6	4.5	1	13.5	2.95	1.6	13.9	3.08	2.2	37.5	NC
P9	6	5	1	14	3.11	1.79	14.4	3.25	2.4	34.07	NC
P10	6	2	0.5	11	2.17	1.09	11	2.14	1.5	37.61	NC
P11	6	0	0.5	9	1.60	0.63	9.4	1.57	0.675	7.142	NC

H_{air}: Mean height of each pier in air, H_{water}: Mean height of water column at each pile, H_{liq}: Mean depth of liquefaction; *NC: No Collapse - The piers did not collapse after full liquefaction; *C: Collapse - The piers collapsed after full liquefaction.

II. Natural period (Tpre, Tpost):

For the estimation of natural period of the piers of Showa bridge before and after the liquefaction, the geometry of the individual pile foundation, length of the pile above the ground surface and depth of liquefied soil have all been taken from available literatures (Bhattacharya et al., 2014). Using the Eqs. (3) and (5) the natural period of different piers have been calculated for both the extreme condition and presented in Table 2.

III. Peak acceleration and displacement (A_{pre}, A_{post} and D_{pre}, D_{post}):

The response spectrum of Type I (Level 2) earthquake as prescribed by JRA (2002) is used for the analysis. The peak acceleration for different piers are estimated from the expressions proposed in Table 6.4.1 of *Standard Values of the Design Horizontal Seismic Coefficient for Level 2 Earthquake Ground Motion (Type I)*, k_{hcO} of DESIGN SPECIFICATIONS FOR HIGHWAY BRIDGES of JRA (2002). This response spectrum was considered to take the worst case scenario into account while doing a generalised design. T_{pre}, T_{post} are determined for different piers, the same can be used to estimate the peak accelerations. These acceleration values are used to find out the peak displacements at the head of each of piers using Eqs. (7) and (8). All these values are mentioned in the Table 2.

4.1.4 Results and Discussions

The following points may be noted based on the Table 2.

1. It can be seen that the natural period of different piers increases due to the liquefaction. For instance; the natural period of the P6 becomes almost 5.8 s at full liquefaction, whereas before the liquefaction it was only 2.47 s. As this resultant increased natural period falls in the displacement sensitive zone of the response spectra, the lateral displacement of the pile top may increase multiple times.
2. Secondly, the peak displacement prior to liquefaction (D_{pre}) for the different piers can range from 0.9 to 1.7 m. This value increases up to a value of 4 m as the liquefaction sets in. It can also be noticed from the Fig. 5, as the natural period of the P6 at full liquefaction increases to almost 2.3 times of its pre-liquefaction value, its spectral displacement increases by almost 200%. As piers have been simplified to be SDOF (single degree of freedom) systems and the liquefied soil still offers some resistance, the actual displacement can be lesser than this.
3. Thirdly, when the degree of increase in the peak displacement of the pile for the different piers before and after liquefaction are compared, it can be seen that for pier no P3, P4, P5, P6, the margin of increase is around 100–200%. The margin of increases of displacement for other piles was at most or less than 50%.

4.2 Failure of Panshan Bridge in Liquefiable Soil

The Haicheng earthquake occurred at around 7.36 pm local time on February 4, 1975 in the north-eastern region of China with a magnitude of Ms = 7.3. The hypocenter depth was around 12 km. Many bridges, buildings, embankments and others were seriously damaged due to soil liquefaction. The overall length of the bridge was 315.64 m and the length of each span was around 22.2 m. The width of the bridge was around 7 m. The bridge was supported on piles of 0.9 m diameter and of 30 m length with the pile cap thickness of 1.7 m. The diameter of piers was 1 m with a length of 7 m. The present literatures are silent about the strength of soil around the piers as well as the number of piles beneath each support. There is no information regarding the SPT N values of the surrounding soil. Hence, it was quite difficult to predict the superstructure load acting on the piles for this bridge. But by looking at the superstructure and substructure similarity between this bridge and the Hangu Bridge, which got damaged during the 1976 Tangshan earthquake, the superstructure load acting from each of the span was taken around 170 ton with 3 columns and piles beneath each support. The schematic diagram of the undamaged and the damaged Panshan Bridge is shown in Fig. 9a.

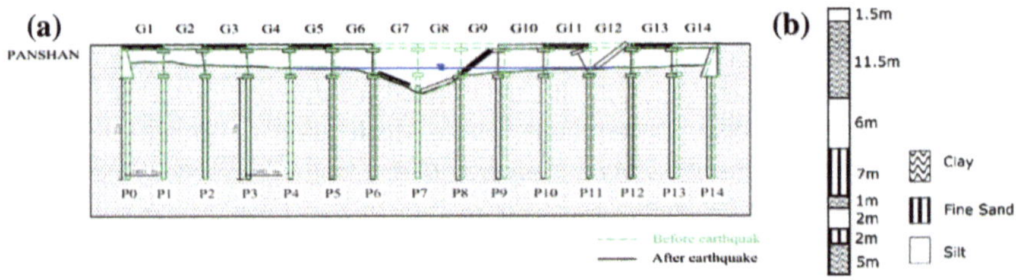

Fig. 9. (a) Schematic diagram of Panshan Bridge (b) Soil Stratigraphy near P4 (derived from (Shengcong and Tatsuoka 1984))

The soil stratigraphy near the Panshan Bridge has been shown in Fig. 9b. It can be seen the soil is highly stratified with clay, fine sand and silty soil. It can be observed that there is a shallow layer of silty soil of around 1.5 m near the ground surface, which had liquefied. But for the pier no P7, this silty layer is absent as the water depth is more as compared to other piers. For the pier P7, the water depth is around 8 m. The soil layer arrangement with respect to depth from the ground surface around all the piers has been assumed as per the information furnished in the work by Shengcong and Tatsuoka (1984). So necessary modification has been done to ascertain the soil stratigraphy near P7.

4.2.1 Liquefaction Profile

The depth of liquefaction near the bridge was around 3 m (Liu et al. 1991). So it is assumed that the top silty soil would have liquefied along with the top portion of the clay, which would have suffered from cyclic mobility.

But for piers P6, P7, P8 and P9 the water depth is more. For the pier P7, the water depth is around 8 m while for other piers it is around 1 m as can be observed from Fig. 9a. Hence, the depth of liquefaction is taken to be around 11 m (up to the interface of silt and fine sand).

4.2.2 Estimation of Pertinent Dynamic Design Parameters of Panshan Bridge

The unsupported length of the pile in pre-liquefaction and post liquefaction stage is determined with the appropriate depth of fixity as per study given by Bhattacharya and Goda (2013). The unsupported length for the piles is estimated as per the data available from the available literatures (Shengcong and Tatsuoka 1984) and is presented in Table 3. For the estimation of natural period of the piers of Panshan bridge before and after the liquefaction, the geometry of the individual pile foundation, length of the pile above the ground surface and depth of liquefied soil have all been taken from available literatures (Shengcong and Tatsuoka 1984). Using the Eqs. (3) and (5) the natural period of different piers have been calculated for both the extreme condition and presented in Table 3. The response spectrum as prescribed by GB50011:2010 (2010) is used for the analysis. The peak acceleration for different piers are estimated from the expressions proposed in **Fig. 5.1.5 Seismic Influence Coefficient Curve** of GB50011:2010 (2010). T_{pre}, T_{post} are determined for different piers, the same can be used to estimate the peak accelerations. These acceleration values are used to find out the peak displacements at the head of each of piers using Eqs. (7) and (8) and are mentioned in the Table 3. The analytical calculation has been carried out taking into account the support systems of the bridges, earthquake, and depth of liquefaction. The values are given in Table 3.

Table 3. Analysis for Panshan Bridge (Haicheng earthquake 1975)

Pier No.	H_{air}	H_{water}	H_{liq}	$L_{0\text{-}pre}$	$L_{0\text{-}post}$	T_{pre}	T_{post}	T_{post}/T_{pre}	D_{pre}	D_{post}	% increase in D	Remarks
1	7	1	2	12.5	13.6	2.2	2.5	1.134	0.127	0.146	15	NC
2	6	2	2	12.5	13.6	2.2	2.5	1.134	0.127	0.146	15	NC
3	6	2	3	12.5	14.6	2.2	2.8	1.262	0.127	0.164	29	NC
4	5	2	3	11.5	13.6	1.9	2.5	1.286	0.111	0.146	32	NC
5	5	2	3	11.5	13.6	1.9	2.5	1.286	0.111	0.146	32	NC
6	4.11	2.89	3	11.5	13.6	1.9	2.5	1.286	0.111	0.122	10	NC
7	7	8	11	19.5	29.6	4.3	8	1.9	0.265	0.709	167	C
8	4.5	3.31	3	12.3	14.4	2.1	2.7	1.266	0.124	0.143	15	NC
9	4.73	2.27	3	11.5	13.6	1.9	2.5	1.286	0.111	0.122	10	NC
10	5.25	1.75	3	11.5	13.6	1.9	2.5	1.286	0.111	0.146	32	NC
11	5.35	1.65	2	11.5	12.6	1.9	2.2	1.146	0.111	0.129	16	C
12	5.87	1.13	2	11.5	12.6	1.9	2.2	1.146	0.111	0.129	16	NC
13	6	0.31	1	10.8	10.9	1.8	1.8	1.013	0.100	0.101	1	NC

H_{air}: Mean height of each pier in air; H_{water}: Mean height of water column at each pile; H_{liq}: Mean depth of liquefaction; *NC: No Collapse - The piers did not collapse after full liquefaction; *C: Collapse - The piers collapsed after full liquefaction.

4.2.3 Results and Discussions

The following points may be noted based on Table 3:

1. It can be seen that the natural period of different piers increases due to the liquefaction; e.g. the natural period of the P7 becomes almost 8 s at full liquefaction, whereas before the liquefaction it was only 4 s. This resulting increased natural period of the piers falls in the displacement sensitive zone of the response spectra. So the lateral displacement of the pile top may increase multiple times.
2. Secondly, the peak displacement prior to liquefaction (D_{pre}) for the different piers can range from 0.1 to 0.2 m. It can be seen that this displacement increases for every pier after liquefaction. But the margin of increases of peak displacement for P7 is much more than any other piers.

5 Conclusions

A review of 10 river bridge collapses in liquefiable soils from 7 different earthquakes from 6 countries is carried out where a repeated observation of midspan collapse is noted. The collapse is caused due to excessive bending of the pier and its foundation leading to fall of the decks. It was also noted that foundations close to the abutments were stable despite large lateral spreading being observed. This paper provides an explanation for such observations through well-established analytical calculations.

The mechanism of failure is based on differential elongation of natural period of different piers supporting a river bridge due to subsurface liquefaction. Due to the riverbed profile (i.e. water depth variation along the river width) the increase in natural period for the central piers is more as compared to the adjacent once. Correspondingly, the displacement demand on the central pier also increases as soil progressively liquefies further promoting differential pier-cap displacement. Two well-known examples of bridge failure (Showa Bridge and Panshan Bridge) are taken to validate the proposed mechanism of failure. It was noted that the displacement demand for central piers increased by more than 100% due to seismic liquefaction owing to the enhanced flexibility of the bridge piers and due to insufficient seating length, the collapse occurred. This is in contrast to the piers close to the abutments where the increase in corresponding displacement demand is in the order of 30-50%. This phenomenon is quite interesting as when the natural period of the pile increases to a higher value due to the liquefaction, the equivalent static force acting at the pile top reduces due to reduction in the value of the peak acceleration. Nonetheless, the lateral displacement due to this effect becomes so high that it may unseat the deck.

River bridges in seismic areas are lifeline structures and they must operate even after an earthquake. As codes of practice do not explicitly mention this proposed mechanism, it may have been overlooked in many designs and there remains a risk of such failures.

References

Abdoun, T., Dobry, R.: Evaluation of pile foundation response to lateral spreading. Soil Dynamics Earthquake Eng. **22**(9–12), 1051–1058 (2002). https://doi.org/10.1016/S0267-7261(02)00130-6

Bhattacharya, S., Goda, K.: 'Probabilistic buckling analysis of axially loaded piles in liquefiable soils'. Soil Dynamics Earthquake Eng. **45**, 13–24 (2013). https://doi.org/10.1016/j.soildyn.2012.10.004. Elsevier

Bhattacharya, S., Kappos, A.: On the collapse of bridge foundations in liquefiable soils during earthquakes. In: Second European Conference on Earthquake Engineering and Seismology, pp. 1–11 (2014)

Bhattacharya, S., Madabhushi, S.P.G., Bolton, M.D.: An alternative mechanism of pile failure in liquefiable deposits during earthquakes. Géotechnique **54**(3), 203–213 (2004). https://doi.org/10.1680/geot.2004.54.3.203

Bhattacharya, S., Tokimatsu, K., Goda, K., Sarkar, R., Shadlou, M., Rouholamin, M.: 'Collapse of Showa Bridge during 1964 Niigata earthquake: a quantitative reappraisal on the failure mechanisms. Soil Dynamics Earthquake Eng. **65**, 55–71 (2014). https://doi.org/10.1016/j.soildyn.2014.05.004. Elsevier

Boulanger, R.W., Curras, C.J., Kutter, B.L., Wilson, D.W., Abghari, A.: Seismic soil-pile-structure interaction experiments and analyses. J. Geotech. Geoenviron. Eng. **125**(9), 750–759 (1999). https://doi.org/10.1061/(ASCE)1090-0241(1999)125:9(750)

Brandenberg, S.J., Boulanger, R.W., Kutter, B.L., Chang, D.: Static pushover analyses of pile groups in liquefied and laterally spreading ground in centrifuge tests. J. Geotech. Geoenviron. Eng. **133**(9), 1055–1066 (2007). https://doi.org/10.1061/(ASCE)1090-0241(2007)133:9(1055)

Dash, S., Rouholamin, M., Lombardi, D., Bhattacharya, S.: A practical method for construction of p-y curves for liquefiable soils. Soil Dynamics Earthquake Eng. **97**, 478–481 (2017). https://doi.org/10.1016/j.soildyn.2017.03.002

Eurocode 8: Design of structures for earthquake resistance—Part 2: Bridges, European Standard (2011)

Fukuoka, M.: Damage to civil engineering structures. Soils Found. **6**(2), 45–52 (1966). https://doi.org/10.3208/sandf1960.6.2_45

GB50011:2010: Code for seismic design of buildings. China: National Standard For People's Republic Of China (2010)

Hamada, M.: Performances of foundations against liquefaction-induced permanent ground displacements'. In: Proceedings of the 12th World Conference on Earthquake Engineering. Auckland, NewZealand, Paper no 1754 (2000)

Idriss, I.M., Boulanger, R.W.: Soil liquefaction during earthquakes, vol. 136, Iss. 6, p. 755. (2008). https://doi.org/10.1016/j.ajodo.2009.10.006

Indian Road Congress (IRC): Standard specifications and code of practice for road bridge, IRC:6-2014 (2014)

Iwasaki, T.: Soil liquefaction studies in Japan, State-of-the-art, Technical Memorandum No. 2239 (1986)

Janalizadeh, A., Zahmatkesh, A.: 'Lateral response of pile foundations in liquefiable soils'. J. Rock Mech. Geotechnical Eng. **7**(5), 532–539 (2015). https://doi.org/10.1016/j.jrmge.2015.05.001. Elsevier Ltd

JRA (2002) 'Seismic design specifications for highway bridges'

Kimura, Y., Tokimatsu, K.: Buckling stress of slender piles with rotational restraint at the pile head in liquefied soil. J. Struct. Construction Eng. **74**(638) (2009) https://doi.org/10.3130/aijs.74.721

Liu, H., Wang, C., Wong, L.: Slides of river banks concerning lateral spread of liquefaction. In: Second International Conference on Recent Advances in Geotechnical Earthquake Engineering and Soil Dynamics, pp. 507–514 (1991)

Lombardi, D., Bhattacharya, S.: Modal analysis of pile-supported structures during seismic liquefaction. Earthquake Eng. Struct. Dynam. **43**(1), 119–138 (2014). https://doi.org/10.1002/eqe.2336

Mohanty, P., Dutta, S.C., Bhattacharya, S: Proposed mechanism for mid-span failure of pile supported river bridges during seismic liquefaction. In: Soil Dynamics and Earthquake Engineering, SDEE 4890 (Reference Number) (2017, accepted)

O'Rourke, T.D., Hamada, M.: Case studies of liquefaction and lifeline performance during past earthquakes, vol. 1, National Center for Earthquake Engineering Research, Technical Report NCEER-92-0001 (1992)

Rollins, K.M., Gerber, T.M., Lane, J.D., Ashford, Sa: Lateral resistance of a full-scale pile group in liquefied sand. J. Geotech. Geoenviron. Eng. **131**(1), 115–125 (2005). https://doi.org/10.1061/(ASCE)1090-0241(2005)131:1(115)

Shanker, K., Basudhar, P.K., Patra, N.R.: Buckling of piles under liquefied soil conditions. Geotech. Geol. Eng. **25**(3), 303–313 (2007). https://doi.org/10.1007/s10706-006-9111-6

Shengcong, F., Tatsuoka, F.: Soil liquefaction during Haicheng and Tangshan earthquake in China: a review. Soils Found. **24**(4), 11–29 (1984)

Varun, V.: A Non-linear Dynamic Macroelement for Soil Structure Interaction Analyses of Piles in Liquefiable Soils. Georgia Institute of Technology, Atlanta (2010)

Wilson, D.W., Boulanger, R.W., Kutter, B.L.: Observed seismic lateral resistance of liquefying sand. J. Geotech. Geoenviron. Eng. **126**(10), 898–906 (2000). https://doi.org/10.1061/(ASCE)1090-241(2000)126:10(898)

Experimental Study on Gas Permeability of Intact Loess Under Applied Load with Constant Stress Ratio Paths

Cun-Li Chen[1,2(✉)], Le Zhang[1], Deng-Fei Zhang[1], and Hui Chen[3]

[1] Institute of Geotechnical Engineering, Xi'an University of Technology, Xi'an, Shaanxi, China
chencl@xaut.edu.cn
[2] Shaanxi Provincial Key Laboratory of Loess Mechanics and Engineering, Xi'an University of Technology, Xi'an, China
[3] Powerchina Northwest Engineering Corporation Limited, Xi'an, China

Abstract. Gas permeability tests were performed to research the effects of deviator stress and mean stress as well as water content on gas permeability coefficient of intact loess, using the refitted stress-controlled triaxial equipment for gas permeability. Decreasing gas permeability coefficient, k_a, against increasing saturation, S_r, was found to be obviously influenced by mean stress and stress ratio (deviator stress) during wetting, water content during loading, respectively. The decreasing rate during wetting was significantly less than that during loading. The unique relationship between gas permeability coefficient and saturation approximately existed whatever stress at the same water content. There was a unique relationship between relative gas permeability k_{ra} (the ratio of gas permeability coefficient at certain water content to that at natural water content under the applied stress) and relative saturation, S_{rr} (the ratio of the difference between saturation, S_r, and the saturation at natural water content under different applied stresses, S_{rn}, to (1-S_{rn})) at different deviator stresses and mean stresses. The relationship could be well fitted by the proposed relative saturation exponential law (RSEL) model using saturation as variable. Linear relationships between logarithm of gas permeability coefficient and stress ratio were paralleled at the same mean stress, but different water contents. These relationships could be modeled by the proposed exponential law (SWEL) model using stress and water content as variable. The prediction results of gas permeability coefficient by the two models proposed were in good agreement with the test results. In comparison with the RSEL model, the SWEL model was more convenient for engineering application.

1 Introduction

Gas permeability coefficient is one of the key parameters in performing consolidation analysis of unsaturated soil. It has been extensively studies that gas permeability coefficient is affected by many internal factors including granularity, density, moisture and structure (Wei et al. 2007; Liu et al. 2010; Yao et al. 2012; Chen 2014; Zhan et al. 2014). Gas permeability coefficient is also related to applied stress. It decreases with the

© Springer International Publishing AG, part of Springer Nature 2019
S. Shu et al. (eds.), *New Developments in Materials for Infrastructure Sustainability and the Contemporary Issues in Geo-environmental Engineering*, Sustainable Civil Infrastructures, https://doi.org/10.1007/978-3-319-95774-6_13

increase of deformation caused by the increase of applied stress (Stoltz et al. 2010; Liu et al. 2015; Chen et al. 2017). The influence of stress on gas permeability coefficient was studied only in the conditions of confining stress (Stoltz et al. 2010) and the isotropic stress (Liu et al. 2015; Chen et al. 2017a, b). However, little measured gas permeability data were reported under the condition of triaxial stress. Both compression deformation and shear deformation are caused by the increase of applied mean stress and deviator stress on soil mass with the same moisture. These are also caused by the increase of moisture at the same mean and deviator stress. Both result in the change of gas permeability coefficient. So, the research of gas permeability coefficient with the change of both stress and moisture is necessarily conducted under the condition of triaxial stress.

A typical unsaturated soil-intact loess was chosen as the research object. Gas permeability tests were performed on the intact loess with various water contents under triaxial stress condition, using the refitted triaxial gas permeability equipment. The effects of water content, deviator stress and mean stress on gas permeability coefficient were analyzed. Gas permeability functions indirectly or directly considering the influences of mean stress, deviator stress and water content were put forward. The foundation was laid by the research in this paper to establish the model of seepage-consolidation coupling for unsaturated soil.

2 Experimental Equipment and Program

2.1 Experimental Equipment and Principle of Gas Permeability Test

Figure 1 shows the stress-controlled triaxial equipment for gas permeability. It can conduct gas permeability tests under applied both isotropic stress and deviator stress. The volume change of specimen can be measured by the volume change tube under applied deviator stress. The main parts of the equipment are the same as used previously by Chen et al. (2017a) except the axial loading device mainly consisting of air-powered cylinder and loading ram. The relationship between gas pressure and the axial force is measured before testing. Applying air pressure to air-powered cylinder, the axial force is applied through loading ram to the specimen.

As water flowed from the base of the water tank through the valve No.4#, a certain negative pressure which even called pressure difference, Δp (i.e. gas permeability pressure, measured by vacuum pressure gauge), between top tank and the base of specimen was generated (Chen et al. 2017a). Then the permeability pressure gradient, i_G, could be determined by Eq. (1). The water discharge, Q_w, flowing from the tank at a certain time interval, t, was measured under a given permeability pressure gradient i_G. Gas discharge flowing through the specimen, Q_G, was determined by Eq. (2) obtained from Boyle law under the standard state (where temperature T_0= 293.15 K, atmospheric pressure p_a= 101 kPa). Gas permeability coefficient, k_a, would be calculated by Eq. (3) on the basis of Darcy's law. The specific principle of gas permeability coefficient measured had been reported by Chen et al. (2017a).

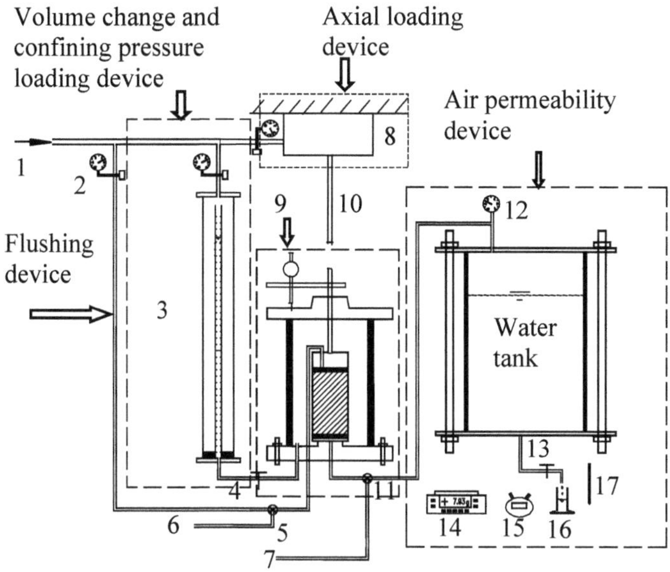

Note: 1. Atmosphere; 2. Pressure gauge; 3. Volume change tube; 4. 2# valve; 5. 1# valve; 6. Atmosphere; 7. Atmosphere; 8. Air pressure cylinder; 9. Pressure room; 10. Force-transfer shaft; 11. 3# valve; 12. Vacuum pressure gauge; 13. 4# valve; 14. Electronic balance; 15. Stopwatch; 16. Graduated cylinder; 17. Thermometer

Fig. 1. Stress-controlled triaxial equipment for gas permeability

$$i_a = \Delta p/\rho_w g L \tag{1}$$

$$Q_a = (p_a - \Delta p) T_0 Q_w / p_a T \tag{2}$$

$$k_a = Q_a/(i_a t A) = Q_a \rho_w g L/(\Delta p A t) \tag{3}$$

Where L, A are the height (cm) and area (cm^2) of the specimen, respectively. ρ_w is water density(g cm^{-3}); g is acceleration of gravity(chosen as 10 m s^{-2}); T is the temperature of the water during testing(*K*) and *t* is interval of time (s).

2.2 Materials and Specimen Preparation

The Q_3 intact loess generated at Late quaternary used in this experimental study was taken from the northern suburb of Xi'an, China, which was the same as that used by Chen et al. (2017a). The sampling depth was 3–4 m, and the main physical indices of the intact loess are illustrated in Table 1. The triaxial specimens with 3.91 cm in diameter and 8 cm in height were prepared by using a special cutting apparatus for gas permeability tests.

The specimens with the natural water content w_n (=15.2%) were wetted by spraying demineralized water to reach a target initial water content w_0 checked by weighing.

Table 1. Physical properties of intact Q_3 loess

Specific gravity G_s	Water content w_n (%)	Dry density ρ_{dn} (g/cm^3)	Liquid limit w_L (%)	Plastic limit w_P (%)	Particle size fraction (%)		
					>0.075 mm	0.075 ~ 0.005 mm	<0.005 mm
2.70	15.2	1.28	30.9	19.8	4	73	23

Note ρ_{dn}, natural dry density.

After reaching the target water content, the specimens were stored in moisturizing cylinder for at least 72 h to allow moisture equalization. The specimen's mass weighed before and after moisturizing equalization was almost same due to water put in moisturizing cylinder.

2.3 Testing Program and Method

A laboratory experimental program was designed whose main objective was to study the effects of deviator stress q ($=\sigma_1 - \sigma_3$, σ_1 and σ_3 are the major principal stress and the minor principal stress, respectively) and mean stress p ($=(\sigma_1 + 2\sigma_3)/3$) on gas permeability of wetting intact loess. Axial stress and confining pressure are equal to σ_1 and σ_3 under triaxial stress, respectively. Stresses were applied under constant stress ratio η ($=q/p$) paths. Gas permeability tests were carried out on intact loess with four initial water contents w_0 including natural water content (=15.2%) and three wetting water contents (=16.6, 19.8, 21.8%) under the two stress conditions. One was the same mean stress p, but increase of deviator stress q (stress ratios η), which was shown in Fig. 2 with dash dot line "①". Another was the same stress ratio η, but increase of mean stresses p, which was represented in Fig. 2 with soild line "②". Figure 2 shows the whole stress conditions for gas permeability tests. The values of deviator stress q and mean stress p were presented in Table 2.

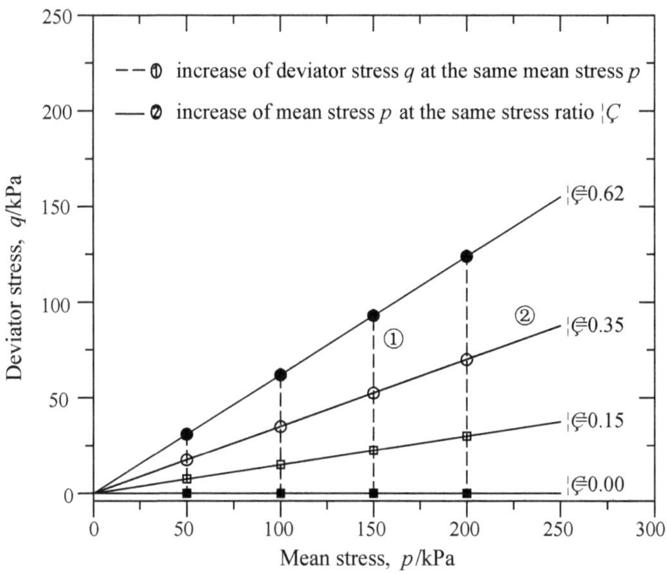

Fig. 2. Triaxial stress conditions of gas permeability tests

Gas permeability test includes two stages: triaxial compression under constant stress ratio path, and gas permeability measurement.

Table 2. Programs and results of gas permeability tests at triaxial stress

η	p/kPa	q/kPa	$w_0 = 1.5\%$		$w_0 = 15.2\%$		$w_0 = 16.6\%$		$w_0 = 19.8\%$		$w_0 = 21.8\%$	
			e	w/%	e	w/%	e	w/%	e	w/%	e	w/%
0	0	0	1.12	1.5	1.12	15.2	1.12	16.6	1.12	19.8	1.12	21.8
	50	0	1.10	1.5	1.08	15.1	1.09	16.5	1.08	19.7	1.08	21.6
	100	0	1.10	1.5	1.08	15.0	1.08	16.4	1.07	19.6	1.06	21.4
	200	0	1.09	1.5	1.03	14.8	1.00	16.1	0.99	19.2	0.99	21.2
0.15	50	8	1.11	1.5	1.08	15.2	1.08	16.6	1.08	19.8	1.07	21.7
	100	15	1.09	1.5	1.07	15.0	1.07	16.5	1.06	19.6	1.06	21.5
	150	23	1.09	1.5	1.07	14.9	1.06	16.4	1.05	19.6	1.04	21.4
	200	30	1.05	1.5	1.00	14.8	1.00	16.3	0.98	19.5	0.97	21.2
0.35	50	18	1.10	1.5	1.08	15.1	1.08	16.5	1.07	19.7	1.07	21.6
	100	35	1.09	1.5	1.07	15.0	1.07	16.5	1.06	19.6	1.05	21.5
	150	53	1.08	1.5	1.06	14.9	1.05	16.3	1.05	19.5	1.03	21.4
	200	70	1.05	1.5	1.04	14.8	0.99	16.2	0.98	19.4	0.97	21.2
0.62	50	31	1.10	1.5	1.08	15.0	1.07	16.4	1.07	19.6	1.07	21.5
	100	62	1.07	1.5	1.06	14.9	1.05	16.4	1.05	19.5	1.03	21.4
	150	93	1.06	1.5	1.06	14.8	1.05	16.3	1.05	19.4	1.02	21.3
	200	124	1.04	1.5	0.99	14.7	0.98	16.2	0.97	19.3	0.97	21.1

In the triaxial compression stage under constant stress ratio path, the mean stress p and the deviator stress q were applied simultaneously to the target stress state (p, q). The stage ended when the increments of volume change and axial deformation were less than 0.01 cm^2/2 h and 0.005 mm/h, respectively. The void ratio, e, of the specimen under triaxial stress was calculated with the following formula.

$$e = e_{\mathrm{n}} - (1 + e_{\mathrm{n}})\frac{\Delta V}{V_0} \quad (4)$$

where ΔV is the volume change of the specimen induced by applying the mean stress and the deviator stress, e_{n} and V_0 are the natural void ratio and the initial volume of the specimen, respectively. The values of the void ratio e were given in Table 2.

In gas permeability measurement stage, the stress (p, q) remains unchanged. Gas permeability tests were performed at the four various pressure differences Δp. The water discharge Q_w (>5 ml) flowing out from the water tank was collected with a graduated cylinder at a certain time t. The test procedure in detail was the same as that reported by Chen et al. (2017b). Gas permeability coefficient k_{a} was determined by Eq. (3) on the basis of the water discharge Q_w and the corresponding time t at the certain gas pressure differences Δp.

After gas permeability tests were completed under different test conditions, the water content, w, of specimen was measured. The values of the water content w were presented in Table 2. It could be seen that there were no much difference ($\leq 1.1\%$) between the initial water content w_0 and the water content w under applying different triaxial stresses, for $p \leq$ 200 kPa, $q \leq$ 124 kPa, $w_0 \leq$ 21.8%. Thus, the initial water content w_0 could be approximately taken as the water content w in the following text. The saturation, S_{r} ($=G_{\mathrm{s}}w/e$), was calculated from the water content w and the void ratio e.

3 Result and Analysis

There were similar characteristics of gas permeability coefficient k_{a} under different stresses for the intact loess with various water contents. Thus, the typical test results were discussed in this paper.

3.1 Modelling the Relationship Between Gas Permeability Coefficient and Saturation

Effect of stress on the relationship between gas permeability coefficient and saturation during wetting

The typical relations between gas permeability coefficient k_{a} and saturation S_{r} are presented in Fig. 3 at the same mean stress p, but different stress ratios η, and at the same stress ratio η, but different mean stresses p during wetting. Gas permeability coefficient k_{a} decreased nonlinearly with increase of the saturation S_{r} induced by the increase of water content at different stress states (q,p). Furthermore, the reduction rate decreased with the increase in saturation S_{r} at a given stress state. The decreasing degree was related to the value of applied stress state.

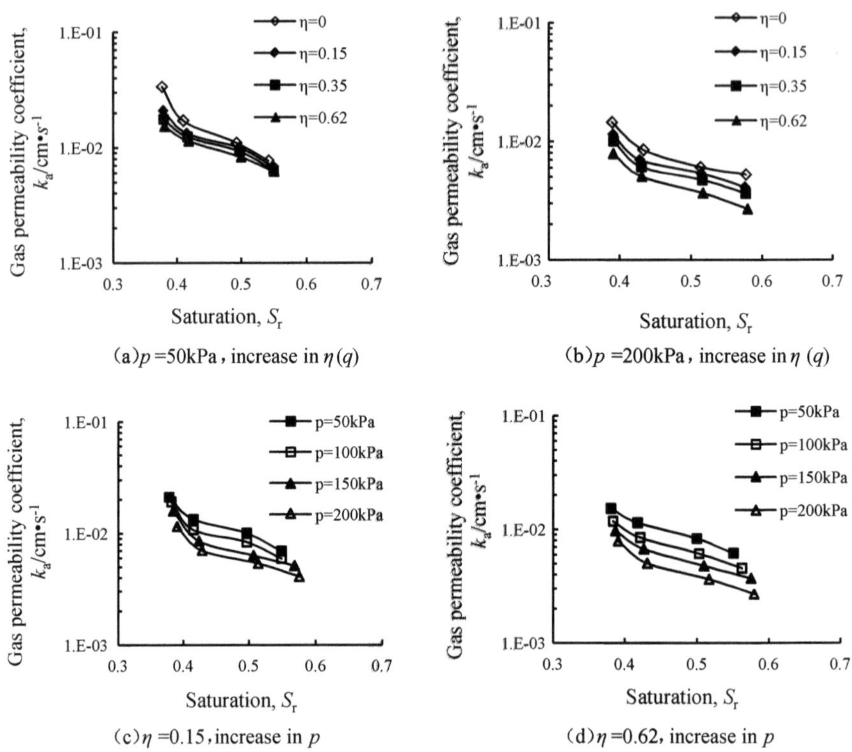

Fig. 3. Influences of deviatoric stress and mean stress on gas permeability coefficient k_a against saturation S_r during wetting

Figure 3a, b indicated that the k_a-S_r curve dropped to the lower part with increase in stress ratio η at the same mean stress p. That is to say, the larger the stress ratio η, the smaller gas permeability coefficient k_a for the specimen with the same saturation S_r. The larger the stress ratio η, the smaller the slope of k_a-S_r curve at the smaller value of mean stress p (=50 kPa), i.e., the smaller the effect of wetting on gas permeability coefficient k_a. The slopes of curve almost unchanged with the increase in stress ratio η at the larger value of mean stress p (=200 kPa), i.e., the influence of wetting on gas permeability coefficient k_a was independent on stress ratio η. It was concluded that the stress ratio (deviator stress) not only had strong influence on the k_a-S_r curve of intact loess wetted at a certain mean stress p, but the influence was dependent on mean stress.

Figure 3c, d revealed that the k_a-S_r curve dropped with increase in mean stress p at the same stress ratio η. In other words, gas permeability coefficient k_a decreased with increase of mean stress p for the specimen with the same saturation S_r. The effect of mean stress on k_a-S_r relationships was independent of stress ratio.

Effect of water content on the relationship between gas permeability coefficient and saturation during loading

Figure 4 shows gas permeability coefficient k_a against saturation S_r while applying stress under different constant stress ratios η for the specimens with different water contents w. Gas permeability coefficient k_a rapidly decreased with increase of saturation

S_r as a result of increase in mean stress p and deviator stress q at different constant stress ratio η paths. The slopes of k_a-S_r curves was obviously higher than that during wetting at the constant stress (Fig. 3). The change in saturation had a stronger influence on gas permeability coefficient at increase in mean stress and deviator stress than in water content. It indicated that different values of gas permeability coefficient were achieved for the same saturation resulted from different mechanism (i.e., a reduction in the void ratio e during loading at nearly constant water content, both a reduction in the void ratio e and an increase in water content during wetting at constant stress (Table 2), respectively).

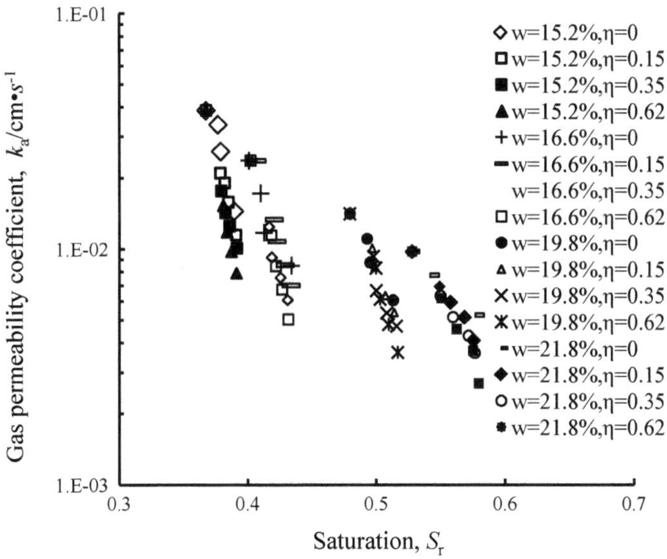

Fig. 4. Influences of water content on gas permeability coefficient k_a against saturation S_r during loading under constant stress ratio path

Inspection of Fig. 4 also suggested that the k_a-S_r data could be merged together at the same water content w, but different constant stress ratios η. The unique relationship between gas permeability coefficient and saturation approximately existed at whatever stress. The k_a-S_r curves dropped to the lower right and the slopes decreased with the increase in water content during loading. It implied that, the higher the water content, the smaller the effect of saturation induced by applied stress on gas permeability.

Model of the normalized relationship between gas permeability coefficient and saturation under different conditions

The aforementioned analysis has shown that, the relationship between gas permeability coefficient k_a and saturation S_r during loading along constant stress ratio η at the same water content was obviously different from that during wetting at the same stress (q, p). Stress (q, p) and water content had pronounced effects on the k_a-S_r curves during wetting and loading, respectively. The influence of saturation resulted from the change in both applied stress and water content on gas permeability coefficient should be taken into account while the k_a-S_r relationship was modeled.

The P-vG-M model (Parker et al. 1987) was commonly used to describe the relationship between the relative gas permeability coefficient, k_{ra}, ($=k_a/k_d$, k_d is gas permeability of dry soil) and effective saturation, S_{re}, at constant void ratio:

$$k_{ra} = (1 - S_{re})^{\gamma}(1 - S_{re}^{1/m})^{2m} \tag{5}$$

where γ is a soil parameter representing the effect of tortuosity of porous medium with a statistically defined value of 0.5, and m is a soil parameter of the capillary model of van Genuchten used to model water retention curve.

It took a long time to obtain the dry specimen by air-drying from natural water content. Shrink and crack may be produced during the air-dried process. Thus, gas permeability coefficient k_d would be more difficultly obtained. For this reason, choosing gas permeability coefficient at natural water content, k_{an}, as reference point values, the relative gas permeability k_{ra} ($=k_a/k_{an}$) was determined.

By means of the data in Fig. 4, relative gas permeability k_{ra} is plotted against saturation S_r during wetting at different stresses in Fig. 5. The data of relative gas permeability and saturation approximately tend to lie on a curve. It was reflected that the k_{ra}-S_r relationships were independent of deviator stress q and mean stress p. So, the k_{ra}-S_r curves could be described by a unique function, during loading and wetting at different water contents and stresses, respectively.

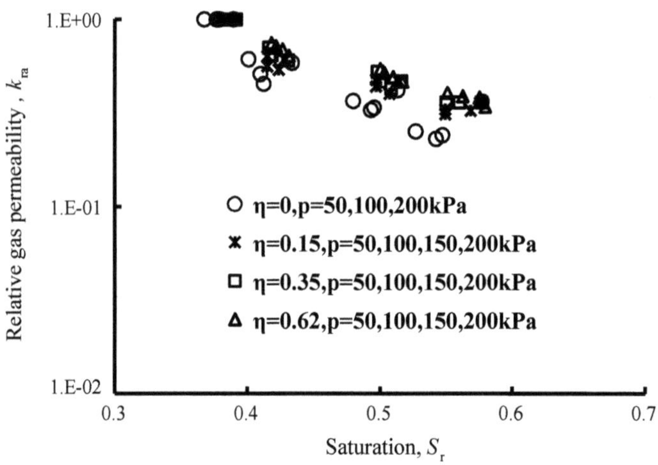

Fig. 5. Relative gas permeability k_{ra} against saturation S_r during wetting at different stress ratios and mean stresses

In order to simply model the normalized relationship between relative gas permeability k_{ra} and saturation S_r, the relative saturation, $S_{rr} = ((S_r - S_{rn})/(1 - S_{rn})$, where S_{rn} is the saturation at natural water content w_n under different applied stresses), was introduced to reflect the change of saturation relative to saturation condition. By means of data presented in Fig. 5, the relative gas permeability k_{ra} is plotted against relative saturation S_{rr} in Fig. 6. The data at different stress ratios η (≤ 0.62) and mean stresses p (≤ 200 kPa) appeared to lie on a curve, and then k_{ra}-S_{rr} relationships were described by a single gas permeability function. These could be fitted by the relative saturation exponential law (RSEL)model.

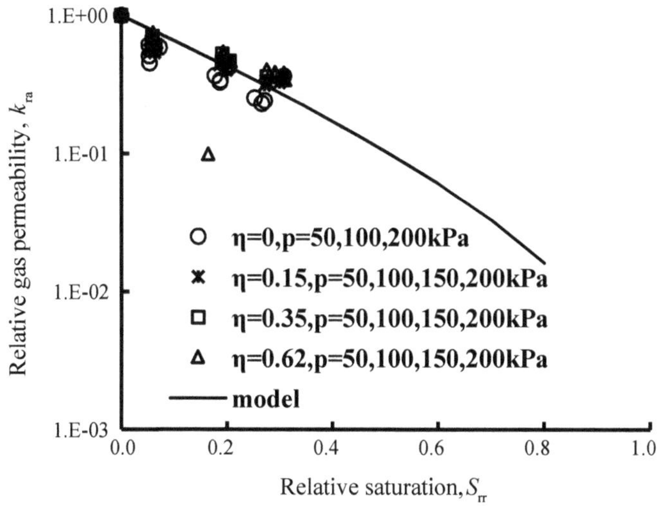

Fig. 6. Relative gas permeability k_{ra} against relative saturation S_{rr} during wetting at different stress ratios and mean stresses

$$k_{ra} = k_a/k_{an} = (1 - S_{rr})\exp(-b_1 S_{rr}) \tag{6}$$

where b_1 is a soil parameter. The boundary condition, $k_{ra} = 1(k_a = k_{an})$, $0(k_a = 0)$ as $S_{rr} = 0(S_r = S_{rn})$, $1(S_r = 1)$, respectively, could be meant by the RSEL model. Regression of the measured data (Fig. 6) to the RSEL model (Eq. 6) obtained a solid curve and $b_1 = 3.15$. In comparison with the P-vG-M model (Eq. 5) reported by Parker et al. (1987), the proposed RSEL model (Eq. 6) was more appropriate to predict gas permeability of intact loess for three reasons. Firstly, its number of parameters was one less. Secondly, relative gas permeability was more conveniently and accurately determined based on gas permeability coefficient k_{an}. Thirdly, the test to determine water retention curve would not be carried out.

To determine gas permeability coefficient k_{an} against the applied stress, change in stress (q, p) was replaced with change in void ratio, e_n (Table 2), which resulted from the applied stress for the specimens with natural water content w_n. Inspection of Fig. 7 revealed that the measured data of k_{an}-e_n relationships could be merged together at different stresses (q, p) applied at different stress ratios. Gas permeability coefficients k_{an} at the same void ratio e_n were almost equal at different stress ratios. The k_{an}-e_n relationships were described by a unique power-law function (Eq. 7).

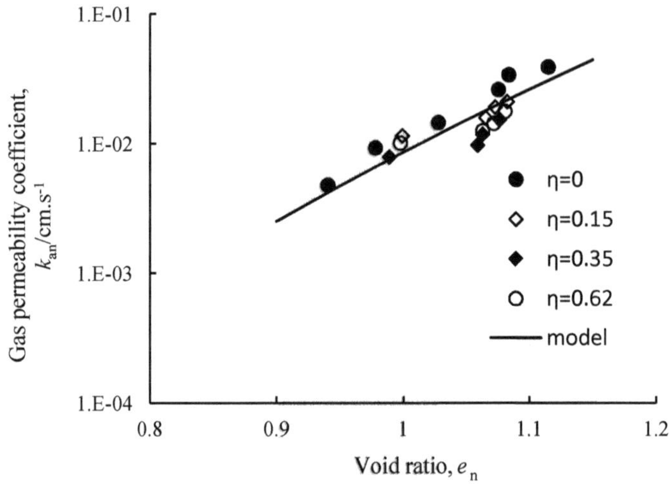

Fig. 7. Gas permeability coefficient k_{an} against void ratio e_{n} during loading under constant stress ratio path

$$k_{\mathrm{an}} = b_2 e_{\mathrm{n}}^{b_3} \tag{7}$$

where b_2 and b_3 are soil parameters. b_2 is permeability coefficient at void ratio $e = 1$ and b_3 represents the decreasing rate with the void ratio resulted from the applied stress for the specimen with natural water content. A regression applied to the measured data resulted in $b_2 = 8.6 \times 10^{-3}$cm s^{-1}, $b_3 = 11.7$.

Substituting Eq. (7) into Eq. (6), the RSEL model was derived as follows.

$$k_{\mathrm{a}} = b_2 e_{\mathrm{n}}^{b_3}(1 - S_{\mathrm{rr}})\exp(-b_1 S_{\mathrm{rr}}) \tag{8}$$

Gas permeability coefficient k_{a} was modeled by the RSEL model for intact loess with various water contents at different triaxial stresses.

Gas permeability coefficient k_{a} was predicted by the RSEL model (Eq. (8)) during wetting at different stresses. The typical relationships between the predicted gas permeability coefficient and the measured gas permeability coefficient are shown in Fig. 8. It was concluded that the predicted results were in good agreement with the measured results.

3.2 Modelling the Relationship Between Gas Permeability Coefficient and Applied Stress

The typical relationships between the logarithm of gas permeability coefficient, ln k_{a}, and stress ratio η at the same mean stress p, but different water contents are shown in Fig. 9. The ln k_{a}-η relationships could be simulated with line. Namely, the k_{a}-η curves at different mean stresses and water contents could be described by exponential law model.

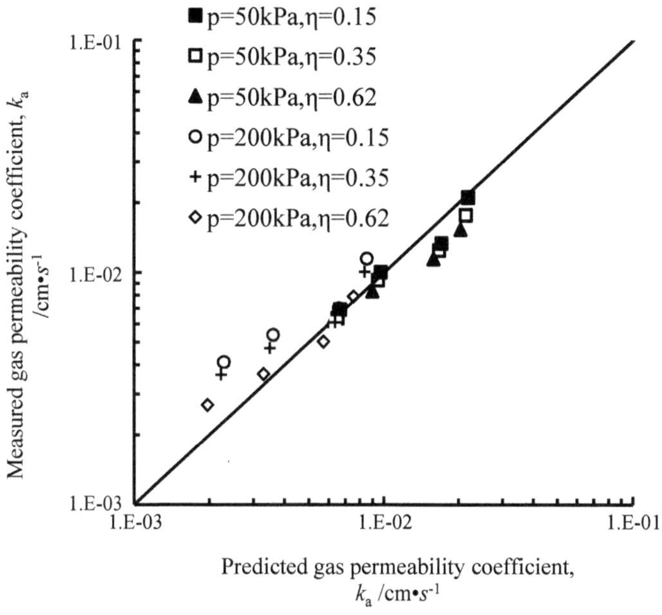

Fig. 8. Scatter plot comparison of predicted and measured gas permeability coefficient during wetting at different stresses

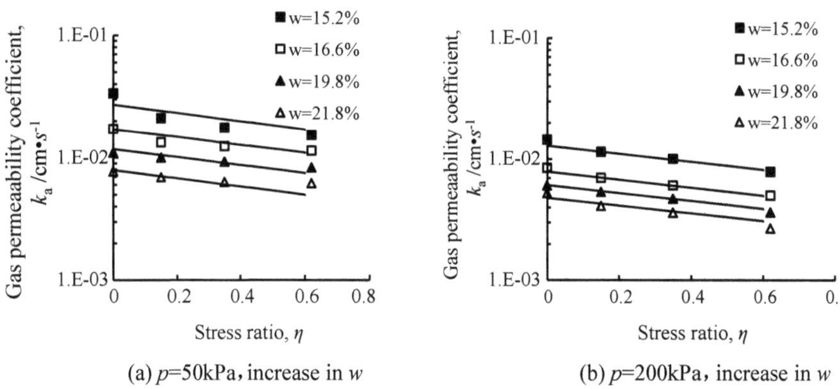

Fig. 9. Influence of water content on gas permeability coefficient k_a against stress ratio η during loading at the same mean stress

$$k_a = k_{ap} \exp(-c_1 \eta) \tag{9}$$

where k_{ap} and c_1 are soil parameters, and $k_a = k_{ap}$ as $\eta = 0$. These equal to the intercept and slope of the ln k_a-η line, respectively, and all have explicit physical meanings. The parameter k_{ap} is gas permeability coefficient corresponding to isotropic stress for a certain mean stress p as stress ratio $\eta = 0$. The parameter c_1 reflects the decreasing rate of gas permeability coefficient with the increase in stress ratio.

Figure 9 also revealed that the ln k_a-η lines at different applied stresses and water contents were approximately paralleled. In other words, the parameters c_1 were nearly equal as the stress ratio increase at different mean stresses for intact loess with various water contents. Regression of the data to the model (Eq. 9) resulted in $c_1 = 0.755$.

Inspection of Fig. 9 also suggested that the ln k_a-S_r curves were dropped to the lower with the increase of water content. That is to say, gas permeability coefficient decreased with increase in water content corresponding to the same stress ratio at the same mean stress.

For the intact loess in this paper, Chen et al. (2017a) found that the linear relationships between the logarithm of gas permeability coefficient, lnk_{ap}, and normalized mean stress p/p_a were approximately paralleled at isotropic stress at various water contents. These could be modeled by the following exponential function

$$k_{ap} = k_{a0} \exp(-c_2 p/p_a) \tag{10}$$

where k_{a0} and c_2 are soil parameters, and are the intercept and slope of the ln k_{ap}-p/p_a lines, respectively. The parameter k_{a0} is gas permeability coefficient at null stress, which decreased with increase of water content w. The parameter c_2 represents the decreasing rate of gas permeability coefficient with increase in normalized mean stress p/p_a, which is independent of water content, and equal to 0.346.

In order to establish the k_{a0}-w relationship, the relative moisture index, w_r (=$(w - w_n)/(w_s - w_n)$, where w_s (=$(G_s - \rho_{dn})/\rho_{dn}G_s$) is the saturated water content of specimen at zero stress) was introduced to reflect the change of water content relative to the saturation condition during wetting. The relative gas permeability at null stress, k_{a0}/k_{an0} (k_{an0} is the gas permeability coefficient at null stress and natural water content w_n), is plotted against relative moisture index w_r in Fig. 10. The k_{a0}/k_{an0} decreased with

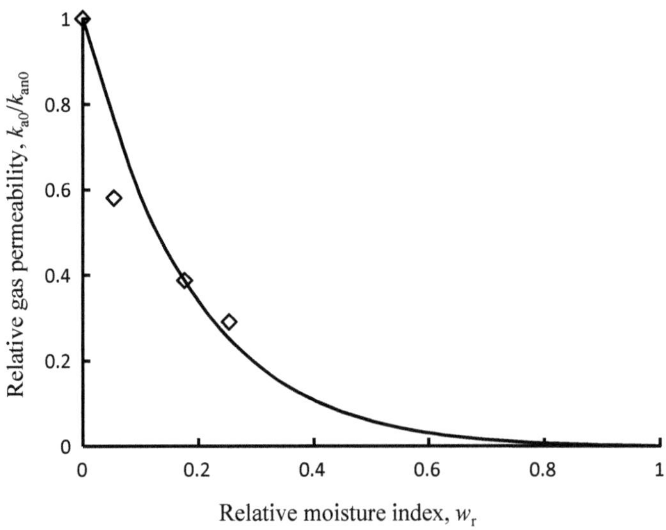

Fig. 10. Relative gas permeability k_{a0}/k_{an0} against relative moisture index w_r

increase in relative moisture index w_r. The k_{a0}/k_{an0} as a function of w_r was expressed by an exponential model

$$k_{a0}/k_{an0} = (1 - w_r)\exp(-c_3 w_r) \tag{11}$$

where c_3 is a soil parameter representing the decreasing rate of gas permeability coefficient k_{a0} with the increase in water content, and the $k_{a0}/k_{an0} = 1, 0$ as $w_r = 0$ ($w = w_n$), 1 ($w = w_s$), respectively. Regression of k_{a0}/k_{an0}-w_r to the Eq. (11) resulted in $c_3 = 4.3$.

Substituting Eqs. (11) and (10) into Eq. (9) yielded

$$k_a = k_{an0}(1 - w_r)\exp[-(c_1\eta + c_2 p/p_a + c_3 w_r)] \tag{12}$$

where the boundary conditions, i.e., $k_a = k_{an}$, 0 as $w = w_n(w_r = 0)$, w_s ($w_r = 1$), respectively, could be satisfied. Equation (12) using stress and water content as dependent variables was called the SWEL model. Gas permeability coefficient k_a at different applied stress (p, η) could be directly calculated using the SWEL model for intact loess with different water contents.

Figure 11 shows the predictions of gas permeability coefficient by the Eq. (12) against the measurements as stress ratio increases at different mean stresses and water contents. The predictions were found to be close to the measurements.

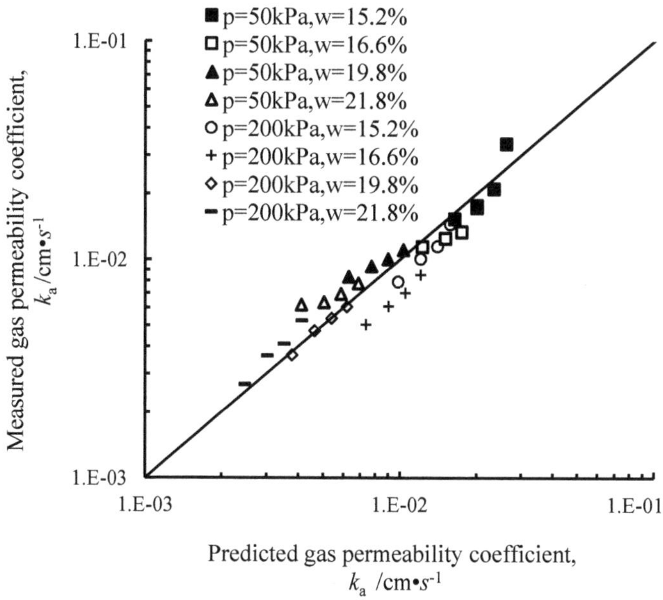

Fig. 11. Scatter plot comparison of predicted and measured gas permeability coefficient as increasing stress ratio at the same mean stress

The aforementioned analysis has shown that, the two gas permeability models (Eqs. 8 and 12) proposed with the same number of parameters all could give good predictions. The SWEL model (Eq. 12) and the RSEL model (Eq. 8) proposed in this study were an explicit and implicit function of applied stress and water content, respectively. Their influences on gas permeability could be directly taken into account by the SWEL model (Eq. 12) rather than the RSEL model (Eq. 8). The stress-strain constitutive relation was necessarily introduced to determine the saturation concerned with void ratio resulted from applied stress using the RSEL model (Eq. 8). Therefore, In comparison with the RSEL model (Eq. 8), the SWEL model (Eq. 12) was more convenient for engineering application.

4 Conclusion

A series of gas permeability tests were conducted to investigate the effects of water content, mean stress and deviator stress on gas permeability for the intact loess from Xi'an, China, using the refitted triaxial equipment for gas permeability. The conclusions drawn from this study were as follows.

Gas permeability coefficient nonlinearly decreased with increase in saturation during wetting at constant mean stress and deviator stress. The stress had strong influence on the relationship between gas permeability coefficient and saturation. The influence was dependent on the stress ratio at constant mean stress.

Applying stress under constant stress ratio path at the same water content, gas permeability coefficient rapidly decreased with the increase in saturation. The decreasing rate was subjected to the water content. The relationship between gas permeability coefficient and saturation could be merged together regardless of applied stresses.

The decreasing rate of gas permeability coefficient with the increase in saturation was related to the mechanism of increase in saturation. It was greater due to the increase of stress at the same water content than that of the increase of water content at the same applied stress.

The relationships between relative gas permeability versus saturation and relative saturation were all well merged together during wetting at different stresses. Gas permeability coefficient as function against relative saturation could be fitted by the proposed RSEL model.

The linear relationships between the logarithm of gas permeability coefficient and deviatoric stress ratio were paralleled at the same mean stress, but different water contents. It could be described by the proposed SWEL model.

Two models proposed all contained three parameters. But the effect of stress and water content on gas permeability coefficient could be directly considered in the proposed SWEL model. Therefore, the SWEL model is more convenient for the application of practical engineering.

References

Chen, Z.: On basic theories of unsaturated soils and special soils. Chin. J. Geotech. Eng. **36**(2), 201–272 (2014)

Chen, C., Zhang, D., Zhang, J., et al.: Gas permeability of intact Q_3loess under isotropic stress. Chin. J. Geotech. Eng. **39**(2), 287–294 (2017a)

Chen, C., Zhang, D., Zhang, J.: Influence of stress and water content on air permeability of intact loess. Can. Geotech. J. **54**(9), 1221–1230 (2017b)

Liu, F., Zhang, Z., Zhou, D.: Density-saturation-dependent gas-water permeability function of unsaturated loess. Chin. J. Rock Mech. Eng. **29**(9), 1907–1914 (2010)

Liu, J.F., Skoczylas, F., Talandier, J.: Gas permeability of a compacted bentonite-sand mixture: coupled effects of water content, dry density, and confining pressure. Can. Geotech. J. **52**(2), 1159–1167 (2015)

Parker, J.C., Lenhard, J.C., Kuppusamy, T.: A parametric model for constitutive properties governing multiphase flow in porous media. Water Resour. Res. **23**(4), 618–624 (1987)

Stoltz, G., Goure, J.P., Oxarango, L.: Liquid and gas permeabilities of unsaturated municipal solid waste under compression. J. Contam. Hydrol. **118**, 27–42 (2010)

Wei, H., Zhan, L., Chen, Y.: Experimental study on gas permeability of municipal solid waste. Chin. J. Rock Mechan. Eng. **26**(7), 1408–1415 (2007)

Yao, Z., Chen Z., Huang, X., et al.: Expermental research on gas permeability of unsaturated Q3 loess. Chin. J. Rock Mech. Eng. **31**(6), 1264–1273 (2012)

Zhan, T.L.T., Yang, Y.B., Chen, R., et al.: Influence of clod size and water content on gas permeability of a compacted loess. Can. Geotech. J. **51**(11), 1468–1474 (2014)

Weathered Swelling Mudstone Landslide and Mitigation Measures in the Yanji Basin: A Case Study

Zhixiong Zeng[1(✉)], Lingwei Kong[1], Min Wang[2], and Juzhao Li[1]

[1] State Key Laboratory of Geomechanics and Geotechnical Engineering, Institute of Rock and Soil Mechanics, Chinese Academy of Sciences, Wuhan, Hubei 430071, China
zzhxhp@163.com
[2] Rockfield Software Limited, SA18AS Swansea Wales, UK

Abstract. Some national expressways and railways were routed through the Yanji basin with weak rock foothills. To achieve acceptable grades, a lot of excavations in swelling mudstones were performed, eventually forming many swelling mudstone slopes. After heavy rainfalls, slope failures occurred which were often attributed to the presence of swelling mudstones. Field and experimental investigations were carried out on a representative landslide with emphasis on the landslide causes and the higher landslide incidence in south-facing slopes in the Yanji Basin was also discussed. On this basis, a multipath mitigation method including drainage and reinforcement measures was employed for the landslide case. Considering the effect of slope orientation, distinctive vegetation was distinguishingly planted on the south-facing and north-facing slopes. The later field survey showed that the undertaken mitigation measures were efficient to prevent the landslide development and deserved to be adopted in other swelling rock and soil regions.

1 Introduction

The Yanji Basin is located on the eastern foot of Changbai Mountain, northeast of China and located in the eastern part of Tianshan-Xinganling fold belt and southeastern part of hercynian fold belt. This region covers an area of about 980 km^2 and measures approximately 400 km from east to west and 35 km from north to south. The elevation in this region ranges from 155 to 1149 m. Rock type in the Yanji Basin mainly consist of silty mudstone and argillaceous siltstone in the Dalazi Formation of the Lower Cretaceous (K_1d_2), Longjing Formation of the Upper Cretaceous (K_2l) and Tertiary Hunchun Formation ($E_{2\text{-}3}h$) (He et al. 2003). In recent years, some national expressways and railways were routed in this region with foothills, creating cut slopes, and many of them failed during and after construction, which not only affected the day to day activities of the inhabitants but also caused great economic losses (Table 1).

In order to minimum the impact of landslides, it is important to indentify the key factors and adopt effective measures based on detailed integrated geological and geotechnical investigations. Previous studies reveal that weak mudstones can be easily

© Springer International Publishing AG, part of Springer Nature 2019
S. Shu et al. (eds.), *New Developments in Materials for Infrastructure Sustainability and the Contemporary Issues in Geo-environmental Engineering*, Sustainable Civil Infrastructures, https://doi.org/10.1007/978-3-319-95774-6_14

Table 1. List of the main landslide occurrences in the Yanji Basin before 2010

No	Location	Year	Failure mode	Rock type	Slope protection method
1	Tumen North River Bridge	1984	Earth slide	Clasolite; claystone	Gravel blind ditch; surface drainage; gravity retaining wall
2	Maoer Mountain	1998	Earth slide	Silty mudstone; muddy siltstone	Trimming and cutting
3	Zhongli Village	1999	Earth slide	Tuffy sandstone	Anti-slide pile
4	Jihun Expressway (K310)	2000	Earth slide	Mudstone	Gravel blind ditch; geotextile
5	Zhongli Village	1984–2012	Earth slide	Gravel; mudstone	Gravity retaining wall; anti-slide pile
6	Yanji Expressway Entrance	2001	Earth slump	Mudstone	Trimming and cutting; shotcrete with wire mesh
7	Taiyan Mountain	2001	Soil creep; cracking	Mudstone	Surface drainage; anchored pile; grille beam
8	Hezuo Village	2001	Soil creep, cracking	Alternating sandstone and mudstone	Artificial digging pile
9	Piyan Mountain	2002	Debris Slump	Mudstones, shaly sandstone	Sand bag barriers

softened and slaked when soaked and weathered, and pose a significant challenge to slope stability (He et al. 2005; Kong et al. 2017). Most of the landslide studies in the Yanji Basin have not paid adequate attention to the slope protect methods due to the lack of detailed investigations in the region. This paper records a representative landslide event during the construction of the Jilin-Hunchun high-speed railway in the Yanji basin and presents an effective and slope protection method for its control.

2 General Site Condition

2.1 Landslide Event

A new high-speed railway line connecting Jilin and Hunchun traverses the Yanji Basin from west to east. To achieve acceptable grades, the construction of this line required numerous excavations in the hillsides and created many north-facing and south-facing cut slopes. In the afternoon of October 30th, 2012 after a continuous 3-day rainfall, the K275 south-facing slope slid between markers km 275 + 945 to km 276 + 205 (Fig. 1).

The total volume of the released rock mass was estimated to be about 503000 m^3. The elevation of the front edge of the landslide was 248 m, and the elevation of the back edge was 274 m; thus, the relative elevation difference 26 m. The length of the slide was approximately 185 m and the width at the toe was around 260 m.

Fig. 1. Site photograph illustrating the K275 landslide

2.2 Geology

An engineering geological survey at the landslide site shows that local geology comprises a mantle of superficial deposits overlying two weathered mudstones in Longjing Formation of the Upper Cretaceous. Figure 2 illustrates a typical profile of the geology in the region. The lower deeper part consists of yellow-brown mudstones, while the upper shallower part is made up of magenta mudstones.

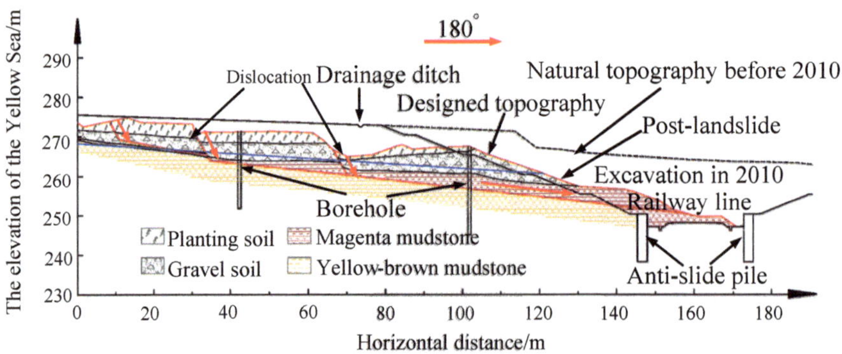

Fig. 2. Typical profile of the K275 landslide

2.3 Climate

The Yanji Basin has a sub-humid continental monsoon climate with dry cold winters and wet hot summers. Based on the data from National Meteorological Information Center, China Meteorological Administration, the temperature in the Yanji Basin changes from being positive to negative approximately in December, and returns from being negative to positive in April of the next year with a maximum frozen depth of about 160 cm. The annual precipitation is about 529.8 mm, more than 80% of the total precipitation is received during the monsoon season from May and September (Fig. 3).

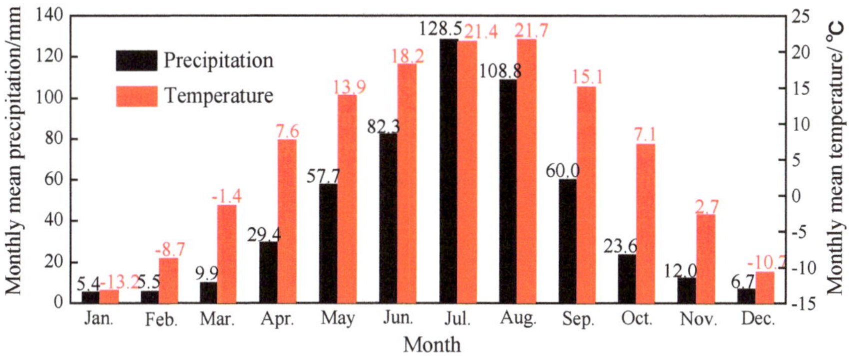

Fig. 3. Monthly mean precipitation and temperature in the Yanji Basin

3 Characteristics of the Mudstones

Several mudstone blocks selected from different positions were crashed into pieces and well mixed for physical and mineralogical analyses. Randomly selected pieces (about 5 g) were oven-dried and then ground again to pass through a 74 μm sieve for X-ray diffraction. Primary mineral components of the yellow-brown mudstone include montmorillonite (23.0%), illite (5.2%) and quartz (23.5%), whereas the magenta mudstone also mainly composes of montmorillonite (29.3%), illite (4.8%) and quartz (32.0%). Table 2 lists detailed physical properties of two weathered mudstones. According to the Unified Soil Classification System (USCS), the yellow-brown and magenta mudstones in the Yanji Basin are respectively a low-plasticity clay (CL) and a high-plasticity silty clay (MH). From the plasticity index and free swelling ratio, it can be deduced that the yellow-brown mudstone has low-swell potential, while the magenta mudstone has medium-swell potential. The zero-swell swelling pressures of yellow-brown and magenta mudstones are 208 kPa and 201 kPa, respectively.

Table 2. Physical properties of the mudstones

Sample	w (%)	ρ_d (g/cm^3)	G_S	e	PL (%)	PI (%)	δ_f (%)	P_s (kPa)	e_s (%)
Yellow-brown	20.0	1.67	2.72	0.63	25.6	5.8	54	208	14.0
Magenta	28.1	1.48	2.79	0.88	34.6	9.4	64	201	20.6

w: natural water content; ρ_d: dry density; G_S: specific gravity; e: void ratio; PL: plastic limit; PI: plasticity index; δ_f: free swelling ratio; e_s: volume shrinkage ratio; P_s: swelling pressure determined by zero-swell method.

From the Table 2, the magenta mudstone is looser with a low density and large void ratio compared with the yellow-brown mudstone. While rainfalls infiltrated into the slope, the yellow-brown mudstone acted as an impervious or relatively impermeable bottom layer that led to the formation of a perched ground water table in the magenta mudstone. The large seepage pressure induced by the flow after heavy rainfalls increased the sliding force and the increase in pore water pressure on the sliding surface led to the decrease in the resistance force, both of which consequently accelerated the slope failure (Wang et al. 2016; Kong et al. 2017).

The shear strength parameters were also obtained from direct shear tests on the disturbed samples. The mudstones exhibit a great decrease in the shear strength once they absorb water (Table 3). During or immediately after heavy rainfall in summer, the groundwater table quickly rose to the ground surface, though it was usually located at a considerable depth below the ground surface (Yue and Lee 2002; Kwong et al. 2004; Kim et al. 2004). Consequently, a sharp reduction in shear strength occurred in the mudstones. In addition, due to the repeated drying, wetting, freezing and thawing conditions in the Yanji Basin, the mudstones exhibited a rapidly decline in the shear strength (Kong et al. 2017).

Table 3. Shear strength parameters of the mudstones

Sample	c_1(kPa)	ϕ_1(°)	c_2(kPa)	ϕ_2(°)	Remarks
Yellow-brown	113.8	30.9	32.1	24.7	Disturbed sample
Magenta	109.4	27.3	31.9	24.5	Disturbed sample

c_1 and ϕ_1: cohesion and internal frictional angle before absorbing water; c_2 and ϕ_2: cohesion and internal frictional angle after absorbing water.

4 Effects of Slope Orientation

Along the Yanji section of the Jilin–Hunchun railway line, it had also been manifested that landslides always occurred in south-facing slopes, the stability of which was distinctly lower than north-facing slopes. A similar phenomenon was also reported by Zheng (2005), Cheng et al. (2011) and Ren (2014), who studied swelling rock or soil landslides during and after the construction of railways or expressways in Nanning, Nanyang and Hanzhong Basins of China. This is due to the microclimatic conditions on the slopes of different orientation will vary dramatically, despite located only a few hundred meters apart and sharing the same macroclimatic zone (Auslander et al. 2003).

The Yanji Basin is located in the middle latitudes of the Northern Hemisphere, and the sun rises in the southeast and sets in the southwest every day. With respect to incident angle and daylight hour, it is easy to explain that south-facing slopes received more solar radiation than north-facing ones. Accordingly, the temperatures in south-facing slopes were higher than those in north-facing slopes, which mean that the south-facing slopes had a more xeric environment that was a warmer, drier and more variable microclimate (Segal et al. 1987; Auslander et al. 2003). On the one hand, the higher temperatures in south-facing slopes caused the development of extensive shrinkage and cracking, which facilitated rainfall infiltration and water evaporation. Under rainfalls following the high temperature, drier mudstones in the south-facing slopes could exhibit larger swelling deformations and lose substantial portions of shear strengths after absorbing water. On the other hand, the mudstones underwent more intensive drying-wetting-freezing-thawing processes, causing larger decreases in the shear strengths of mudstones. Both of them decreased the stability of the south-facing slopes (Kong et al. 2017). Simultaneously, such microclimatic conditions on south-facing slopes also accelerated the erosion of mudstone slopes and reduced plant developmental stability (Auslander et al. 2003), resulting in a higher percentage of bald south-facing slopes (Chang et al. 1996; Lee et al. 2007).

5 Landslide Mitigation

According to the aforementioned analysis, the mitigation measures for landslide hazards in the Yanji Basin should be based on the following essential principles: reduce water seepage and enhance groundwater discharge in slope body; slow down the weathering of the mudstones. Thereby, water-proofing and moisturizing measures must be adopted on the newly formed cut slope surfaces after the excavation to prevent rainfall infiltration into the slope body and reduce evaporative water loss. For example, drainage ditches and catchment wells could be excavated to lower the groundwater table, and vegetation cover can be established on the slope surfaces to maintain the water content of the slope body. With consideration of the effect of slope orientation, vegetation cover should be established on the basis of microclimatic conditions in south-facing and north-facing slopes. Besides, a number of reinforcement measures, such as reinforced concrete grids, anti-slide piles, and retaining walls with or without anchors, can also be employed to control the swelling deformations of mudstones and increase their shear strengths, and the reinforcement facilities must have adequate strengths to resist the swelling pressures of the mudstones. Additionally, the reinforcement facilities in south-facing slope are supposed to have higher strengths to recover the greater loss in shear strength of slope body under drying-wetting-freezing-thawing cycles.

As for the K275 landslide, a slope protection method was proposed (see Fig. 4). Specifically, drainage, transverse and filter ditches, as well as catchment wells were excavated to drain groundwater and surface water. Arch grille beams with vegetation and double-row anti-slide piles were constructed to prevent the swelling deformations of mudstones. The vegetation on the south-facing and north-facing slopes differed considerably. The south-facing slopes were planted with Mongolian pine, a heliophilic, deep-rooted, and evergreen conifer that adapts well to various climatic regions and is widely used for afforestation in the northeast region of China. By contrast, the north-facing slopes were covered with Amorpha fruticosa Linn., a perennial leguminous shrub with good resistance to shady, cold and dry conditions.

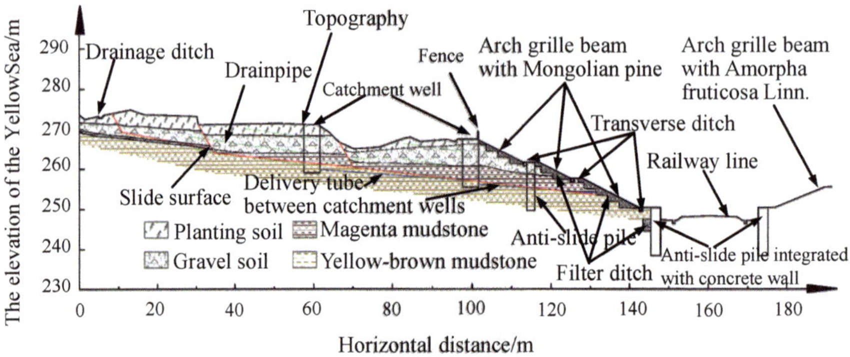

Fig. 4. Slope protection method used in the K275 landslide

The Yanji section of Jilin-Hunchun high-speed railway line has been exploited commercially since September 2015. On September 23, 2016, we returned to the K275 site. There was no damage observed on the surface of the cut slopes (see Fig. 5), and thus we deduced that the slope protection method was effective.

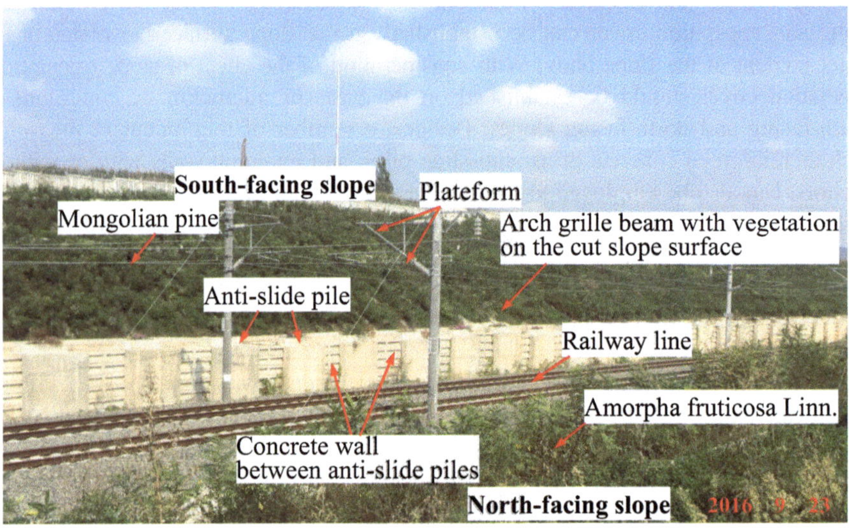

Fig. 5. Site photograph illustrating the K275 landslide a year after the opening of the railway line to traffic

6 Conclusions

Landslides occurred frequently in the Yanji Basin, northeast of China. The landslides were caused by the combined effect of the water seepage due to heavy rainfalls and the weathering of mudstones induced by fluctuating climatic conditions.

It was also inferred that slope orientation differences significantly affected the stability of swelling mudstone cut slopes. In contrast to north-facing slopes, south-facing slopes had a more xeric and variable microclimate that not only accelerated the weathering of mudstones, but also influenced the plant growth on the slopes, eventually increasing the susceptibility to slope failure.

A multipath mitigation method for the landslides in the Yanji Basin mainly included construction of drainage and reinforcement measures, and planting of vegetation. With the consideration of the effect of slope orientation, Mongolian pine and Amorpha fruticosa were planted on the south-facing and north-facing slopes, respectively.

Acknowledgments. This work was financially supported by the National Natural Science Foundation of China (Grant Nos. 41430634) and the State Key Laboratory of Geomechanics and Geotechnical Engineering (Grant No. Y11002). The authors are very grateful to their financial support.

References

Auslander, M., Nevo, E., Inbar, M.: The effects of slope orientation on plant growth, developmental instability and susceptibility to herbivores. J. Arid Environ. **55**(3), 405–416 (2003)

Chang, H.W., Tien, T.M., Juang, C.H.: Formation of south-facing bald mudstone slopes in southwestern Taiwan. Eng. Geol. **42**(1), 37–49 (1996)

Cheng, Z.L., Li, Q.Y., Guo, X.L., Gong, B.W.: Study on the stability of expansive soil slope. J. Yangtze River Sci. Res. Inst. **28**(10), 102–111 (2011)

He, M.C., Leng, X.C., Heng, C.Y., Wu, X., Zhang, M.F.: Research on the expansive soft rock in Yanbian region. Chin. J. Rock Mechan. Eng. **22**(7), 1151–1155 (2003)

He, M.C., Liu, C.Y., Wu, X.: A case study on the reinforcement countermeasures for strongly swelling soft rock llope in Yanji Basin. J. Jiling University **35**(4), 496–500 (2005)

Kim, J., Jeong, S., Park, S., Sharma, J.: Influence of rainfall-induced wetting on the stability of slopes in weathered soils. Eng. Geol. **75**(3), 251–262 (2004)

Kong L.W., Zeng Z.X., Bai W., Wang M.: Engineering geological properties of weathered swelling mudstones and their effects on the landslides occurrence in the yanji section of the jilin-hunchun high-speed railway. Bull. Eng. Geol. Environ. 1–13 (2017)

Kwong, A.K.L., Wang, M., Lee, C.F., Law, K.T.: A review of landslide problems and mitigation measures in Chongqing and Hong Kong: similarities and differences. Eng. Geol. **76**(1), 27–39 (2004)

Lee, D.H., Yang, Y.E., Lin, H.M.: Assessing slope protection methods for weak rock slopes in Southwestern Taiwan. Eng. Geol. **91**(2), 100–116 (2007)

Ren, Q.: The Research of characteristics and failures mechanism on typical landslide systems along Shiyan-Tianshui Expressway West Hanzhong Section. Ph.D. thesis, Changsha University of Science and Technology, Changsha, China (2014)

Segal, M., Ookouchi, Y., Plelke, R.A.: On the effect of steep slope orientation on the intensity of daytime upslope flow. J. Atmos. Sci. **44**(44), 3587–3592 (1987)

Wang, J., Su, A., Xiang, W., Yeh, H.F., Xiong, C., Zou, Z., Zhong, C., Liu, Q.B.: New data and interpretations of the shallow and deep deformation of Huangtupo no.1 riverside sliding mass during seasonal rainfall and water level fluctuation. Landslides 1–10 (2016)

Yue, Z.Q., Lee, C.F.: A plane slide that occurred during construction of a national expressway in Chongqing, SW China. Q. J. Eng. Geol.Hydrogeol. **35**(4), 309–316 (2002)

Zheng, P.: The stability analysis and treatment of technical research for expensive soils cut slope in Nanning-Youyiguan Expressway. Master thesis, Changsha University of Science and Technology, Changsha, China (2005)

Soils Improvement by PVD in a Harbor Storage Area

Alexandre Santos-Ferreira[1(✉)], Joana Lemos[2], and Paula F. da Silva[3]

[1] DGRM, Lisbon, Portugal
asantos.ferreira1954@gmail.com
[2] Faculdade de Ciências e Tecnologia, Universidade Nova de Lisboa, Lisbon, Portugal
j.lemos@campus.fct.unl.pt
[3] Departamento de Ciências da Terra, Faculty of Sciences and Technology, NOVA University of Lisbon, Lisbon, Portugal
apfs@fct.unl.pt

Abstract. In the south bank of the Tagus River, in the Lisbon harbor, it was necessary to create a large raw mineral storage area for a steel plant adjacent to it, as well as a new quay for discharge of those materials, some years ago. The occurring formations in the area included very soft soils layers, up to 17 m thick. The inherent geotechnical problems to the development were of two types: those soils did not have enough strength to support a walking man, least the storage piles, without a substantial improvement of their strength; and the equipment used, on rail, to move the raw minerals could not be submitted to differential settlements greater than 15 cm, over 500 m. The total area of the storage facility to improve was 25 ha. This paper introduces the comparison between the monitoring data of the experimental embankment built on top of the soil improved area by PVD with a recent 3D modelling of the stresses evolution. A special attention is given to the behavior, under loading, of the soft soils foundation, especially by comparing it with the first results of the trial embankment.

1 Introduction

The need of a new raw mineral storage area (NRMS) for a steel plant adjacent to the River Coina, located at the south bank of the Tagus River (Lisbon harbor), led to the execution of major earthworks, some years ago, covering a total area of 25 ha, which 546,000 m^2 presented soft to very soft clays, with 17 m of maximum identified thickness in some places – Fig. 1.

This particular situation was studied by an extended geotechnical exploration plan, including field and laboratory tests, complemented by an instrumentation plan for an initial general soil treatment, of an area of 1500 × 150 m^2, by a 2.5 m embankment surcharge – Fig. 1. Afterwards, an experimental embankment was design to have a better insight of the soft soil behavior and the PVD improvement of the area under operation with the future raw materials piles – 12 m high ore piles with a specific weight of 23 kN/m^3; this was achieved by a detailed monitoring plan of the experimental embankment. Due to all this, it was achieved a very complete and full scaled field test.

© Springer International Publishing AG, part of Springer Nature 2019
S. Shu et al. (eds.), *New Developments in Materials for Infrastructure Sustainability and the Contemporary Issues in Geo-environmental Engineering*, Sustainable Civil Infrastructures, https://doi.org/10.1007/978-3-319-95774-6_15

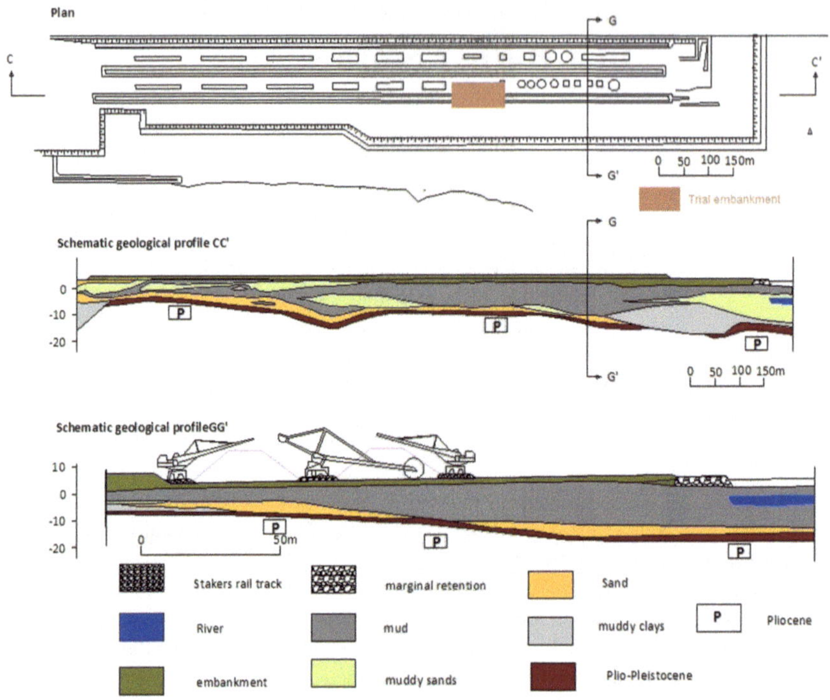

Fig. 1. NRMS: general plan and two geological profiles

In this paper, the geotechnical problems of NRMS are summarized, with special attention to their description as well as the behavior of the soft soils foundation, and the comparison between the monitoring data of the trial embankment built with a recent 3D modelling of the stresses is also discussed.

2 Geological and Geotechnical Setting

The location of the NRMS, in a floodplain, next to an important river, has largely conditioned the geological settings at the site. The complexity of the geological conditions, both the vertical and the lateral variability of the sedimentary layers, were anticipated due to the site location, in the Tagus River Estuary, and it was necessary to carry out an extensive site investigation plan. This included dynamic probing tests (about 60 across NRMS) and percussion drilling, which were accompanied by regular SPT as well as undisturbed systematic sampling (more than 300 piston samples across the NRMS). The undrained strength characteristics of the alluvium was assessed, at the site, by vane shear tests (VST) in soft to very soft clays (in a total of more than 80 tests), and more than 300 laboratory test (identification and mechanical proprieties) in the unaltered samples – Table 1.

Table 1. Main geotechnical proprieties of the mud alluvium

Geotechnical propriety	Area improved (PVD)		Surrounding areas	
	Limit values	Average	Limit values	Average
W_n [%]	75–104	91	40–110	80
Sr [%]	100	100	100	100
γ [KN/m^3]	1.41–15.5	14.9	1.41–17.0	15.4
γ_d [KN/m^3]	6.9–8.7	7.8	6.7–12.9	8.7
% < #200$_{ASTM}$	75–99	93	500–100	91
W_L [%]	48–72	59	31–67	56
IP [%]	22–36	28	14–33	25
c_u [KPa]	1.00–13	5	0–31	11
φ_u (°)	0–8	1	0–12	0
c_{cu} [KPa]	0–19	3	1–35	2.10
φ_{cu} (°)	12–21	18	3–20	19
c′ [KPa]	0–0	0	0-15	3
φ' (°)	25–38	31	5–20	11
Es [KPa]	1400–11500	8100	1100–12400	5600
e_o	2.1–2.926	2.463	0.958–2.855	2.095
C_c	0.54–1.07	0.71	0.42–0.88	0.65
C_v [m^2/s]	7×10^{-10}–3×10^{-9}	1×10^{-9}	8×10^{-10}–3×10^{-9}	1×10^{-9}
C_h [m^2/s]	2×10^{-9}–6×10^{-9}	4×10^{-9}	1×10^{-9}–9×10^{-9}	5×10^{-9}
C_α	6×10^{-4}–2×10^{-2}	1×10^{-2}	5×10^{-9}–1×10^{-2}	0.9×10^{-4}
σ_p/σ'_o	0.2–1	0.5	0.1–1	0.6
c_{uVST} [KPa]	1.50–38.30	13	1–49	17
S_{tVST}	4–28	–	4–8	–

Figure 1a shows the outline for the NRMS, while Fig. 1b, c present, respectively, the longitudinal and transversal geological profiles C-C′ and G-G′, respectively, with the latter showing the most critical situation in NRMS, from a geotechnical point of view (Soeiro and dos Santos 1983). Those profiles (Fig. 1) represent a set of geological units, deposited between the Pliocene until recent times and encompasses the following lithology, from more recent to the older ones:

(a) Alluvium (Recent):

The top unit is composed by dark organic grey clays, with muddy characteristics and frequent shell fragments. The upper layers of this formation displayed variable consistencies, between soft and very soft ($N_{SPT} < 4$); in deeper layers or in already surcharged layers with a pre-existing embankment of alluvial soils and slags, it was possible to identify a medium to hard consistence ($4 < N_{SPT} < 15$). It's possible to acknowledge that, due to its physical and mechanical characteristics, this first layer is the conditioning stratum from a geotechnical point of view. Its thickness varies from 5 to 20 m, along the site. Table 1 summarizes some of the geotechnical characteristics of these layers across the NRMS area.

Bellow, there was a stratum composed by variable sized light grey sand with shell fragments, moderately ($10 < N_{SPT} < 30$) to very compacted ($N_{SPT} > 60$).

Deeper, it was detected a layer composed by light grey muddy fine to coarse sands with some shell fragments and exhibit moderate ($10 < N_{SPT} < 30$) to high compaction ($N_{SPT} > 60$).

Bellow, there was grey, sometimes greenish, silty clays with some shell fragments; this layer was more dense than the upper stratum, with variable consistence between hard ($8 < N_{SPT} < 15$) and stiff ($N_{SPT} > 30$).

(b) Sands with scattered pebbles (Plio-Pleistocene):

This layer represents the transitions between the Alluvium and the Pliocene strata, and were sometimes eroded and decompressed. In its composition, it was possible to identify coarse to fine sands, sometimes clayey, with scattered pebbles. This layer presents light colors, like white stained yellow and its density varies between moderate to very compacted ($16 < N_{SPT} < 60$). This layer has a quite uniform thickness of 5 m.

(c) Very compacted Sands (Pliocene):

The oldest formation bored are Pliocene sands. This unit presents the highest relative compaction - very compacted ($N_{SPT} > 60$), and were constituted by brownish yellow or whitish yellow sands. In the south zone it was detected some intercalations of hard clay lenses.

Summing up, for the targeted area, the following zoning can be systematized:

(a) The top seventeen meters maximum are constituted by alluvium layers, with variable thickness, overlapping a substratum formed by medium to very high compact sands from Plio-Pleistocene;
(b) In general, the thickness of the alluvium layers increases from South to North, and from West to East;
(c) As usual in this type of formations, the nature and evolution of the different kind of alluvium layers was very irregular;
(d) It is clear the predominance of the muddy beds in the north zone (Fig. 1), where important fossil valleys can be identified, which were eroded in recent geological times.

3 Design Constraints and Premises

As previously mentioned, a total area of 1,900,000 m^2 was subjected to earthworks, with a total volume of controlled fill of about 669,900 m^3. This area is divided into several platforms, at various heights: +2.60 m; +3.40 m; +6.00 m; +7.50 m, imposed by the global project of the NRMS. The NRMS was established by a platform at the height of +3.40 m, having an extended stabilizing +2.60 m height berm. In this particular area, the average ground surface had an average height of 0.00 m, and it is where the higher thickness of soft clays was expected, showing undrained resistance values of about 1 kPa in the top layers, increasing with depth. This situation also occurred in the surrounding areas, totalizing an area of 546,000 m^2.

As already mentioned, the mud alluvium - soft and very soft clays, had the most critical conditions in NRMS, therefore these formations were deeply studied, along with some of the underlying recent clay layers. About the upper alluvium layer, it is important to refer the following facts:

- The analysis of 60 VST results allowed to determine that 16% of the very soft clays are quick clays' type ($s_t > 16$), 34% were extra-sensitive type ($8 < s_t < 16$), 40% were of sensitive type ($4 < s_t < 8$) and only 10% were normal type clays.
- The analysis of the first elements lead to consider that the more unfavorable values to be used in numerical analysis was $C_v = 1 \times 10^{-9}$ m^2/s, and $C_h = 1 \times 10^{-8}$ m^2/s. However, subsequent analysis of the trial embankment allowed considering more favorable values.
- Considering values of $c' = 0$ kPa and $\varphi' = 25°$, the increase of undrained strength for a settlement under a surcharge $\Delta\sigma'_c$ (in terms of effective stresses), follows the equation $c_u = 0.34.U.\Delta\sigma'c$. When considering a value of $\varphi' = 18°$, the equation is $c_u = 0.32.U.\Delta\sigma'c$. These values were confirmed, later, by results of the instrumentation placed on site.

In this context, four distinct solutions to improve the ground were considered, based on the following parameters:

- Execution methods;
- Execution and implementation time;
- Advantages and disadvantages;
- Financial costs.

The geological zoning showed that was impossible to execute the earthworks, not only due to the impossibility of circulating the equipment for filling and ground leveling, even small ones, but also because the lack of strength of the very soft clays compared to the loads imposed by the embankments to be built.

Therefore, it was decided to properly analyze the alternative solutions for the embankments execution, taking in consideration the following initial restrictions:

(a) It was necessary to assure that it would be possible to execute the construction method proposed;
(b) The chosen solution would have to ensure, in useful time, firstly the achievement of the stability characteristics of the embankment, and secondly the global stability of the NRMS in the future, during operation.

Considering that the natural geotechnical characteristics did not guarantee the stability of the embankments that needed to be build, when exposed to loading during exploration phase, it was possible to preliminarily design an adequate treatment to guarantee the necessary improvement to the soil properties of the upper alluvium layers.

4 Design Solutions

It was based on the above mentioned limitations that a final decision for the development was made. In general and in order to improve the soil foundations and to decrease the magnitude of the settlements, or to reduce the speed at which the settlements would proceed, there were two general processes to consider: replacement of the compressible material or pre-consolidation prior to the beginning of operation of the project. This way, four design solutions were prepared and compared, based on the following procedures:

(a) Total replacement of the muds (very soft clays);
(b) Sand vertical drains with a helical brought;
(c) Sand vertical drains made by water jets;
(d) Prefabricated vertical drains (PVD).

The last three mentioned predicted the use of a surcharge to accelerate the muddy alluvium consolidation, increasing its stiffness and shear strength. It was proven that the solutions (a) and (c) were the most expensive; on the other hand, solutions (b) and (d) were practically similar in costs. The solution (d) had the advantage of allowing a greater acceleration in the consolidation time prior the operation phase while also increasing, simultaneously, the stiffness and strength of the muddy formations.

Concerning the constructive problem of the embankment's execution mode, and regarding that the top soils couldn't bear any extra loads, like workers or machines, it was necessary to create in it an advanced platform that would allow the machines to circulate. This platform would also act as a blanket drainage layer, considering the need to drain all the waters to be removed from the muds by PVD. This construction problem was solved by using in a large scale nonwoven geotextile (450 g/cm^2 and 650 g/cm^2). A total area of 540,000 m^2 were covered with this type of geotextile, which allowed the using of light equipment to perform the first layer of the embankment with 0.50 m of thickness. Thereafter, it was possible to use normal equipment to complete the earthworks. The purposes of the geotextile were:

- Initial reinforcement of the soil allowing the circulation of earthwork light equipment;
- Contribution of its tensile strength to the global stability of the embankments;
- Separation between the saturated soft soil foundation and the embankment material, preventing its contamination and consequent reduction of its strength properties and permeability.

The sizing of the PVD mesh to be applied was elaborated according to the traditional theory of Barron (1947). It was considered, for equivalent diameter d of each drain, used in the definition of n (quotient of the radius of each drain), the value of $d = 2\ (b + e)/\pi$, where b is the width of the drain and e its thickness. In the case of the drains used in this project, b = 0.1 m e e = 0.004 m, which results in a d = 0.05 m, taking into account the disturbance induced by the drains setting.

The calculation of the meshes size was based on the following assumptions, needed to assure that the NRMS could become operational on schedule, and in its best conditions:

- Consolidation ratio would have to be at least 60% after 180 days;
- Consolidation ratio would have to be at least 70% after 240 days.

Based on these elements, the area to be improved was divided in three distinct zones, according to the thickness variation of the soft soils, with three different drain meshes:

- Zone A, with average soft soils thickness of 7 m, was defined a square mesh with 1.80 m of length;
- Zone B, with average soft soil thickness of 11 m, was defined a triangular mesh with 1.80 m of length;
- Zone C, with average soft soil thickness of 15 m, was defined a triangular mesh with 1.80 m of length, reinforced with one drain in the center.

The work platform was defined at a height of +2.60 m; after the execution of the drains, this height was covered with 0.80 m of a blanket drainage layer, comprising sandy soils with high permeability. In total, 1,100,000 m of PVD were used, which makes this project one of the largest of this type at the time. The platform final height of +3.40 m was later raised, in two phases, to a final height of +5.00 m, thus constituting the surcharge associated with the vertical drainage system for speeding up the muds consolidation.

The behavior of the foundation was duly followed by observation and interpretation of a general instrumentation plan – settlement plates. Through the analysis of the settlements referenced in these plates, using the Asaoka method (Asaoka 1978), it was possible to confirm the analytical predicted amplitude of the settlements (ranging 1 m for this surcharge of about 5 m of fill, and for a compressible layer thickness of 8–10 m), and it was also possible to conclude that the consolidation was much faster than theoretically predicted. After 180 days, the consolidation ratio was in the order of 70%, after 240 days was about 80–85%, and after 360 days was near 100%. This data confirms the excellent functioning of the vertical drains, as well as the efficiency of the preloading solution, allowing to consider the hypothesis of using this method when the NRMS goes into operation – phase of ore loading – thereby reducing the cost of the treatments to be undertaken. The general strength improvement was assessed by comparing VST results, both before and after the treatment.

An important problem that arose, concerned to the predicted settlements for the rail paths of the overloading machines. The severe limitations imposed by the equipment manufacturers – they could not be submitted to differential settlements greater than 15 cm, over 500 m, made it impossible to solve the problem only by using surcharges. Therefore, it was necessary to analyze other solutions, among which the execution of reinforced concrete piles, supporting a platform, also in reinforced concrete, where the rails of the stackers would be installed. The final solution adopted, as being the most favorable, was reinforcement with stone columns of the foundation soils.

5 Trial Embankment

5.1 Design of the Embankment

Due to the ground characteristics previously described, as well as the uncertainties that still existed, and which could not be answered through laboratory tests and calculations, it was also decided to build an experimental embankment with the purpose of analyzing, in real and full scale, the behavior of the very soft clays after improvement the whole area with PVD and first phase surcharge fill. This embankment would have to be duly instrumented and well observed. The location to install these experimental infrastructure – Fig. 1a, was chosen based on which areas were more unfavorable and which one was treated with the PVD longer time ago. This way, it was chosen an area where the drains had been installed 5 month ago, after the consolidation ratio was more than 90%. The general second height increase was not followed up in the area of the experimental embankment, since this area was intended to be as stable as possible, having even been defined an area for its protection in which the circulation of vehicles was forbidden.

To better represent the loading with ore piles, as well as to have areas far enough away from the limits of the NRMS, to be able to consider flat stages of stress distribution, it was chosen an embankment with dimensions of 40 × 80 m, and 6 m height, to be performed in phases of 2 m each. Afterwards, it was decided to change the total height to 8 m, so that there would be 4 loading stages, but unfortunately only the first two were completed – Fig. 2.

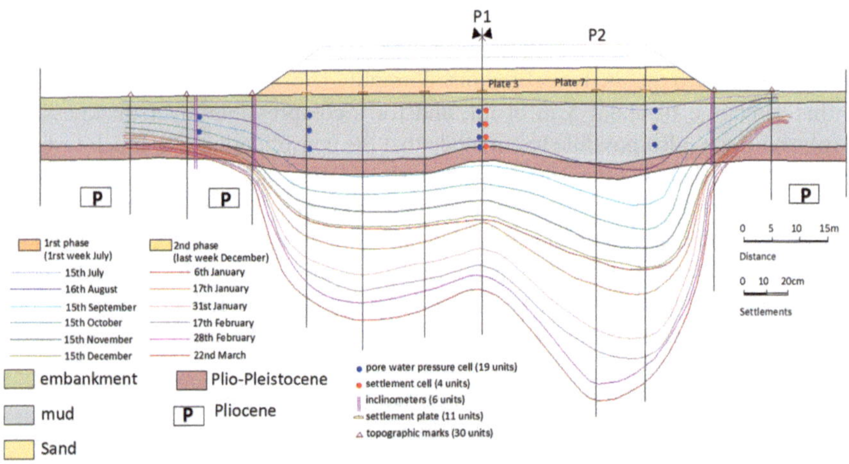

Fig. 2. Longitudinal cross section of the trial embankment, with simplified geology, instrumentation and settlements evolution based on the settlement plates measuring

Summing up, with this experimental embankment it was intended to set the amplitude of the settlements and their speed; to define the lateral deformations of the soils of the foundation; and to verify the efficiency of the improvement with PVD. The embankment was monitored with piezometers, settlement cells and plates (in a total of 4 and 11, respectively), pore water pressure cells (total of 19), topographic marks (in a total of 30) and vertical inclinometers (in a total of 6) – Fig. 2. Due to the size of the embankment, two different profiles were instrumented, a transverse one and a longitudinal one, intersecting in the center of the embankment. Each loading phase was executed as fast as possible to allow an acceptable interpretation of the observations.

Figure 2 presents the cross section of the embankment, including the soil foundation and the registered settlements along time. The frequency of observation of the instrumentation, for each loading phase, was stablished as follows:

i. during the construction of the embankment, two daily readings (at the beginning and at the end of the works);
ii. in the next 15 days following the construction of the embankment, one daily reading;
iii. in the next 15 days, readings every two days;
iv. in the next 3 months, weekly readings;
v. in the next 3 months, readings every two weeks.

5.2 Instrumentation: Results and Discussion

The observation of the general embankment allowed, since it was constructed and monitored, to predict, at the time of the construction, a probable maximum value of 0.57 m of settlement, for a 2 m height load, and a consolidation ratio between 80–90%, after 5 months of the surcharge implementation. From the observation of the trial embankment, it was observed that the consolidation ratio ranges between 80–86%, with settlements of around 0.49–0.50 m, after 5 months. Therefore, there was an excellent agreement of the results with the empirical methods then used. Figure 3 presents the observation results of the pore water cells, under the center of the trial embankment.

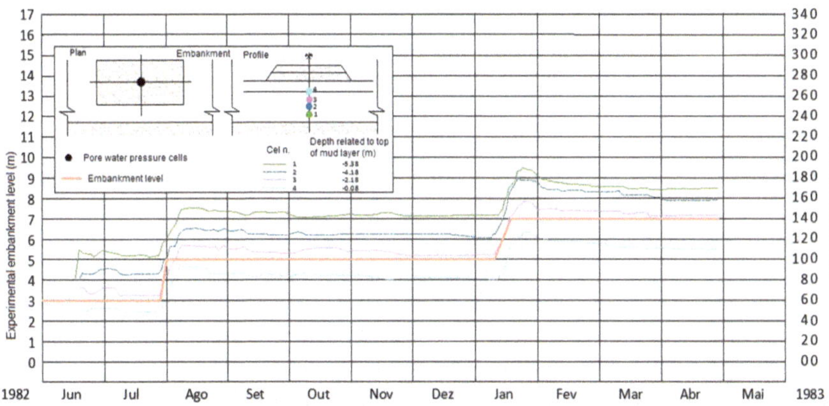

Fig. 3. Monitoring of the four pore water pressure cells under the center of the trial embankment, and evolution of the embankment height

As expected, the raise of the pore water pressures, directly follows the exact values of the embankment's height (Fig. 3). However, the dissipation of these pressures is very small, as it was generally observed in this type of soils (muddy soft clays) for the first upload, and there is still no justification for this type of behavior.

Related to the horizontal displacements, these were reasonably within the values predicted (0.15–0.24 m for the shoulder), displaying a fast decrease in time.

6 3D Modelling: Analytical Versus Experimental Results

Settle3D is a three-dimensional (3D) software from Rocscience®, which allows to analyze foundations consolidation and settlements or surface excavations. This program can quickly and accurately help solve settlement problems considering three dimension stresses, create complex soil profiles and loading conditions, staged modelling and perform time-dependent consolidation analysis. Settlements, stresses and pore pressures are computed throughout the 3D volume, and results can be contoured along any horizontal or vertical plane, or plotted along any line.

This program was chosen to be used in this case, not only because of its features, but also because it is user friendly, with an intuitive interface.

To create the model, some assumptions were made. This was a probable situation in which it was needed to use both immediate settlement and primary consolidation. Immediate settlement is generally assumed to be the settlement that occurs quickly after the construction of the embankment. The primary consolidation settlement is the gradual re-arrangement of grains as water is expelled. This component of settlement is usually dominant in saturated soft clays.

To calculate the settlements, in this case, it was considered that immediate settlements occur instantly after the load application and it was assumed to be linear elastic. Bowles (1996) suggests that immediate settlements are those that take place within 7 days of loading. Settle3D also uses elastic theory to calculate it. Thus, by turning it on, the software calculates essentially the volume distortion due to stress changes.

Settlement depends on effective stress, which is the total stress of the ground due to gravity and any external loads minus the pore water pressure. Vertical stress due to three-dimensional loads are estimated using elastic analyses such as the Boussinesq method.

Settle3D uses one-dimensional consolidation theory to calculate the primary consolidation settlement. However, most loading scenarios are actually three-dimensional. Therefore, even with an incompressible fluid, some volume distortion settlements will occur, immediately, in a saturated clay.

In Settle3D, these 3D effects can be more easily accounted for if, in the Project Settings, the Mean 3D Stress is chosen instead of using vertical stress only. Using the mean stress instead of the vertical stress, means that when a load is applied, part of it is instantly carried by the grain skeleton. This results in some immediate settlement. Since the calculated mean stress is a function of the load geometry, there is no need to apply a fudge factor to account for the loads' size.

After the application of the preload it was possible to verify that after 1 year the settlement values were, in average, about 1.20 m, with a consolidation coefficient estimate of 95%. After 400 days, the first stage load of the experimental embankment was applied. Although the emplacement of the trial embankment was carefully chosen, the muddy layers do not have a homogeneous thickness. So, it was considered for this discussion the observation of the trial embankment profiles P1 and P2 (Fig. 2), and the results, for the same points, of the numeric modelling with Settle3D. Profile P1, at the center of the trial embankment, corresponds to 7 m thickness of muddy layers, and at profile P2 there was a 10 m thickness for the same layers.

In Fig. 4, the total settlements modelled for those two profiles, P1 and P2 (Fig. 2) are presented, as well as the measured settlement in Plates 3 and 7, respectively (Fig. 2), located immediately underneath the trial embankment. The not so good correspondence between the modelling and observed values in the second charge stage can be due to the time that was necessary to execute the trial embankment and, also, owed to the fact that the observation of the trial embankment was stopped too early, only some 60 days after the surcharging.

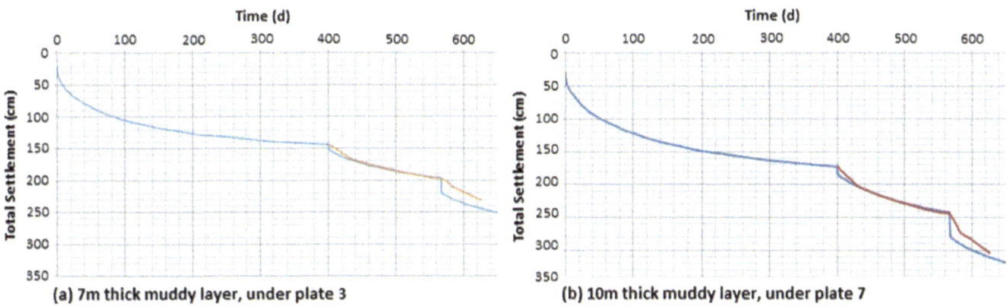

Fig. 4. Modelling of settlements under plates 3 and 7 (blue) versus registered settlements (red)

As for the consolidation achieved, the results obtained with the modelling for the same two profiles are presented in Fig. 5.

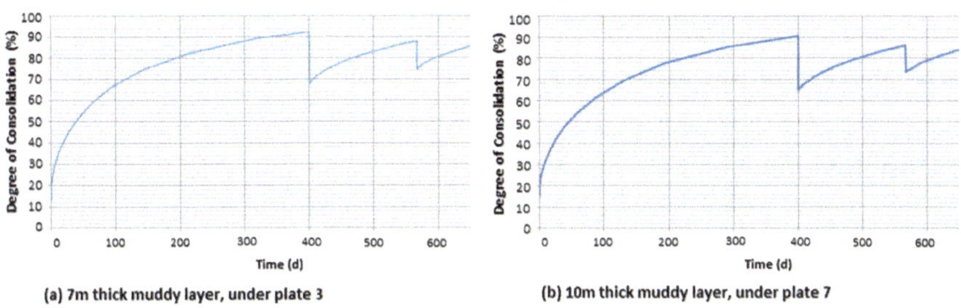

Fig. 5. Modelling of degree of consolidation beneath plates 3 and 7 (in Fig. 2)

As expected, in profile P2, crossing a thicker layer of soft soils, the degree of consolidation obtained at the same time is slightly inferior to the one obtained in the other profile, P1. In both situations, the trial embankment, constructed 400 days after the general embankment, has reached more than 90% of consolidation, as defined in project. The second stage of the trial embankment was built, according the situations, with respectively 87 or 89% of degree of consolidation accomplished. The project defined, for the second layer of the trial embankment, a minimum of consolidation under the first stage of charge of 80%. These values are in accordance with the values obtained by the Asaoka method.

For a comparison of the pore water pressure evolution, only the profile P1, across a 7 m thick muddy layer, and the pore water pressure cell 1, located 5 m below the soft soil top layer, were considered. The corresponding computed hydrostatic pore water pressures with Settle 3D is presented in Fig. 6. Related to the readings from this cell (see Fig. 3), the increase of pore water pressure with the embankment construction is higher in the simulation, but the dissipation of the excess pore water pressure is also faster than what is observed in the cell. Anyway, the dissipation of the pore water pressure in the 3D modelling is more in accordance with the settlements calculated and observed, and with the degree of consolidation calculated, as well as the one estimated by the Asaoka method.

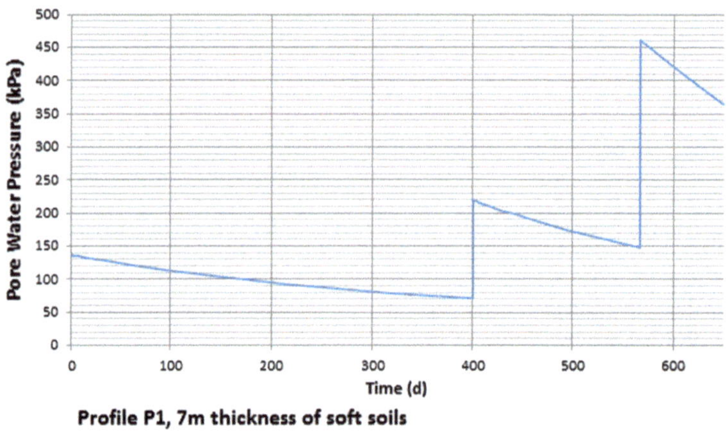

Fig. 6. Modelling of pore water pressure in profile 1 at pore water pressure cell 1 (in Fig. 2)

7 Conclusions

A case study of a ground improvement of a muddy alluvium by PVD and a surcharge fill was discussed. The presence of very soft and very sensible clays, with thickness of up to 17 m, in the selected implementation area of a new raw materials storage park, led to the execution of an extensive geotechnical exploration plan, with a large amount of prospection methods and laboratory and in situ tests, as well as the construction of an experimental embankment.

The response of the muddy alluvium to the staged embankment loading was observed by a set of instruments, including settlements plates and cells and pore water pressure cells, among many others. This allow to the monitoring of the settlements as well as the pore water pressure built up and, therefore, the assessment of the degree of consolidation evolution, during the staged construction of the trial embankment.

The monitoring results at the bottom of the fill were compared with a 3D model of the settlements (progress of the consolidation degree) and excess pore water pressure evolution using Settle3D. This analysis allowed for the simulation of the nonlinear and time dependent behavior of soft soils underneath the trial embankment. The results obtained by these two methods, one experimental and other analytical, were compared for two settlements plates: plate 3 (in Fig. 2), located at the center of the embankment and over 7 m thick muddy layer, and Plate 7 (in Fig. 2), positioned south of the previous and resting on top of 10 m thick muddy layer. There was an effective simulation by the 3D modelling of the real behavior of the soft soils.

The 3D modelling implemented shows to be a worthy tool for the design of similar future earthworks, as long as there is a good definition of the foundation geotechnical scheme, complemented by an adequate characterization of the engineering proprieties of the foundation soils. The model used allows, in a very easy way, to compare limit situations, according to the variability of the characteristics of the soft soils present. you should directly and clearly present your findings.

References

Asaoka, A.: Observational procedure of settlement prediction. Soils Found. **18**(4), 87–101 (1978)

Barron, R.A.: Consolidation of fine-grained soils by drain Wells. Jl. of SMFED, ASCE, 73 (SM6), 811–835 (1947)

Bowles, J.E.: Foundations Analysis and Design. The McGraw-Hill Companies, Inc. (1996)

Soeiro, F.A., Santos, M.P. dos.: Design and behavior of an iron ore storage yard on soft and sensitive clay foundation. VIII E.C.S.M.F.E., Helsinki (1983)

Research of Geopolymer Deal with the Strength of Soft Soil and Microstructure Test

Huicong Wang[1(✉)], Jialiang Yao[2], Yi Lin[1], and Hua He[1]

[1] School of Traffic and Transportation Engineering, Changsha University of Science and Technology, Changsha, Hunan 410114, China
{770692963,812782481,490706532}@qq.com

[2] Key Laboratory of Road Structure and Material of Ministry of Transport (Changsha), School of Traffic and Transportation Engineering, Changsha University of Science and Technology, Changsha, Hunan 410114, China
yao26402@126.com

Abstract. This paper aims at a highway that adopted geopolymer processing the strength of soft soil foundation engineering in Wenzhou city, Zhejiang province. The paper discussed the influencing factor of geopolymer soil's mechanical strength by orthogonal optimization test and microstructure test, and obtained the primary factors that affect change laws of geopolymer soil strength as well as the reasonable blending ratio of geopolymer variation. It considered the best ratio of geopolymer to soil of geopolymer reinforced soft soil is 14%, at the same time this paper found that the unconfined compressive strength of geopolymer is reduced with the increasing of soft soil's plasticity index through different plastic index of geopolymer soft soil strength test. Experiments also showed that the relation of reinforced soil sample's shear strength and compressive strength expressed as positive relationship. internal cohesion and angle of internal friction are enhanced with the increasing of compressive strength within limits, the relation of shear strength and compressive strength is $\tau = 0.25fcu$. Scanning electron microscope (SEM) sweeping test indicated that geopolymer produced cementation and filling to soil after reacting; Computed tomography (CT) sweeping test indicated that geopolymer could decreased the pore of soil and reduced porosity. This paper provided scientific evidence for substantial project that adopted geopolymer to treat soft soil foundation.

1 Introduction

Geopolymer is a kind of silicon aluminum material that mainly composed of metakaolin or other silicon and aluminum elements, and generating three-dimensional network structure's polymeric cementing material composed of silicon tetrahedron and aluminum-oxide tetrahedron under the excitation of strong-base group. It has a wide range of raw materials, low energy consumption, small environmental pollution, easy preparation, at the same time with good mechanical properties and corrosion resistance, it is a kind of environment-friendly new building materials and its application prospects are very widely [1].

© Springer International Publishing AG, part of Springer Nature 2019
S. Shu et al. (eds.), *New Developments in Materials for Infrastructure Sustainability and the Contemporary Issues in Geo-environmental Engineering*, Sustainable Civil Infrastructures, https://doi.org/10.1007/978-3-319-95774-6_16

The geopolymer material has higher mechanical properties than high polymer material, cement, ceramics and metals in many respects. On the other hand, the energy consumption and discharge amount of "three wastes" in the process of polymer materials are very low, the materials are eco-friendly and belong to sustainable development of green materials. Geopolymer materials used as soft soil foundation, and can improves soft soil subgrade's bearing capacity. However, there are few studies about treating soft soil foundation through adopting the materials of geopolymer, and there is no design and construction control standard. Based on these, this paper systematical developed the indoor test of geopolymer processing soft soil foundation combined with physical engineering, and provided scientific evidence for substantial project that adopted geopolymer to treat soft soil foundation.

2 The Research of Orthogonal Experimental Test

2.1 Experimental Materials

This paper's experiment mainly relying on soft soil of engineering as soil samples to carry out research by adopting geopolymer used as curing agent. The primary properties of raw materials are as follows:

The geopolymer is composed of steel slag, cinder, high calcium fly ash, metakaolin and alkali activator, ect mixed preparation according to a certain proportion.

The soil samples be used in this test are taken from the soft soil of a site in Wenzhou city, Zhejiang Province, and a paddy field in the Xiang River, Wangcheng County. The three soil samples are separately referred to bog soil, riverside soil and field soil blow. The sample of riverside soil and field that taked from construction site crushed and paved in a dry environment soil, and sieving after air drying, the sample of soil after sieving are stored in sealed barrels.

2.2 Experiment Scheme

The orthogonal experiment is proposed to use the geopolymer and two categories of soft soil to carry out the crossing contrast test, and to verify the feasibility of the geopolymer improving the soft soil foundation. In this test, factor A is content of geopolymer, factor B is moisture content of soft soil, factor C is mixing time of geopolymer and soil, factor D is content of organic matter. These four factors separately take three levels to analysis that the effect of these factors to strength of geopolymer soil and find the optimal level combination.

The test mixed the samples of soft soil with geopolymer, to achieve the best strength of geopolymer and obtained best mixing quantity ratio of soft soil and geopolymer by testing.

According to the orthogonal experimental test and sample's preparation requirement separately weighed geopolymer, soft soil and water, then mixed them and placed in blender to stir balance. According to the setting moisture of geopolymer soil to weigh water, then mixing them, half of the polymer soil samples were putted into the test model (70.7 mm $\times$ 70.7 mm $\times$ 70.7 mm), the production of test samples according to the requirements. Finally, covering the plastic film on the samples to prevent moisture evaporation.

These test samples formed a certain strength after one day, then removal of shuttering. These test samples setted into the curing box for conservation After removal of shuttering, The curing temperature of samples is 20 ± 2 °C, the humidity is about 75%. the samples was subjected to unconfined compression test after 7 days of health in the curing box.

2.3 The Result and Analysis of Test

The results of riverside soil's orthogonal experimental test is given in Table 1. From the above calculation and analysis, it can be seen that the influencing factor's primary and secondary sequence of 7d unconfined compressive strength about samples that made of geopolymer and riverside soil is A, B, C and D. According to the comparison of average size, the optimal level is A3, B1, C3, D1, and the optimal level combination is A3B1C3D1.

Table 1. The result of riverside soil's orthogonal experimental test

Number	Factor					
	A the mixing ratio of geopolymer (%)	B the moisture content of soft soil ω (%)	C the mixing time t (s)	D the content of organic matter (%)	Y unconfined compressive strength (MPa)	Y^2
1	10	41	240 (one round)	1.1	0.64	0.4096
2	10	44	480 (two rounds)	1.6	0.56	0.3136
3	10	47	720 (three rounds)	2.1	0.51	0.2601
4	12	41	480 (two rounds)	2.1	0.76	0.5776
5	12	44	720 (three rounds)	1.1	0.67	0.4489
6	12	47	240 (one round)	1.6	0.56	0.3136
7	14	41	720 (three rounds)	1.6	0.87	0.7569

Variance analysis table is given in Table 2, the results of variance analysis showed that factors A and B have a significant effect on the results of test, while the factors C and D have no significant effect on the results of test. The primary and secondary order of the factors are A, B, C, D. It is consistent with results of range analysis.

Draw a graph of 7d unconfined compressive strength value and factor level, the figure is given in Fig. 1. As can be seen from Fig. 1:

(1) Factor A is geopolymer content, There is a direct proportion relationship between geopolymer content and strength. The slope of strength curve of geopolymer soil

Table 2. Variance analysis table

Origin of variance	Sum of squares	DOF	Mean	Value of F	Critical value	Significant
Factor A	0.0644	2	0.032183	160.917	F0.01 (2, 2) = 99.0 F0.05 (2, 2) = 19.0 F0.10 (2, 2) = 9.0	***
Factor B	0.0406	2	0.020283	101.417		***
Factor C	0.0020	2	0.000983	4.91667		
Factor D	0.0003	2	0.000150	0.75000		
Error	0.0002	2	0.000100			
Sum	0.1101	10				

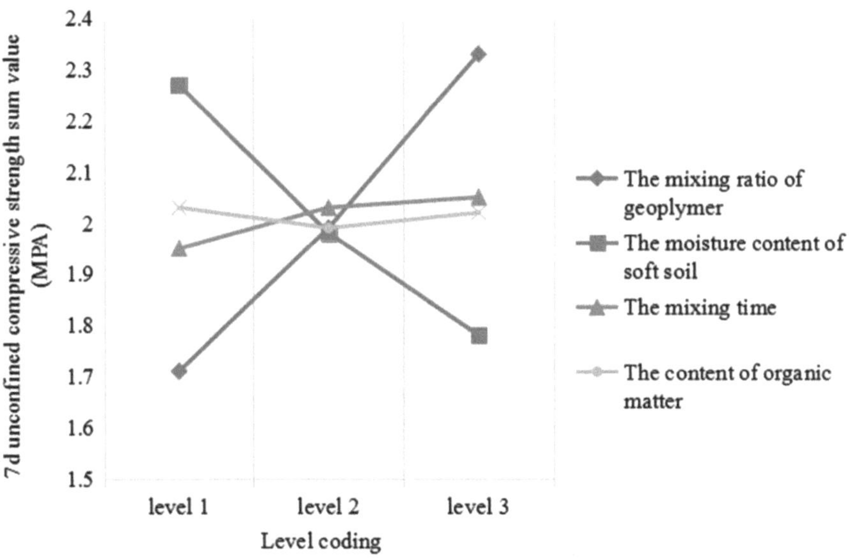

Fig. 1. Graph of 7d unconfined compressive strength value and factor level

is most significant effect which can be seen that geopolymer content has the greatest effect on the strength of geopolymer soil, The size of geopolymer content is the key factor affecting economic efficiency of the soft foundation treatment. Therefore, it is necessary to find the balance of quality and economy of geopolymer. When content is increased to 14%, the strength requirement can be reached. If content is increased to 16% or 18%, the strength will be increased correspondingly, but the cost will be too high. the dosage is 14% more reasonable when considering the requirement of strength and economy.

(2) Factor B is moisture content of soft soil, which is the second only to the influence factor geopolymer content.

(3) Factor C is mixing time which is primary effecting the mixing uniformity of geopolymer and soft soil. Factor D is organic content, It can be seen from the graph that there is no effect on the strength of geopolymer soil when organic content is 1.1–2.1%.

The result and analysis of field soil's orthogonal test. From the above calculation and analysis, it can be seen that the influencing factor's primary and secondary sequence of 7d unconfined compressive strength about samples which made of geopolymer and field soil is A, B, C and D. According to the comparison of the average size, the optimal level is A3, B1, C3, D1, and the optimal level combination is A3B1C3D1. Variance analysis table of field soil orthogonal experimental test is given in Table 3.

Table 3. The orthogonal test result of riverside soil and geopolymer

Origin of variance	Sum of squares	DOF	Mean	Value of F	Critical value	Significant
Factor A	0.052567	2	0.026283	131.4167	F0.01 (2, 2) = 99.0 F0.05 (2, 2) = 19.0 F0.10 (2, 2) = 9.0	***
Factor B	0.036167	2	0.018083	90.41667		**
Factor C	0.001767	2	0.000883	4.416667		
Factor D	0.0003	2	0.00015	0.75		
Error	0.0002	2	0.000100			
Sum	0.091001	10				

The results of variance analysis Table 3 showed that factors A and B have a significant effect on the results of test, while the factors C and D almost have no significant effect on the test results. The primary and secondary order of the factors are A, B, C, D. It is consistent with the results of the range analysis.

Draw a graph of 7d unconfined compressive strength value and factor level, the graph is given in Fig. 2. The variation rule of Fig. 2 is consistent with Fig. 1, the influence factors and change rule of the formation strength after field soil and geopolymer be mixed which is consistent with the influencing factors and change rule of the formation strength of riverside soil and geopolymer.

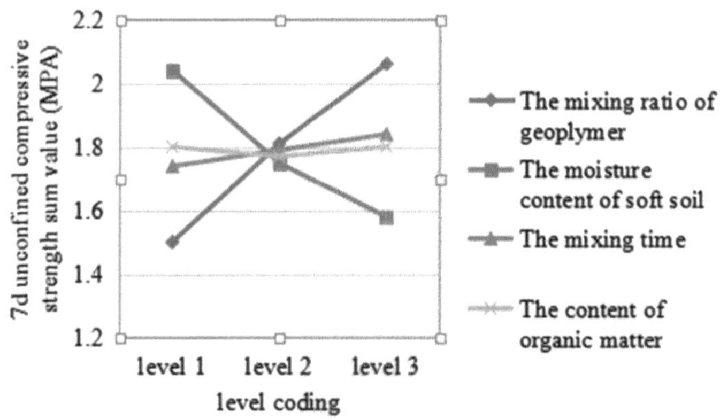

Fig. 2. Graph of 7d unconfined compressive strength value and factor level

3 Experimental Study on Different Plasticity Index of Strength of Geopolymer Soft Soil

3.1 The Materials of Test

In this study, the soft soil of 104 National Highway in Zhejiang Province was used as the test soil. Curing agent used DJZ type of geopolymer, because there are no relevant performance standards currently and referred to the standard of ordinary Portland cement [3].

3.2 Experimental Scheme

The indoor experiment separately adopted different ratio of geopolymer to soil (10%, 12%, 14%) [5] and detected the sample's age are 7d, 28d, 60d of unconfined compressive strength and shear strength according to natural moisture content of 52%.

3.3 Processing of Test

The soft soil be taken in the oven crushed after drying according to the above test program, then via 5 mm sieve and the residue on sieve of soil placed in the sealed barrel to save for sparing. The sample of straight shear test was made up of preparation of ring knife which diameter of 61.8 mm and high of 20 mm; the sample of unconfined compressive strength by the test mode of 70.7 mm × 70.7 mm × 70.7 mm cube preparation. Taking a certain amount of dry soil and geopolymer placed in the mixing pan according to the ratio, and then taking a suitable amount of water into the mixing pan to start the blender stir evenly according to the corresponding water content and water cement ratio of 0.5. Each of test samples' proportions produced three parallel patterns. The inner wall of mold is coated with oil, then soil is divided three times and mixed into test samples which stir well, each layer vibrated 2 min to exclude bubble until filling up and scratched evenness with a scraper. then sealed with plastic film to prevent moisture evaporation too fast. After 48 h, the form removal of mold samples will be placed in the standard curing box (20 ± 2 °C, humidity maintained at about 75%) maintained to 7d, 28d and 60d and other set age. Cutting ring's preparation is similar to the above.

Unconfined compressive strength test and shear strength test's samples are taken out for soaking in water for 24 h when it is cured to the specified age. The test equipment is Longwei LM-02 digital hydraulic dynamometer, to read the maximum pressure, and then calculate the maximum compressive strength of samples, test 3 times and take the average in the results of the analysis. Fast shear test's equipment is strain-controlled direct shear apparatus, recorded the breach value and organized data at the end of experiment.

3.4 The Result and Analysis of Test

The results of test as shown in Tables 4, 5 and 6. Tables 4, 5 and 6 test's results show that: At the same age, the strength of reinforced soil samples increases with the increase of geopolymer content (the ratio of geopolymer to content);

Table 4. Test data of 7 day reinforced soil sample

Strength and dosage	Content of geopolymer (10%)	Content of geopolymer (12%)	Content of geopolymer (14%)
Unconfined compressive strength (kPa)	500	580	780
Internal friction angle (°)	14.05	15.43	17.52
Cohesion (kPa)	83.47	114.34	187.6
The ratio of cohesion and fcu (%)	16.7	19.7	24

Table 5. Test data of 28 day reinforced soil sample

Strength and dosage	Content of geopolymer (10%)	Content of geopolymer (12%)	Content of geopolymer (14%)
Unconfined compressive strength (kPa)	1310	1440	1670
Internal friction angle (°)	16.27	16.92	18.84
Cohesion (kPa)	109.35	147.9	239.57
The ratio of cohesion and fcu (%)	8.3	10.3	14.3

Table 6. Test data of 60 day reinforced soil sample

Strength and dosage	Content of geopolymer (10%)	Content of geopolymer (12%)	Content of geopolymer (14%)
Unconfined compressive strength (kPa)	4250	4390	4730
Internal friction angle (°)	17.35	18.77	20.73
Cohesion (kPa)	221.57	299.12	421.76
The ratio of cohesion and fcu (%)	5.2	6.8	8.9

The 7d's data was taked as an example, the cohesion (c), internal frictional angle (φ), compressive strength (fcu) of geopolymer content apart increased by 37%, 10%, 16%, when geopolymer content from 10% to 12%. c, φ and fcu of soil samples separately increased by 64%, 13.5%, 34% when geopolymer content from 10% to 12%. The strength growth rate of content from 12% to 14% is greater than strength growth rate of content from 10% to 12%. It is considered that the ratio of geopolymer to soil of 14% is the best blending ratio. The same conclusion can be obtained from the 28d of data.

The shear strength (τ) of geopolymer increased with increasing unconfined compressive strength (q_u) and expressed as a linear correlation. When fcu = 0–4730 kPa, c = 83.47–421.76 kPa, φ = 14.05–20.73°. Fcu is roughly 1/10–1/4 of no lateral compressive strength when fcu = 0–4730 kPa. That is τ = (1/10–1/4) fcu. And with the increase of fcu, the ratio of q_u to fcu ratio has a tendency that gradually become smaller and taking into account the strength of discrete, shear strength value τ = 0.25 fcu.

4 Research of Microstructure Test

4.1 Materials of Experiment

A small amount of soil samples which collected in the sample of prepared during the unconfined compressive strength test and the direct shear test were dried in an oven at 105 °C and crushed. And sieving larger particles, to select size of a rice grain and the surface smooth of soil samples [9–11].

4.2 Testing Program

SEM scanning electron test used the experiment equipment for the Japanese Hitachi S-3000N-type scanning electron microscopy, Its maximum voltage is 30 kV, the maximum magnification times can reach 300,000 times and the maximum resolution is 5 nm.

The equipment of CT scanning test used by the German company YXLON 20105348-type industrial CT, its maximum acceleration voltage is 320 kV and the minimum resolution is 83 μm.

4.3 The Result and Analysis of SEM Scanning Electron Test

Sample to be tested separately named a, b, c, d and carried out electron microscopy scanning test, the result is given in Fig. 3, figure (a)–(d) is SEM scanning electron microscopy image of part of samples.

(1) Figure (a) is Wenzhou original soil pictures, we can see that structural distribution has more dense pores and more loose skeleton, which is consist of different shapes clay mineral particles cluttered accumulation from the figure,
(2) Figure (b) is 10% proportion of geopolymer reinforced soil samples of 28d age, It can be seen that original loose skeleton and dense pores became no longer obvious, the surface of soil particles appear at some long strip, flocculent white crystals which adhere to the part of pores to enhance interaction between soil and improving strength.
(3) Figure (c) is 12% proportion of geopolymer reinforced soil samples of 28d age, It can be seen that strip and threadiness of substance's distribution range is greater than figure b, These materials are filled in the pores which making the structure of soil sample densely and connected the surrounding of loose particles. And also improving the overall strength of samples [4].

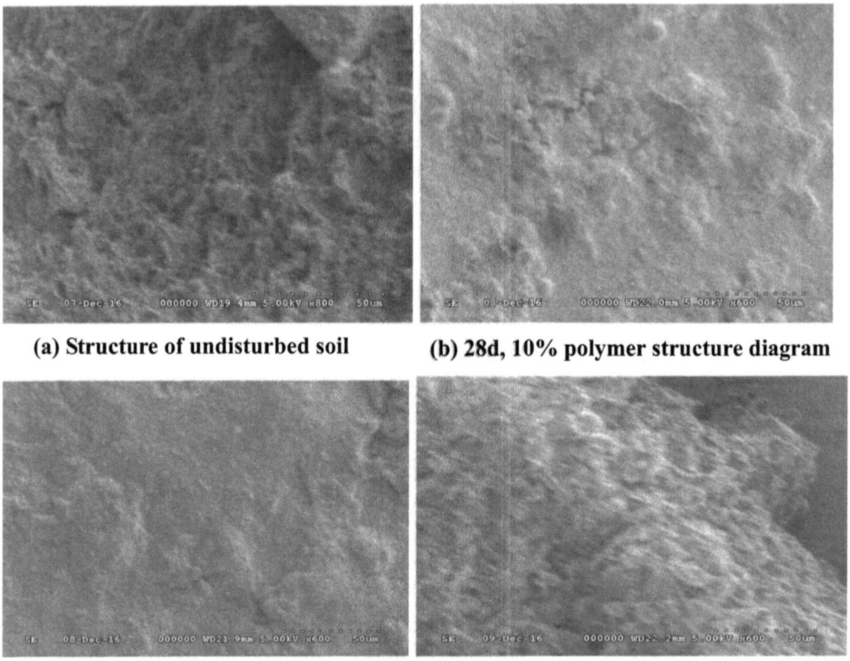

(a) Structure of undisturbed soil

(b) 28d, 10% polymer structure diagram

(c) 28d,12% polymer structure diagram

(d) 28d,14% polymer structure diagram

Fig. 3. SEM scan test piece

(4) Figure (d) is 14% proportion of geopolymer reinforced soil samples of 28d age, It is clear that surface of soil can be seen a large number of spider web-like white crystals, which firmly tied to cover the surface of soil and let soil particles and its around glued to become a complete overall. these zeolite precursors that are produced by hydration of geopolymer after dewatering and to form amorphous phase material with soil particles combined to form a whole, so the structural features of soil from friability messy accumulation into a granular glue, The porosity is drastically reduced, shear, compression and impermeability significantly improved.

4.4 The Result and Analysis of CT Scanning Test

The results of test is shown in Figs. 4 and 5. The content of geopolymer from left to right was 10%, 12%, 14% in picture, the age of test samples are 28d, Figs. 4 and 5 are part of sample picture and which are different cross-section cut by same batch of samples, the height of section respectively is 60 cm and 70 cm. It can be seen that the amount of porosity of the samples decreased with the increasing of the amount of geopolymer in the same water content from the following figure, that is porosity decreased and the ratio of macrovoid decreased. This also illustrated that the filling effect of geopolymer to soil and the results of SEM scanning test are also verified.

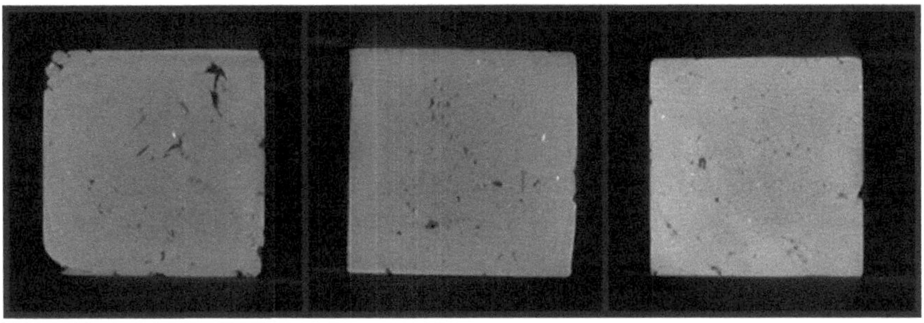

Fig. 4. Section height 60 cm specimen

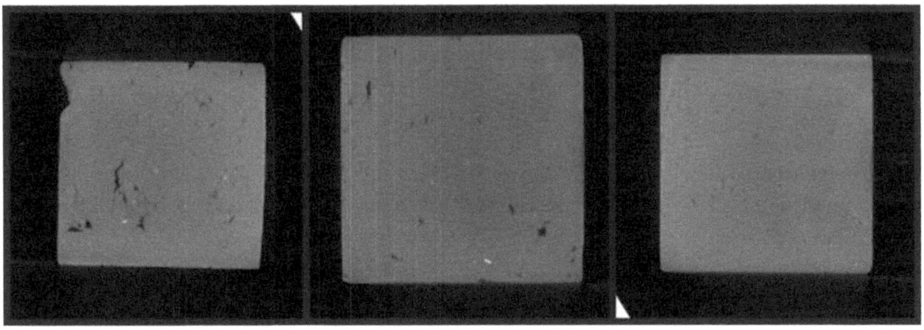

Fig. 5. Section height 70 cm specimen

The results of CT scanning test further showed that geopolymer to soil produced a cementation and filling effect. Soil is denser and compact and the strength of soil is improved in the same mixing amount and age, especially macrovoid reduced in the late period of reaction, this is consistent with the previous macroscopic shear strength, compressive strength and the results of SEM test.

5 Conclusions

This paper first obtained that the influencing factor of geopolymer soil's mechanical strength by analyzing orthogonal test of field soil and geopolymer, orthogonal test of riverside soil and geopolymer, then the effect of geopolymer to strength of Wenzhou city's soft soil was discussed by a mass of unconfined compression test and direct shear test. Finally to analyze that effect of geopolymer reinforced soft soil on compressive strength and shear strength from a micro perspective by SEM electron test and CT scanning test. Conclusions of experimental study are as follows:

(1) The main factors affecting the strength of geopolymer soil are the amount of geopolymer and the moisture content of soft soil. The secondary factors are the stirring time and the content of organic matter.

(2) The strength increasing rule of geopolymer reinforced the strength of soft soil expressed as with age and the amount of geopolymer increased with increasing, the content of geopolymer in soft soil is 10–14% can make the treatment of soft soil foundation to achieve good results. When the content is increased to 14%, the strength requirement can be reached. If content is increased to 16% or 18%, the strength will be increased correspondingly, but the cost will be too high. The best ratio of geopolymer to soil is 14% when comprehensive considering the requirement of strength and the economy.
(3) Experiments also showed that the relation of reinforced of soil sample's shear strength and compressive strength expressed as positive relationship. internal cohesion and angle of internal friction are enhanced with the increase of compressive strength within a certain range, the relation of shear strength and compressive strength is $\tau = 0.25$ fcu.
(4) SEM sweeping test indicated that geopolymer produced cementation and filling to soil after reacting; CT sweeping test indicated that geopolymer could decreased the pore of soil and reduced porosity.

Acknowledgements. This research was supported by the National Natural Science Foundation of China under Grant 51578080, Grant 51178064 and the Transportation Department of Hunan Province, No. 2013-01.

References

1. Li, C.: Preparation and Performance of Na-GeoPolymer Cementitious Material Based on Fly Ash. Chang'an University (2015)
2. Ruijiang, L., Yewang, Z., Chongwei, W., Jian, T.: Study on the design and analysis methods of orthogonal experiment. Exp. Technol. Manag. **09**, 52–55 (2010)
3. GB 175-2007, common Portland cement. General Administration of Quality Supervision, Inspection and Quarantine, Beijing (2007)
4. Nie, Y.-M., Ma, H.-W., Giudice, L.C. et al.: The mineral aggregate material polymerization reaction mechanism in the process of solidification research. Modern Mass **11**:340–346 (2006)
5. JGJ/T 233-2011. cement-soil mixture ratio design procedures
6. Li, K.: Silicon powder—metakaolin base polymer microstructure analysis. J. Build. Mater. **15**(4), 553–556 (2012)
7. JTG E40-2007 highway geotechnical testing standards. The Ministry of Communications of the People's Republic of China, Beijing (2007)
8. DB/T 904-2013, 33 highway soft soil embankment design specification. Zhejiang Province Quality Technology Supervision Bureau, Zhejiang (2013)
9. Duan, P., Yan, C., Zhou, W.: Influence of partial replacement of fly ash by metakaolin on mechanical properties and microstructure of fly ash geopolymer paste exposed to sulfate attack. Ceram. Int. (2016). https://doi.org/10.1016/j.ceramint.2015.10.154
10. Schmücker, M., MacKenzie, K.J.D.: Microstructure of sodium polysialate siloxo geopolymer. Ceram. Int. (2005) https://doi.org/10.1016/j.ceramint.2004.06.006
11. Tailby, J., MacKenzie, K.J.D.: Structure and mechanical properties of aluminosilicate geopolymer composites with Portland cement and its constituent minerals. Cem. Concr. Res. (2010). https://doi.org/10.1016/j.cemconres.2009.12.003

Mechanical Properties of Geopolymers Cured in Saline Water

Xiaonan Ge(✉) and Guoping Zhang

Department of Civil and Environmental Engineering,
University of Massachusetts Amherst, Amherst, MA 01003, USA
xge@umass.edu

Abstract. An experimental study of the mechanical properties of fly ash-based geopolymer cured in saline water is presented. The resistance of building materials to saline water is of key important to the design and safety of civil infrastructure and constructions. The studied geopolymers were synthesized from the mixture of Class C fly ash and metakaolin with alkali activation, and cured immediately in saline water at ambient temperature. The ordinary Portland cement (e.g., even used as oil well cement) usually have poor performance when cured in saline water. However, fly ash-based geopolymer cements cured in saline water showed an increased unconfined compressive strength than the cured in fresh water. Chemical and compositional analyses, including the pH of the curing water, salinity or electrical conductivity of the curing water, scanning electron microscopy, X-ray diffraction, and X-ray fluorescence spectroscopy, were conducted to investigate the mechanisms for the improved performance of geopolymers cured in saline water. Results show that the increase in the strength of geopolymers cured in saline water is related to the leaching of reactants from the material and infiltration of saline water into the material.

Keywords: Geopolymer · Fly ash · Metakaolin · Saline water
Unconfined compressive strength · Ambient temperature

1 Instruction

The soil with high content of soluble salts is located at the arid and semi-arid regions of China, mainly including the Inner Mongolia, Ningxia Hui, and Xinjiang Uygur Autonomous Region, Heibei and Gansu provinces. Due to inwelling, high-salinity environment also exists in the soil of coastal regions. Chemical composition of salts is complicated, such as chlorate in the coastal region, sulfate in some inland areas, carbonate in Shanxi province and some regions of northeast. The corrosive characteristic of these salts to different types of civil infrastructure would be obvious, which would also bring about economic loss.

The research of resistance of geopolymers to salts corrosion mainly focuses on chlorate, sulfate, and carbonate and so on. Preceding research showed that the performance of concrete formed from ordinary Portland cement would be degrade in the saltness condition [1–3]. There is a complicated chemical reaction between hydrating products of cement and different kinds of salts, which refers to CH, C-S-H, aluminate,

© Springer International Publishing AG, part of Springer Nature 2019
S. Shu et al. (eds.), *New Developments in Materials for Infrastructure Sustainability and the Contemporary Issues in Geo-environmental Engineering*, Sustainable Civil Infrastructures, https://doi.org/10.1007/978-3-319-95774-6_17

and so on [4, 5]. For example, the reaction between sulfate (Including magnesium sulfate, sodium sulfate, calcium sulfate) and calcium hydroxide in cement would generate gypsum, whose crystal expand volume. Then hydrated calcium sulfate would be produced by the reaction between gypsum and hydrated calcium aluminate. Due to amounts of crystal water in hydrated calcium sulfate, volume can be expanded to about 2.5 times than that of hydrated calcium aluminate, which would do great damage to hardened cement paste [4–6]. This process mainly consumes calcium in the C-S-H [7, 8].

Davidovits [9] indicated that geopolymers, a class of inorganic polymeric materials, possess three-dimensional aluminosilicate structures consisting of linked SiO_4 and AlO_4 tetrahedral, by sharing all the oxygen atoms [10, 11]. A general formula for the chemical composition of geopolymers is as follows:

$$M_n^+ \{-(SiO_2)_z-(AlO_2)-\}_n$$

Where M^+ is an alkali cation (K^+ or Na^+) for balancing the negative charge for Al, n is the degree of polymerization, and z is the Si/Al molar ratio, ranging from 1 to 15, and up to 300 [12, 13]. Geopolymerization can be divided into three major steps that interact with each other along the reaction: (1) the amorphous aluminosilicate materials, dissolved by alkali hydroxide solution to form reactive silica and alumina; (2) reorientation, which is the transportation, orientation, or condensation of dissolved precursor ions into monomers; (3) the dissolved species polycondense into amorphous or semi-crystalline aluminosilicate polymers. The formation of geopolymers contains two following reactions [14, 15].

This experimental study mainly investigates the fly ash-based geopolymer's properties of resistance to slats by comparing their unconfined compressive strength after 28 days curing in saline water with different salinity with those of Class G oil well cement (GOWC), which is type of cement with high sulfate resistance. Meanwhile several methods are conducted to analyze the physicochemical characteristic of cured samples and solution.

2 Materials

In this research, geopolymers were synthesized using class C fly ash [16] (Headwater Resources, Inc., USA), metakaolin (Advanced Cement Technologies, LLC, USA), solid sodium hydroxide (>95% purity quotient, Fisher Scientific of Thermo Fisher Scientific Inc., USA), sodium silicate solution with specific gravity 1.4 consisting of 9.1 wt% Na_2O, 29.2 wt% SiO_2, 61.7 wt% H_2O (Fisher Scientific of Thermo Fisher Scientific Inc., USA) and reverse-osmosis (RO) water. Table 1 presents the chemical compositions of dry fly ash and metakaolin powders, while Fig. 1 shows the particle size distributions [17, 18]. Saline water was made from synthetic sea salts (Instant

Ocean Company) [19], whose chemical composition showed in Table 2 is close to ocean water. For benchmarking purposes, GOWC (Lafarge, USA) was studied at the same conditions.

Table 1. Chemical composition of fly ash and metakaolin (mass%)

	Fly ash	Metakaolin
SiO_2	38.56	56.61
Al_2O_3	19.8	39.16
Fe_2O_3	6.26	1.87
MgO	5.15	0.09
CaO	22.28	0.05
SO_3	1.72	0.05
K_2O	0.61	0.3
Na_2O	1.58	0.01
Moisture	0.11	0.13
LOI	0.41	0.19
Total	96.48	98.46

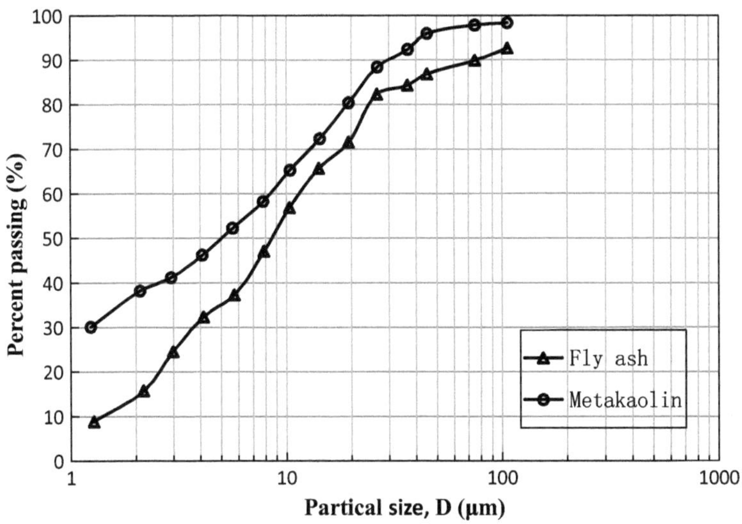

Fig. 1. Particle size distribution of raw materials used for geopolymers syntheses

Table 2. Chemical composition of seasalts (from Instant Ocean Company)

	Seawater	Instant ocean
ppt	35	29.65
Major cations (mmol kg^{-1})		
Na^+	470	462
K^+	10.2	9.4
Mg^{+2}	53	52
Ca^{+2}	10.3	9.4
Sr^{+2}	0.09	0.19
Sum	607	594
Major anions (mmol kg^{-1})		
Cl^-	550	521
SO_4^{-2}	28	23
TCO_2 (total CO_2)	1.90	1.90
TB (total B)	0.42	0.44
Sum	608	569

3 Sample Preparation

Geopolymers were synthesized from mixture of fly ash and metakaolin in mass proportion of 1:1. They were prepared with Na/Si molar ratio of 1.0 and Si/Al molar ratio of 2.0. Water to mixture powder mass ratio of 0.5 was used to prepare slurry. The NaOH and RO water were firstly mixed by a glass stirring rod and left to cool down to the room temperature. Then the sodium silicate solution was added to the NaOH solution, followed by mixing for 5 min at least and allowing resting overnight with covering in order to fully blend. The blended solution was added to the dry powder mixture, followed by mixing for minimum of 15 min [20] to ensure complete mixing between the powder and solution, resulting in the formation of geopolymer precursor in the form of slurry.

To make regularly shaped specimens for mechanical testing, one part of prepared geopolymer precursors were poured into preassembled cylindrical porous molds with an inner diameter of 2.54 cm and height of 6.35 cm (i.e., an aspect ratio of 2.5 to minimize the end effects), vibrated by hand to remove the trapped air, followed by totally immersing and curing in the saline water with different salinity (0, 15, 35 ppt; ppt - part per thousand) in a laboratory ambient environment for 28 day. The other slurry was poured into plastic split mold (Air-curing) with inner diameter of 2 cm and height of 5 cm, followed by fixing with steel clamp, where no exchange of chemicals in or out of samples were allowed. Prior to assembling, the interior of molds were coated with a thin layer of vacuum grease for easy removal of specimens after 28 day curing duration. For ensure repeatability, typically 4 specimens were prepared for each type of geopolymer. For GOWC, the same method of sample preparation for geopolymer was adopted with liquid to solid ratio of 0.5. After 28 day curing, each specimen was de-molded and then allowed to air dry for a minimum of one week.

4 Characterization

Cured samples were tested for their unconfined compression strength (UCS) [21] using Geotest instruction strain-controlled loading frame with 10,000 lb load cell at a constant strain rate of 0.5%/min for GOWC and geopolymer samples. The process of strength testing needs 5–7 min for one sample. In addition, scanning electron microscope (SEM), X-ray diffraction (XRD), X-ray fluorescence spectroscopy (XRF), the pH and electrical conductivity testing were conducted to analyze the mechanism for performance of geopolymers cured in saline water. With regarding to samples for SEM, XRD and XRF, small pieces selected from the interior and exterior of specimens in failed unconfined compression testing were examined separately.

5 Results and Discussion

5.1 Mechanical Properties

Figure 2 presents the average unconfined compressive strength (UCS) for all cured samples in different curing environment after 28 day curing duration. Geopolymers showed an increase in UCS of 50.28 $\pm$ 2.96 MPa for 0 ppt, 61.34 $\pm$ 3.02 MPa for 15 ppt, and 64.27 $\pm$ 2.99 MPa for 35 ppt. Such result presents that the higher salinity of salt water has a positive effect on the development of mechanical properties, which is directly related to the level of polymerization. According to the reaction principle of formation of geopolymers, two main parts of reactant, concentration of alkali cations and sodium hydroxide in solutions would influence on the degree of chemical reaction, which is also changed by curing solution with different salinity. Therefore, variation of UCS of geopolymers reflects the existing difference among three types of solutions for 0, 15, 35 ppt. The concrete change of concentration of alkali cations and sodium hydroxide in solutions were tested through electrical conductivity and pH methods, whose results are presented in the later sections. For geopolymers, the difference of strength between air cured and the saline water cured samples clearly shows that ion exchange really has influence on the UCS. However, the opposite trend occurred to GOWC cured in different saline water. Lots of previous articles had given reasonable explanation about the degradation of Portland cement in different kinds of saline solutions [22]. Class G Oil Well Cement, as one of basic oil well cement, is one of special hydraulic binding materials. Its main content is clinker of Portland cement, which would be degraded by saline solutions in same mechanism compared to Portland cement. Moreover, leaching of row materials from cement paste is one of the reason that UCS of GOWC decreases with the increasing salinity of saline water.

5.2 Microstructure and X-Ray Diffraction Analysis

Figure 3 presents the SEM micrographs for geopolymers cured in 0, 15, 35 ppt saline water. Generally speaking, un-reacted and partially reacted components in the exterior are more than that in the interior. SEM images for the interior show fairly homogeneous and continuous matrix with more reactive components. Meanwhile, among

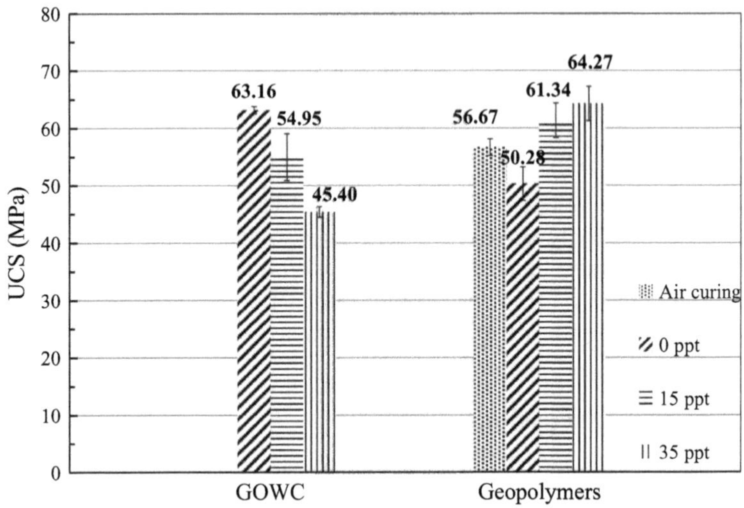

Fig. 2. Unconfined compressive strength of cured geopolymers and oil well cement after 28 days (the error bars represent one standard deviation, which applies to figures as well)

samples in different salinity, it seems that geopolymers cured in 15 or 35 ppt saline water are more homogeneous and have larger part of reactive matrix than that cured in 0 ppt saline water. Overall the curing water would have more effect on the exterior of geopolymer specimens than the interior. The main reason is more leaching of reactant in the exterior from precursor leading to decline of extent of chemical reaction.

The XRD patterns of geopolymers cured in different salinity are shown in Fig. 4, respectively. A pronounced broad hump between 20–40° 2θ with a few sharp peaks is present for all samples. This is the characteristic reflection of amorphous geopolymers [23]. The crystalline phase include Quartz, Sodalite, Mullite, Calcite, Halite, and Chabazite-Ca.

5.3 Ion Exchange and Chemical Composition

In order to verify the ion exchange during curing duration, values of pH [24, 25] and salinity [26] would be tested before and after curing process, as shown in Figs. 5 and 6 ("plus" means increase in value; "minus" means decrease in value).

According to the records, the pH values don not change anymore after the first 24 h, which is consistent to the results from Haider [27]. Figure 5 illustrates that the increasing salinity would reduce the degree of ion exchange. In other words, amount of the alkali metal ion and hydroxide ion cut down due to the raising salinity. In addition to this, the rates of ΔpH for GOWC and geopolymers are also different. The decreasing rate of ΔpH for GOWC grows faster (13.3% for 15 ppt; 43.6% for 35%), while that for geopolymers becomes slower (50.8% for 15 ppt; 74.9% for 35 ppt). Therefor, high salinity has larger influence on GOWC than geopolymers. The rate of change of strength in Fig. 2 is also consistent to that of ΔpH.

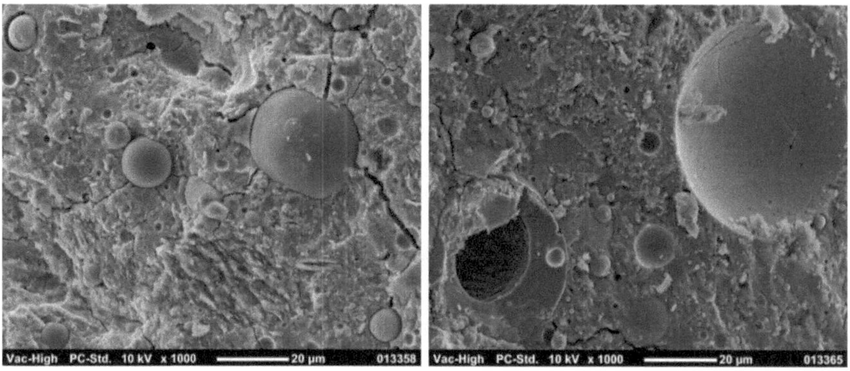

(a) 0 ppt

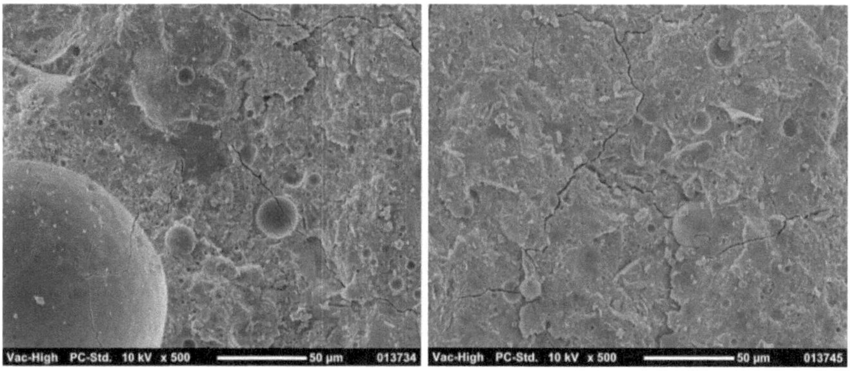

(b) 15 ppt

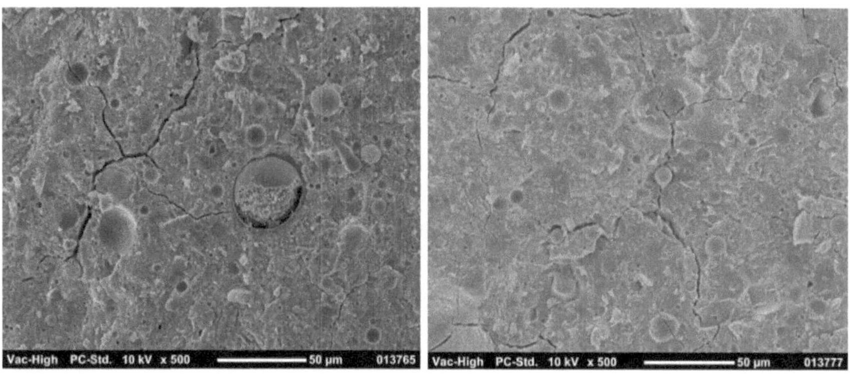

(c) 35 ppt

Fig. 3. SEM results of geopolymer samples, interior (left) and exterior (right)

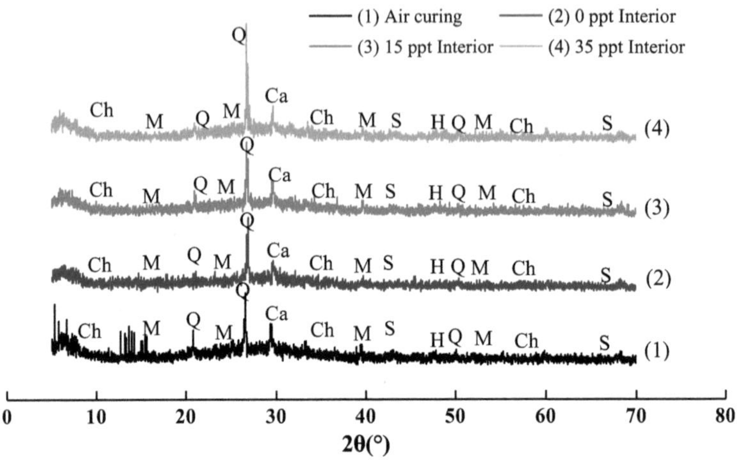

Fig. 4. XRD patterns of geopolymer samples (Q: Quartz, M: Mullite, Ch: Chabazite-Ca, Ca: Calcite, S: Sodalite, H: Halite)

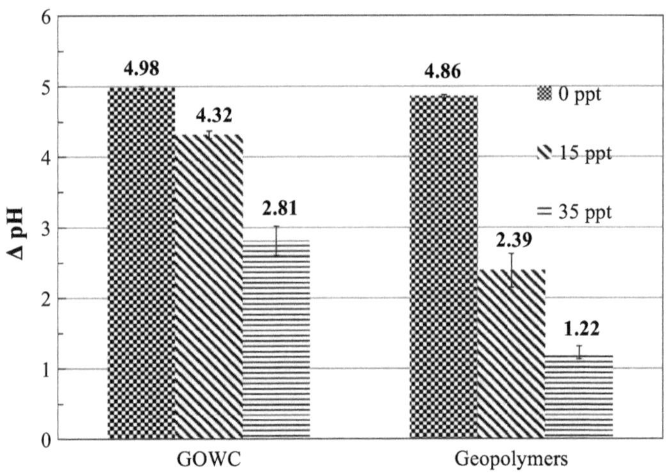

Fig. 5. The change of pH of curing water before and after 28 days

The change of salt concentration of the curing water before and after 28 day is shown in Fig. 6. The value still decrease with the increasing salinity. But there are some negative values meaning part of saline ions go into samples. For GOWC, intrusion of saline ions leads to a series of chemical reaction destroying its structure and dropping the strength.

XRF supplies the mass percentage of a series of oxides including SiO_2, Al_2O_3, Fe_2O_3, CaO, Na_2O, and so on. With comparison, it is found that only Na_2O has relatively obvious alter among data, while others almost keep unchanged. Therefore, in

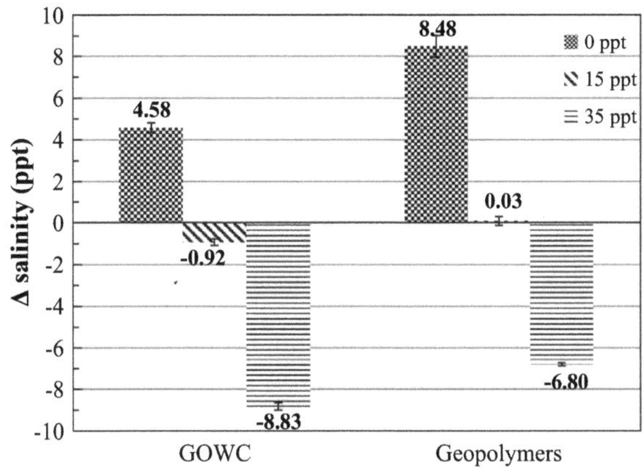

Fig. 6. The change of salt concentration of curing water before and after 28 days

this study, only Na_2O was analyzed among all data as an indicator to present the condition of ions exchange, as shown in Fig. 7. Since the chemical composition of air curing geopolymers from XRF is same to that of pre-designed, air curing sample can be reference standard.

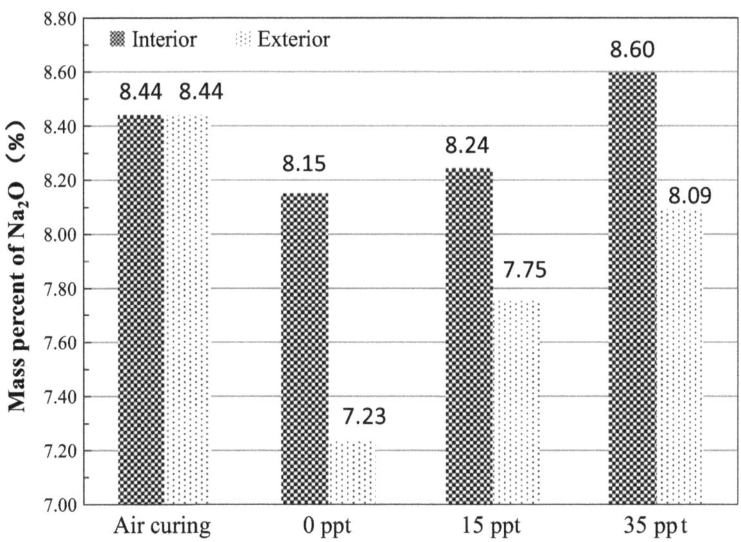

Fig. 7. The XRF results of geopolymer

For air curing sample, the percentage of Na_2O in exterior is same to that in interior, which indicates the chemical composition is homogeneous. As for all other three samples, the percentage of Na_2O in exterior is less than that in interior and than that of air curing at the same time. This record illustrates Na^+ is dissolved out from sample during curing process, especially from the exterior. The phenomenon that the percentage of Na_2O increases with the increasing salinity means high salinity would control the leakage of Na^+.

By comparison among all XRF results, it is found that only the percentage of Na_2O of interior curing in 35 ppt is larger than that of air curing, which shows Na^+ from curing water intrudes into the body of sample. This variation trend exactly correspond to those showed in Figs. 2 and 6. Coincidentally, UCS for 35 ppt cured sample is higher than that of air curing. Generally speaking, consistency of change between UCS and percentage of Na_2O presents that interaction between geoopolymers and salt water really has directly effect on the mechanical property. In other words, the increasing Na^+ within a certain range could promotes the process of polymerization which improves the development of strength.

6 Conclusions

This study investigates certain characteristics of geopolymers cured in the salt water by taking tests such as unconfined compression test, SEM, XRD, XRF, pH test and electrical conductivity. And based on the limited data presented above, specific conclusions are showed as followed:

(1) For this geopolymer with such design mix, its UCS after curing 28 day increases with the raise of salinity. And the average strengths cured in the 15, 35 ppt condition exceed the values of sample cured in sealed condition. To the contrary, the UCS of GOWC decreases with the increasing salinity, whose largest decline is up to 28%.
(2) All salt water with different salinity have more effect on the exterior of geopolymers than interior. The result is that degree of polymerization of exterior would be weakened.
(3) The change of strength of geopolymers is related to variation of the metal ions. ① High salinity prevents the leakage of the metal ions from body of samples (such as Na^+); ② High salinity also intrudes the metal ions into interior of sample which could improve the polymerization within certain range.
(4) Geopolymers' performance of resistance to salt corrosion is better than GOWC. For samples cured in the 35 ppt condition, the mechanical properties can be improved.

References

1. Mehta, P.K.: Sulfate attack on concrete: a critical review. In: Villarreal, R.R. (ed.) Concrete Durability, pp. 107–132. Universidad Autonoma de Nuevo Leon (1993)
2. Wakely, L.D., Poole, T.S., Ernzen, J.J., Neeley, B.D.: Salt saturated mass concrete under chemical attack. Am. Concr. Inst. SP **140**, 239–267 (1993)

3. Irassar, E.F., Di Maio, A., Batic, O.R.: Sulfate attack on concrete with mineral admixtures. Cem. Concr. Res. **26**(1), 113–123 (1996)
4. Ferraris, C.F., Clifton, J.R., Stutzman, P.E., Garbocsi, E.J.: Mechanisms of degradation of Portland cement-based systems by sulfate attack. In: Scrivener, K.L., Young, J.F. (eds.) Mechanisms of Chemical Degradation of Cement-Based Systems, pp. 185–192. E and FN Spon, London (1997)
5. Santhanam M., Cohen, M.D., Olek, J.: Mechanism of sulfate attack: a fresh look: Part 2. Proposed mechanisms. Cem. Concr. Res. **33**(3), 341–346 (2003)
6. Taylor, H.F.W., Gollop, R.S.: Some chemical and microstructural aspects of concrete durability. In: Scrivener, K.L., Young, J.F. (eds.) Mechanisms of Chemical Degradation of Cement-Based Systems, pp. 177–184. E and FN Spon, London (1997)
7. Bonen, D., Cohen, M.: Magnesium sulfate attack on Portland cement paste: I. Microstructural analysis. Cem. Concr. Res. **22**(1), 169–180 (1992)
8. Bonen, D., Cohen, M.: Magnesium sulfate attack on Portland cement paste: II. Chemical and mineralogical analyses. Cem. Concr. Res. **22**(4), 707–718 (1992)
9. Davidovits, J.: Chemistry of geopolymeric systems. In: Proceedings of 99 Geopolymer Conference, pp. 9–40 1999
10. Thakur, R., Ghosh, S.: Fly ash based geopolymer composites: manufacturing and engineering properties. Lambert Academic Publishing (2011)
11. Shi, C.J., Jimenez, A.F., Palomo, A.: New cements for the 21st century: the pursuit of an alternative to Portland cement. Cem. Concr. Res. **41**(7), 750–763 (2011)
12. Davidovits, J.: Properties of geopolymer cements. In: First International Conference on Alkaline Cements and Concretes, vo. 1, pp. 131–149 (1994)
13. Zhang, G.P., He, J.A., Gambrell, R.P.: Synthesis, characterization, and mechanical properties of red mud-based geopolymers. Transp. Res. Rec. **2167**, 1–9 (2010)
14. Khale, D., Chaudhary, R.: Mechanism of geopolymerization and factors influencing its development: a review. J. Mater. Sci. **42**(3), 729–746 (2007)
15. Duxson, P., et al.: Geopolymer technology: the current state of the art. J. Mater. Sci. **42**(9), 2917–2933 (2007)
16. ASTM D618-15. Standard Specification for Coal Fly Ash and Raw of Calcined Natural Pozzolan for Use in Concrete. ASTM International, USA (1994)
17. ASTM D422-63. Standard Test Method for Particle-Size Analysis of Soil. ASTM International, USA (2007)
18. ASTM D854-05 Standard Test Methods for Specific Gravity of Soil Solids by Water Pycnometer. ASTM International, USA (2005)
19. Atkinson, M.J., Bingman, C.: Elemental composition of commercial seasalts. J. Aquariculture Aquatic Sci. **8**(2) (1997)
20. He, J., Jie, Y.X., Zhang, J.H., Yu, Y.Z., Zhang, G.P.: Synthesis and characterization of red mud and rice husk ash-based geopolymer composites. Cem. Concr. Compos. **37**, 108–118 (2013)
21. ASTM C39/C39 M-15a. Standard Test Method for Compressive Strength of Cylindrical Concrete specimens. ASTM International, USA (2015)
22. Gollop, R.S., Taylor, H.F.W.: Microstructural and microanalytical studies of sulfate attack. IV. Reactions of a slag cement paste with sodium and magnesium sulfate solutions. Cem. Concr. Res. **26**(7), 1013–1028 (1996)
23. He, J., Zhang, J.H., Yu, Y.Z., Zhang, G.P.: The strength and microstructure of two geopolymers derived from metakaolin and red mud-fly ash admixture: a comparative study. Constr. Build. Mater. **30**, 80–91 (2012)

24. ASTM D1293-12. Standard Test Methods for pH of Water. ASTM International, USA (2012)
25. ASTM D5464-11. Standard Test Methods for pH of Measurement of Water of Low Conductivity. ASTM International, USA (2011)
26. ASTM D5391. Standard Test Method for Electrical Conductivity and Resistivity of a Flowing High Purity Water Sample. ASTM International, USA (2014)
27. Giasuddin, H.M., Sanjayan, J.G., Ranjith, P.G.: Strength of geopolymer cured in saline water in ambient conditions. Fuel **107**, 34–39 (2013)

Water Permeability Reduction in THF Hydrate-Bearing Sediments

Nariman Mahabadi[1], Tae Sup Yun[2], and Jaewon Jang[3(✉)]

[1] Arizona State University, Tempe, AZ, USA
[2] Yonsei University, Seoul, South Korea
[3] Hanyang University, Seoul, South Korea
jwj@hanyang.ac.kr

Abstract. Water permeability in hydrate-bearing sediments is a key parameter in gas production affecting the effective depressurization boundary from a wellbore and contributing heat transport associated with fluid flow. The experimental measurement of water permeability in the presence of hydrates is associated with many difficulties such as dynamic hydrate dissolution and formation during fluid flow and long induction time. In this study, we formed tetrahydrofuran (THF) hydrates in a core-scale chamber to explore water permeability as a function of hydrates. Wave velocities during the permeability measurement were also measured. The results show that water permeability decreases as hydrate saturation increases. Shear and compression wave velocities increase with increasing hydrate saturation, but the velocity decreases a little during the repetitive permeability measurement at a given hydrate saturation.

1 Introduction

Permeability of hydrate-bearing sediments is critically important for gas production because it governs the propagation of depressurization from a well bore and heat transport from far field associated with water flow. If the permeability of the hydrate-bearing reservoir is very low (e.g., clayey sediment), the dissociation of hydrates occurs at very short interface between the free gas zone and the hydrate layer. But if the hydrate-bearing sediment has high permeability (e.g., sandy sediment), the dissociation of hydrate occurs over wider area (Jang and Santamarina 2016; Kumar et al. 2010; Santamarina and Jang 2009).

Hydrate pore habit in addition to hydrate saturation is important to govern permeability reduction pattern and acoustic properties of the hydrate-bearing sediment (Kleinberg et al. 2003; Mahabadi and Jang 2014; Mahabadi et al. 2016a, b). Different permeability values can be obtained even at the same hydrate saturation depending on hydrate pore habit (Kang et al. 2016; Liang et al. 2011). Therefore, more efforts are needed to verify permeability in hydrate-bearing sediments formed by different ways for in-situ gas production (Kneafsey et al. 2011).

The studies on in-situ hydrate formation show pore-filling hydrate habit, based on in-situ seismic analysis for Black Ridge (Jakobsen et al. 2000) and Mallik site (Lee and Collett 2001; Sakai 2000), and compressional (P) and shear (S) wave velocity measurement for the pressurized cores recovered from Nankai Trough (Santamarina et al.

© Springer International Publishing AG, part of Springer Nature 2019
S. Shu et al. (eds.), *New Developments in Materials for Infrastructure Sustainability and the Contemporary Issues in Geo-environmental Engineering*, Sustainable Civil Infrastructures, https://doi.org/10.1007/978-3-319-95774-6_18

2015). However, in most laboratory experiments to measure permeability, gas (either CH_4 or CO_2) hydrates have been formed under unsaturated condition (or gas-excess condition) because the gas-excess method reduces the experimental time dramatically for hydrate formation. The gas-excess method results in hydrates cementing predominantly grain contacts (cementation) or on grain surface, which is not in-situ hydrate pore habit (pore-filling). Therefore, the laboratory experimental results may not be able to represent water permeability of in-situ hydrate-bearing sediments.

It is known that the pore habit of THF hydrates is pore-filling based on seismic measurement (Kunerth et al. 2001), X-ray computed micro-tomography (Kerkar et al. 2009), and transparent microfluidic chip test (Mahabadi et al. 2016a). In this study, THF hydrate was formed to study permeability reduction. In addition, P and S wave velocities were monitored to capture any hydrate dissolution during the test.

2 Permeability Model and Experimental Data

The analytical models for water permeability k_{rw} as a function of hydrate saturation S_h have been developed depending on the shape of pore space (Capillary tube model and Kozeny grain model) and the location of hydrates in pore space (coating and pore-filling) (Kleinberg et al. 2003).

Capillary tube model

$$\text{Coating} \quad k_{rw} = (1 - S_h)^2 \tag{1}$$

$$\text{Pore - filling} \quad k_{rw} = 1 - S_h^2 + 2(1 - S_h)^2/\log(S_h) \tag{2}$$

Kozeny grain model

$$\text{Grain coating} \quad k_{rw} = (1 - S_h)^{n+1} \tag{3}$$

where n = 1.5 for $0 < S_h < 0.8$ and $n > 1$ for $S_h > 0.8$

$$\text{Pore - filling} \quad k_{rw} = (1 - S_h)^{n+2}/\left(1 + S_h^{0.5}\right)^2 \tag{4}$$

where $n = 0.7S_h + 0.3$

Permeability as a function of hydrate saturation based on capillary tube model and Kozeny grain model is shown in Fig. 1. Capillary tube model shows a wider range of permeability values at a given hydrate saturation compared to the Kozeny grain model.

The empirical models available in the literature include Tokyo model, $k_{rw} = (1 - S_h)^N$ where N is the permeability reduction exponent (Masuda et al. 2002) and hybrid model considering both coating and pore-filling hydrate pore habits (Delli and Grozic 2014). Moreover, the permeability reduction curves are also suggested by considering capillary effect through numerical simulation using the Lattice Boltzmann method (Kang et al. 2016).

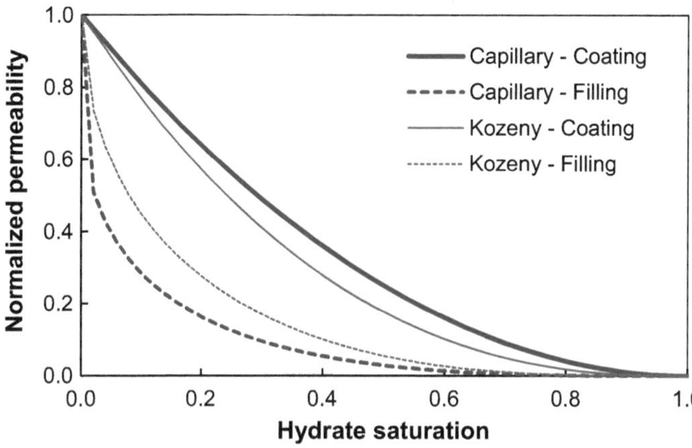

Fig. 1. Water permeability reduction as a function of hydrate saturation – effect of hydrate pore habit.

Table 1 summarizes the experimental results of permeability results. It includes the various formation and measurement methods such as gas type (methane, carbon dioxide), circulating fluid (water and gas), formation method (excess-gas, excess-water, and ice seeding), and sample type (laboratory and pressurized-core sample).

3 Experimental Details

The falling head conductivity cell (Humboldt Mfg. Co., Model: HM-5982) was modified to measure the permeability as a function of hydrate saturation (Fig. 2a). The acrylic cylinder (15.2 cm in height and 8.9 cm in diameter) of the cell was replaced by an aluminum cylinder in the same dimension. Bender elements (Piezo Systems, item #: T2LCH) and piezo-crystals (APC International, Ltd., MFT-15T-6.7A1 RoHS) for P- and S-wave velocity measurement were installed in the cylinder. A function generator (Agilent, 33210A), oscilloscope (Keysight, DSOX-3014A), and filter/amplifier (Krohn-Hite, model: 3364) were connected to the bender elements and piezo-crystals to measure P- and S-wave velocities (Fig. 2b). Thermocouples (Conax Technologies, K-type) connected to the data-logger (Agilent, 34972A) were placed in the cell to monitor the temperature profile during the hydrate formation and permeability measurement.

Sands (passing through #10 sieve and retained on #200 sieve, ASTM C778) were compacted into the cylinder by three layers (the porosity of the sediment was n = 0.39). Two wire meshes were placed in the bottom and top of the sediment to prevent the migration of sands during the permeability test. Once the sediment was compacted, the water/THF solution with various water-THF mass ratio including from 92:8 for S_h = 0.4 to 87:13 for S_h = 0.7 was injected into the sample. The temperature of the environmental chamber was decreased to 0.5 °C. After hydrate formation was detected by temperature peak, the chamber was remained at the temperature for 3 days for complete

Table 1. Summary of permeability tests in hydrate-bearing sediments.

Type		Measure method	Formation method	Sample	References
Artificially-formed hydrates	CH_4 hydrate	CH_4 gas flow	Excess gas	N/A	Liang et al. (2011)
			Ice seeding	Clayey	Liu et al. (2016)
			Ice thawing	Ksand/F110/F110 + Silt	Kneafsey et al. (2011)
				F110	Gupta (2007)
		Water flow	Excess gas	Sands	Sakamoto et al. (2003, 2004)
				Core recovered (Mt. Elbert)	Johnson et al. (2011)
				Toyoura sand No. 7 & 8 sand Mallik simulated sand	Minagawa et al. (2005)
				Toyoura sand	Minagawa et al. (2007)
		NMR/water flow	Excess gas (gas bubbling)	Silica sand	Minagawa et al. (2008a, b, 2009)
			Excess gas/excess water	Silica sand	Minagawa et al. (2012)
		NMR	Gas bubbling	Berea sandstone	Kleinberg et al. (2003)
		PNM	Not described	Uniform-sized quartz	Wang et al. (2015)
	CO_2 hydrate	Water flow	Excess gas	Ottawa 20/30	Delli and Grozic (2014)
				Silty fine sands	Liu et al. (2017)
		CO_2 gas flow	Excess gas	Glass beads	Kumar et al. (2010)
		Gas flushing		Sand + kaolin	Chuvilin and Grebenkin (2015)
	R11	Liquid flow	Excess gas	Sand $\mu = 280\ \mu m$	Berge et al. (1999)

(*continued*)

Table 1. *(continued)*

Type		Measure method	Formation method	Sample	References
Naturally-formed hydrates	Pressurized core	NMR	In-situ	Silt (70%) and clay (15–30%)	Li et al. (2014)
		Water flow		Nankai Trough (silty sand)	Santamarina et al. (2015)
					Priest et al. (2015)
					Konno et al. (2015)
		Consolidation			Yoneda et al. (2015)
	Well-logging	NMR		Mallik 5L-38	Lee (2008)
					Kleinberg et al. (2005)
				Nankai Trough	Murray et al. (2006)

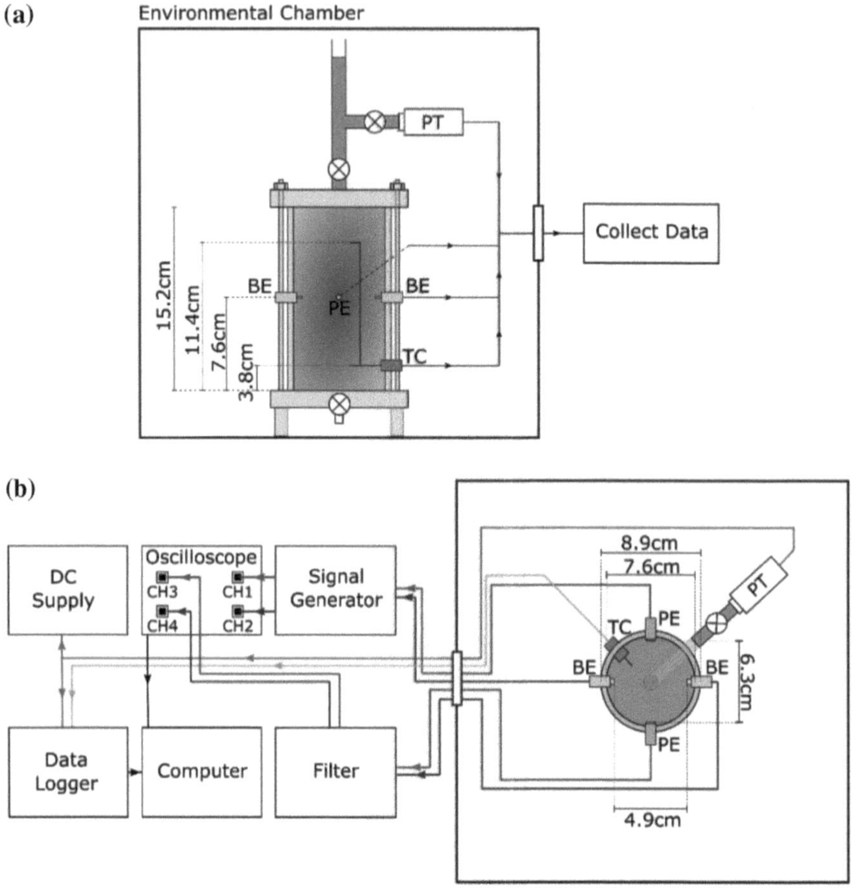

Fig. 2. (a) Modified falling head permeability test cell (Pressure Transducer PT, Bender Element BE, Piezo-crystal PC, Thermocouple TC). (b) Top view of the falling head cell and the details for the connections.

conversion to THF hydrates (until there was no change in temperature and wave velocity measurement). After 3 days, the valve of the water column was opened and the falling head permeability test was performed repeated three times at a given hydrate saturation. P- and S-wave velocities were measured after each permeability test.

4 Results and Analyses

Water permeability as a function of hydrate saturation. Water permeability test was performed for the hydrate saturations including S_h = 0, 0.4, 0.5, 0.6 and 0.7. The water permeability values measured at each hydrate saturation are normalized by the water permeability at S_h = 0 (Fig. 3). The normalized permeability decreases as the hydrate saturation increases, which is in agreement with reported permeability measurements in the literature for in-situ data, pressurized core test, and lab tests. The measured normalized permeability is 0.026 for $S_h = 0.7$. However, hydrate dissociation caused by repetitive permeability test results in 4–7% increase in normalized permeability during three permeability tests at a given hydrate saturation. Figure 3 shows the permeability measurements for different hydrate saturations. The measured permeability from the first conductivity test are 143.8 mD for $S_h = 0$, 33.8 mD for $S_h = 0.4$, 15.8 mD for $S_h = 0.5$, 5.8 mD for $S_h = 0.6$ and 3.8 mD for $S_h = 0.7$. These values are increased to 36.3 mD for $S_h = 0.4$, 17.0 mD for $S_h = 0.5$, 6.1 mD for $S_h = 0.6$ and 4.1 mD for $S_h = 0.7$. The measured permeability values are distributed within the boundaries between Kozeny coating and pore-filling models.

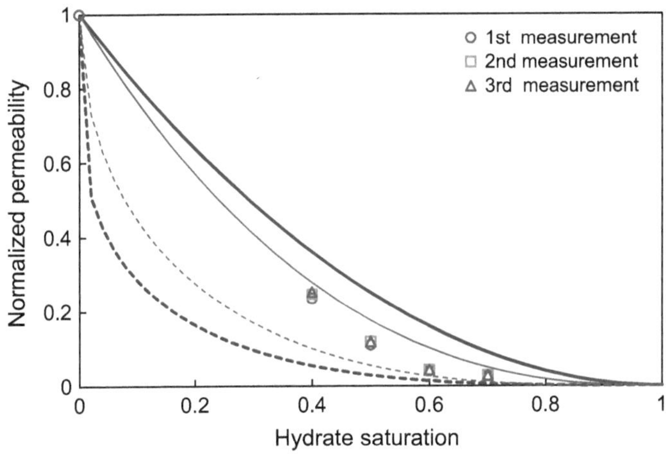

Fig. 3. Normalized water permeability as a function of hydrate saturation. Three permeability values at a given hydrate saturation are shown (Absolute water permeability measured at $S_h = 0$ is 143.84 mD).

Compressional and shear wave velocity. Figure 4 shows the compressional and shear wave velocity measurements for different hydrate saturations. The results show V_s and V_p increasing with hydrate saturation. The shear wave velocity increases 106% from $V_s = 0.107$ km/s for $S_h = 0$ to $V_s = 0.221$ km/s for $S_h = 0.4$, 208% for $S_h = 0.5$ ($V_s = 0.331$ km/s), 516% for $S_h = 0.6$ ($V_s = 0.516$ km/s) and 777% for $S_h = 0.7$ ($V_s = 0.942$ km/s). The percentage increase for V_p is almost 10 times lower than the percentage increase for V_s. The compressional wave velocity increases 13% for $S_h = 0.4$ ($V_p = 1.611$ km/s for $S_h = 0$ and $V_p = 1.826$ km/s for $S_h = 0.4$), 24% for $S_h = 0.5$, 48% for $S_h = 0.6$ and 78% for $S_h = 0.7$.

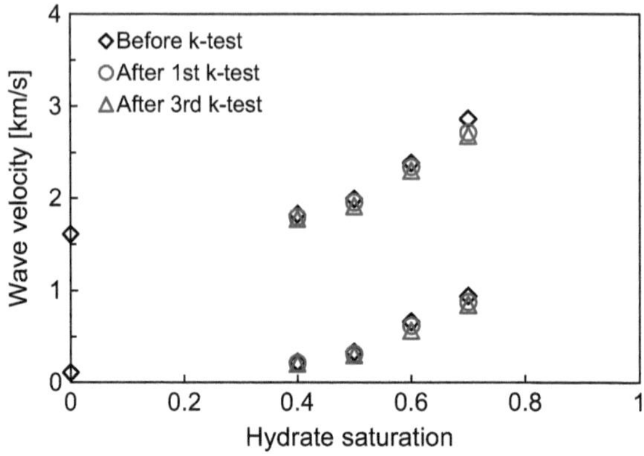

Fig. 4. Compressional and shear wave velocity measurements for different hydrate saturations, before and after hydraulic conductivity test.

5 Conclusions

Water permeability is a critical parameter to govern fluid flow that is also associated with heat transfer and pressure diffusion. But the laboratory test to mimic the hydrate pore-habit of in-situ hydrate is very difficult in terms of experimental time and hydrate properties. In this study, THF hydrate was selected to form hydrates due to pore-habit similarity to in-situ hydrates. Water permeability and wave velocities are measured as a function THF hydrate saturation. The water permeability decreases as hydrate saturation increases. And at a given hydrate saturation, water permeability decreases during the three repetitive permeability measurements. Compression and shear wave velocity results also follow the trend of permeability results.

Acknowledgments. This work was supported by the research fund of Hanyang University (HY-201700000002411).

References

Berge, L.I., Jacobsen, K.A., Solstad, A.: Measured acoustic wave velocities of R11 (CCl3F) hydrate samples with and without sand as a function of hydrate concentration. J. Geophys. Res. **104**(B7), 15415–15424 (1999)

Chuvilin, E.M., Grebenkin, S.I.: Gas permeability variations in gas-filled soils upon hydrate formation and freezing: an experimental study. Sci. J. Earth's Cryosphere **19**(2):59–64 (2015)

Chuvilin, E.M., Ebinuma, T., Kamata, Y., Uchida, T., Takeya, S., Nagao, J., Narita, H.: Effects of temperature cycling on the phase transition of water in gas-saturated sediments. Can. J. Phys. **81**(1–2):343–350 (2003) https://doi.org/10.1139/p03-028

Delli, M.L., Grozic, J.L.H.: Experimental determination of permeability of porous media in the presence of gas hydrates. J. Pet. Sci. Eng. **120**, 1–9 (2014). https://doi.org/10.1016/j.petrol.2014.05.011

Gupta, A.: Methane hydrate dissociation measurements and modeling: the role of heat transfer and reaction kinetics. Colorado School of Mines, Golden, CO (2007)

Jakobsen, M., Hudson, J.A., Minshull, T.A., Singh, S.C.: Elastic properties of hydrate-bearing sediments using effective medium theory. J. Geophys. Res. Solid Earth **105**(B1), 561–577 (2000). https://doi.org/10.1029/1999jb900190

Jang, J., Santamarina, J.C.: Hydrate bearing clayey sediments: formation and gas production concepts. Mar. Pet. Geol. **77**, 235–246 (2016). https://doi.org/10.1016/j.marpetgeo.2016.06.013

Johnson, A., Patil, S., Dandekar, A.: Experimental investigation of gas-water relative permeability for gas-hydrate-bearing sediments from the Mount Elbert Gas Hydrate Stratigraphic Test Well, Alaska North Slope. Marine Pet. Geol. **28**(2), 419–426 (2011). https://doi.org/10.1016/j.marpetgeo.2009.10.013

Kang, D.H., Yun, T.S., Kim, K.Y., Jang, J.: Effect of hydrate nucleation mechanisms and capillarity on permeability reduction in granular media. Geophys. Res. Lett. **43**(17), 9018–9025 (2016). https://doi.org/10.1002/2016gl070511

Kerkar, P., Jones, K.W., Kleinberg, R., Lindquist, W.B., Tomov, S., Feng, H., Mahajan, D.: Direct observations of three dimensional growth of hydrates hosted in porous media. Appl. Phys. Lett. **95**(2), 024102 (2009). https://doi.org/10.1063/1.3120544

Kleinberg, R.L., Flaum, C., Collett, T.S.: Magnetic resonance log of JAPEX.JNOC:GSC et al. Mallike 5L-38 gas hydrate production research well: gas hydrate saturation, growth habit, and relative permeability. In: Dallimore, S.R., Collett, T.S (eds.) Scientific Results from the Mallik 2002 Gas Hydrate Production Research Well Program, Mackenzie Delta, Northwest Territories. Geological Survey of Canada, Canada (2005)

Kleinberg, R.L., Flaum, C., Griffin, D.D., Brewer, P.G., Malby, G.E., Peltzer, E.T., Yesinowski, J.P.: Deep sea NMR: methane hydrate growth habit in porous media and its relationship to hydraulic permeability, deposit accumulation, and submarine slope stability. J. Geophys. Res. **108**(B10), 2508 (2003). https://doi.org/10.1029/2003jb002389

Kneafsey, T.J., Seol, Y., Gupta, A., Tomutsa, L.: Permeability of laboratory-formed methane-hydrate-bearing sand: measurements and observations using X-ray computed tomography. SPE J. SPE **139525**, 78–94 (2011)

Konno, Y., Yoneda, J., Egawa, K., Ito, T., Jin, Y., Kida, M., Suzuki, K., Fujii, T., Nagao, J.: Permeability of sediment cores from methane hydrate deposit in the Eastern Nankai Trough. Mar. Pet. Geol. **66**, 487–495 (2015). https://doi.org/10.1016/j.marpetgeo.2015.02.020

Kumar, A., Maini, B., Bishnoi, P.R., Clarke, M., Zatsepina, O., Srinivasan, S.: Experimental determination of permeability in the presence of hydrates and its effect on the dissociation characteristics of gas hydrates in porous media. J. Pet. Sci. Eng. **70**(1–2), 114–122 (2010). https://doi.org/10.1016/j.petrol.2009.10.005

Kunerth, D.C., Weinberg, D.M., Rector III, J.W., Scott, C.L., Tohnson, J.T.: Acoustic laboratory measurements during the formation of a THF-hydrate in unconsolidated porous media. J. Seismic Explor. **9**, 337–354 (2001)

Lee, M.W.: Models for gas hydrate-bearing sediments inferred from hydrate permeability and elastic velocities. Scientific Investigations Report Rep. 2008-5219, U.S. Geological Survey (2008)

Lee, M.W., Collett, T.: Elastic properties of gas hydrate-bearing sediments. Geophysics **66**(3), 763–771 (2001)

Li, C.-H., Zhao, Q., Xu, H.-J., Feng, K., Liu, X.-W.: Relation between relative permeability and hydrate saturation in Shenhu area, South China Sea. Appl. Geophys. **11**(2), 207–214 (2014). https://doi.org/10.1007/s11770-014-0432-6

Liang, H., Song, Y., Chen, Y., Liu, Y.: The measurement of permeability of porous media with methane hydrate. Pet. Sci. Technol. **29**, 79–87 (2011)

Liu, L., Zhang, H., Liu, C., Li, C., Sun, J., Meng, Q.: Determining the permeability of hydrate-bearing silty-fine sands with water transient flow. In: CGS/SEG International Geophysical Conference, Qingdao, China (2017)

Liu, W., Wu, Z., Li, Y., Song, Y., Ling, Z., Zhao, J., Lv, Q.: Experimental study on the gas phase permeability of methane hydrate-bearing clayey sediments. J. Nat. Gas Sci. Eng. **36**, 378–384 (2016). https://doi.org/10.1016/j.jngse.2016.10.055

Mahabadi, N., Jang, J.: Relative water and gas permeability for gas production from hydrate-bearing sediments. Geochem. Geophys. Geosyst. **15**:2346–2353 (2014). doi:10.1002/

Mahabadi, N., Zheng, X., Jang, J.: The effect of hydrate saturation on water retention curves in hydrate-bearing sediments. Geophys. Res. Lett. **43**(9), 4279–4287 (2016a). https://doi.org/10.1002/2016gl068656

Mahabadi, N., Dai, S., Seol, Y., Yun, T.S., Jang, J.: The water retention curve and relative permeability for gas production from hydrate-bearing sediments: pore-network model simulation. Geochem. Geophys. Geosyst. **17**, 3099–3110 (2016b). https://doi.org/10.1002/2016GC006372

Masuda, Y., Kurihara, M., Ohuchi, H., Sato, T.: A field-scale simulation study on gas productivity of formations containing gas hydrates. Paper Presented at Fourth International Conference on Gas Hydrates, Yokohama, Japan (2002)

Minagawa, H., Sakamoto, Y., Komai, T., Narita, H.: Relation between pore-size distribution and permeability of sediment. Paper Presented at the 19th International Offshore and Polar Engineering Conference, Osaka, Japan (2009)

Minagawa, H., Ohmura, R., Kamata, Y., Ebinuma, T., Narita, H., Masuda, Y.: Water permeability measurements of gas hydrate-bearing sediments. Paper Presented at the 5th International Conference on Gas Hydrates, Trondheim, Norway (2005)

Minagawa, H., Nishikawa, Y., Ikeda, I., Sakamoto, Y., Komai, T., Narita, H.: Measurement of methane hydrate sediment permeability using several chemical solutions as inhibitors. Paper Presented at the 7th ISOPE Ocean Mining Symposium, Lisbon, Portugal (2007)

Minagawa, H., Egawa, K., Sakamoto, Y., Komai, T., Tenma, N., Narita, H.: Characterization of hydraulic permeability of methane-hydrate-bearing sediment estimated by T2-distribution of proton NMR. Paper Presented at The 22nd International Offshore and Polar Engineering Conference, Rhodes, Greece (2012)

Minagawa, H., Nishikawa, Y., Ikeda, I., Miyazaki, K., Takahara, N., Sakamoto, Y., Komai, T., Narita, H.: Relation between permeability and pore-size distribution of methane-hydrate-bearing sediments. Paper Presented at Offshore Technology Conference, Houston, TX (2008a)

Minagawa, H., Nishikawa, Y., Ikeda, I., Miyazaki, K., Takahara, N., Sakamoto, Y., Komai, T., Narita, H.: Measurement of water permeability and pore-size distribution of methane-hydrate-bearing sediments. Paper Presented at the 6th International Conference on Gas Hydrates, Vancouver, Canada (2008b)

Murray, D., Fukuhara, M., Khong, C.K., Namikawa, T., Yamamoto, K.: Permeability estimates in gas hydrate reservoirs of the Nankai Trough. In: SPWLA 47th Annual Logging Symposium, Veracruz, Mexico (2006)

Priest, J.A., Druce, M., Roberts, J., Schultheiss, P., Nakatsuka, Y., Suzuki, K.: PCATS Triaxial: a new geotechnical apparatus for characterizing pressure cores from the Nankai Trough, Japan. Marine Pet. Geol. **66**, 460–470 (2015). https://doi.org/10.1016/j.marpetgeo.2014.12.005

Sakai, A.: Can we estimate the amount of gas hydrates by seismic methods? Ann. N. Y. Acad. Sci. **912**, 374–391 (2000). https://doi.org/10.1111/j.1749-6632.2000.tb06792.x

Sakamoto, Y., Komai, T., Kawabe, Y., Yamaguchi, T.: Properties of multiphase flow in marine sediments with gas hydrate. Paper Presented at 5th ISOPE Ocean Mining Symposium, Tsukuba, Japan (2003)

Sakamoto, Y., Komai, T., Kawabe, Y., Tenma, N., Yamaguchi, T.: Formation and dissociation behavior of methane hydrate in porous media—estimation of permeability in methane hydrate reservoirs, Part 1. J. Min. Mater. Process. Inst. Jpn. **120**(2), 85–90 (2004)

Santamarina, J.C., Jang, J.: Gas production from hydrate bearing sediments: geomechanical implications. Fire Ice **9**, 18–22 (2009)

Santamarina, J.C., et al.: Hydro-bio-geomechanical properties of hydrate-bearing sediments from Nankai Trough. Mar. Pet. Geol. **66**, 434–450 (2015). https://doi.org/10.1016/j.marpetgeo.2015.02.033

Wang, J.-Q., Zhao, J.-F., Yang, M.-J., Li, Y.-H., Liu, W.-G., Song, Y.-C.: Permeability of laboratory-formed porous media containing methane hydrate: observations using X-ray computed tomography and simulations with pore network models. Fuel **145**, 170–179 (2015). https://doi.org/10.1016/j.fuel.2014.12.079

Yoneda, J., Masui, A., Konno, Y., Jin, Y., Egawa, K., Kida, M., Ito, T., Nagao, J., Tenma, N.: Mechanical properties of hydrate-bearing turbidite reservoir in the first gas production test site of the Eastern Nankai Trough. Mar. Pet. Geol. **66**, 471–486 (2015). https://doi.org/10.1016/j.marpetgeo.2015.02.029

Concept of a Geotechnical Solution to Address the Issues of Sea Water Intrusion in Ashtamudi Lake, Kerala

T. G. Sitharam[1(✉)], Sreevalsa Kolathayar[2], Shuqing Yang[3], and Amala Krishnan[2]

[1] Department of Civil Engineering, Indian Institute of Science Bangalore, Bangalore, India
sitharam@civil.iisc.ernet.in
[2] Department of Civil Engineering, Amrita University, Coimbatore, India
[3] International Center for Coastal Reservoir Research, University of Wollongong, Wollongong, Australia

Abstract. This paper presents an innovative concept to transform brackish Ashtamudi Lake into a clean freshwater lake (partially or fully) using advanced geotechnology to construct a dike at appropriate place. This will serve two purposes: to meet water demands of the people and to save tiny islands from the threat of sea level rise. The sea level rise, according to researchers, is of serious concern in the state of Kerala which has a 590 km-long coastline and large expands of backwaters and estuaries and low-lying areas such as filtration ponds. Some tiny islands in Ashtamudi lake like Munroe Thuruthu islands seem to be gradually sinking due to reduced sediment deposits and sea-level rise. Nearby areas are facing frequent and increased tidal flooding throughout the year. The role of proposed project in saving these islands and other low lying areas from tidal flooding and sea level rise are discussed in the paper. Ashtamudi wetland is an estuary filled with brackish water and sewage, which lies in the Kollam district of Kerala. This study aims to conduct a comprehensive feasibility of the concept of conversion of brackish Ashtamudi Lake to a fresh water reservoir by building a barrage at the outlet of lake at the mouth of the Arabian Sea (Neendakara). On one side, it will block freshwater flowing from Kallada River to the lake from being discharged into the sea. On the other side, the barrage will prevent seawater from entering the lake, avoiding the adulteration of the freshwater supply with saltwater. This paper presents three different possible schemes to meet the water demands of Kollam city, Kollam district and entire state of Kerala. The paper also presents the required lengths of dikes for these schemes and suggest different methods of dike construction.

1 Introduction

Kollam district (Fig. 1), is located on the southwest part of Kerala State and has a geographical area of 2491 km^2 which is about 6.48% of the total geographical area of the State. Population of the district is 2629703 as per 2011 census, (8.12% of the total population of the State) and the population density is 1056 persons per km^2 (Census of

© Springer International Publishing AG, part of Springer Nature 2019
S. Shu et al. (eds.), *New Developments in Materials for Infrastructure Sustainability and the Contemporary Issues in Geo-environmental Engineering*, Sustainable Civil Infrastructures, https://doi.org/10.1007/978-3-319-95774-6_19

India 2011). Pathanapuram, Kunnathur, Kottarakkara, Karunagapally and Kollam are the five Taluks in the district. The district is further subdivided into 11 development blocks, 73 Panchayaths and 104 villages. Paravoor, Punalur and Karunagapally are Municipal towns and Kollam has the status of a City Corporation.

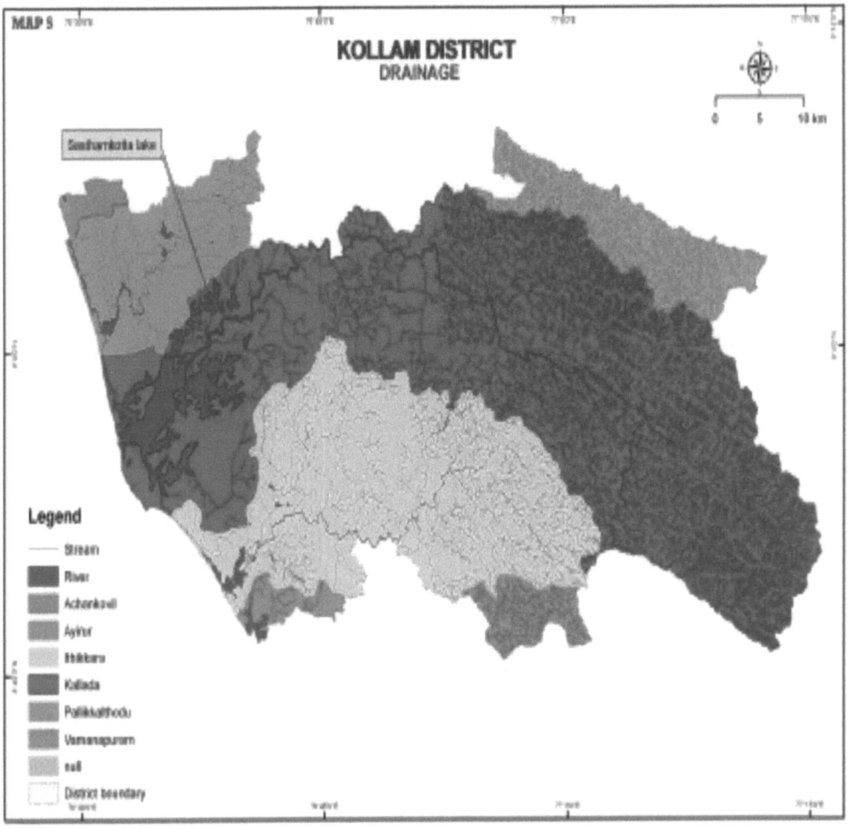

Fig. 1. Kollam district map

Ashtamudi Lake in the Kollam District of Kerala, is the second largest lake, next to Vembanad Lake in the state of Kerala, India. Ashtamudi wetland is an estuary filled with brackish water and sewage. This lake has eight arms and all the arms converge into a single outlet at Neendakara near Kollam, to enter the Arabian Sea. Ashtamudi is the deepest estuary in Kerala with a maximum depth of 6.4 m at the convergence zone. Kallada River is a major river flowing into the Ashtamudi Lake (Fig. 2). The Kallada river which originates from the western Ghats, travels 121 km and enters in to the Ashtamudi basin with an average annual runoff of 2,152 million cubic meters (MCM) of freshwater at the inlet of basin.

Sasthamkotta Lake is rain fed and has no visible fresh water source or tributaries. The ground water recharge to lake is greatly reduced these days. But it is not a best

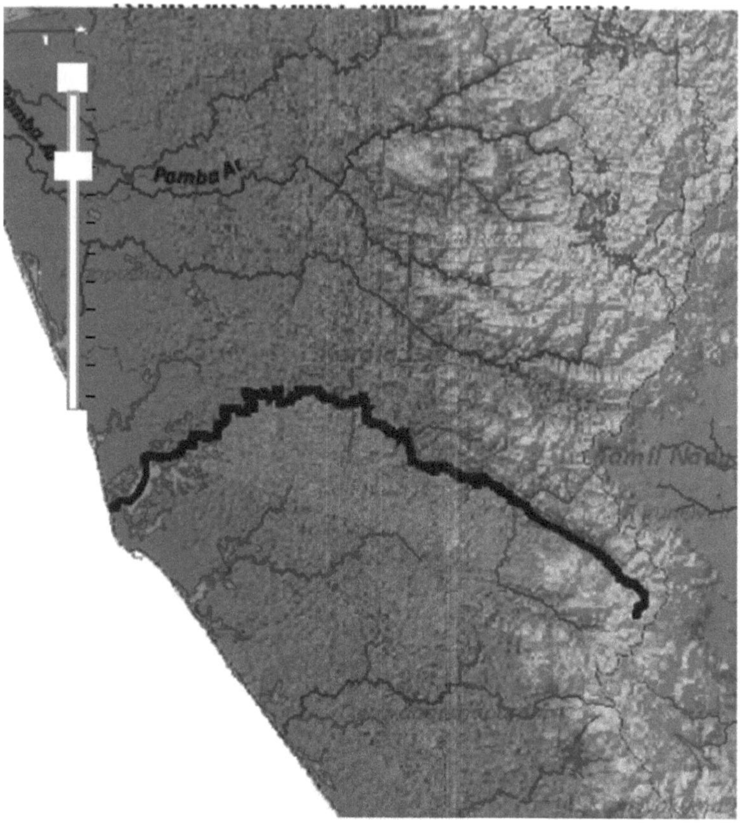

Fig. 2. Kallada river

option to take Kallada water to Sasthamkotta. Because during rainy season it is already fed and there is no point in taking flood water to Sasthamkotta. In contrary, Ashtamudi is a brackish lake and even if one mudi (arm) is replaced with fresh water it can augment the water demand of Kollam and adjacent regions throughout the year which is not possible from Sasthamkotta which has insufficient water in summer.

2 Need of the Study

In-spite of high rainfall in the district, Kollam city and surrounding regions are facing water shortage in summer due to storage shortage. Kollam city has total water demand of 60 MLD of water per day and the current availability is 28 MLD. The total current water requirement in Kollam city is 0.77 TMC ft. The population statistics and water requirement for Kollam city, Kollam district and state of Kerala are shown in Table 1. Kollam Municipal Corporation draws 16 MLD of water from Sasthamkotta Lake, which goes almost dry in summer months. Ashtamudi lake is a much bigger lake (6424 ha) which contains brackish water and hence unqualified for domestic use. This necessitates a feasibility study on the concept of formation of a fresh water reservoir to augment the water demands of Kollam city by converting Ashtamudi Lake to a fresh water lake (partially or fully). Besides this, the sea level rise, according to researchers, is of serious concern in the state of Kerala which has a 590 km-long coastline and large expands of backwaters and estuaries and low-lying areas such as filtration ponds. Some tiny islands in Ashtamudi lake like Munroe Thuruthu islands seem to be gradually sinking due to eroded shoreline, reduced sediment deposits and sea-level rise. Nearby areas are facing frequent and increased tidal flooding through the year. The role of proposed project in saving these islands and other low lying areas from tidal flooding and sea level rise will be investigated and presented. This study proposes the concept of constructing a dike at the mouth of a popular brackish water lake of southern India, to check sea water intrusion into the lake and to save tiny islands in the lake from the after effects of sea level rise and climate change. Saline water intrusion is perceived in the shallow aquifer in the western part of the district which is in hydraulic connection with the back water. There is water scarcity along the eastern hilly areas and Laterite hillocks. Issues of water logging is perceived along the western parts of the district bordering the back water lagoons (Eg: Munroe Thuruthu) during the rainy season. Chavara, Kundara and Pozhikara areas face the issues of groundwater pollution as well (CGWB 2013).

Table 1. Population statistics and water demand

Region	Population (2011 census)	Water requirement (MLD)	Annual water requirement (TMC ft)
Kollam city	346,147	69.22	0.9
Kollam district	2,629,703	525.94	6.7
Kerala	34,523,726	3367.13	44

3 Proposed Fresh Water Storage Strategy

This study aims to conduct a comprehensive feasibility of the concept of conversion of brackish Ashtamudi lake to a fresh water reservoir by building a barrage at the outlet of lake at the mouth of the Arabian Sea (Neendakara) (Fig. 3). On one side, it will block freshwater flowing from Kallada River to the lake from being dispersed into the sea. On the other side, the barrage will prevent seawater from entering the lake, avoiding the contamination of the freshwater supply from saltwater. The project schemes comprises mainly two steps first the construction of the dike at the mouth of Ashtamudi Lake, and second the process of natural replacement of salty water by rainwater and surface runoff to the lake. The dike must be designed to separate fresh water from the salty waters of the Arabian Sea. The quantity of flood water has to be estimated and possibility to alleviate flooding during high tides in ocean.

Three schemes are proposed here as shown in Fig. 4. Scheme 1 is planned in Chittumala Chira which is already connected to Ashtamudi through a canal. If we can have a canal from Kallada to Chittumala Chira, fresh water from Kallada can be drawn

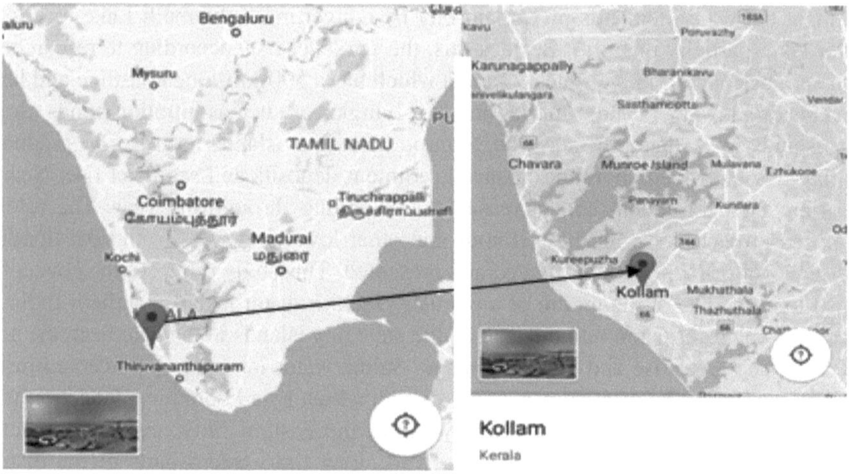

Fig. 3. Location of proposed fresh water reservoir

for Scheme 1. Scheme 2 envisages to convert one arm of the lake into complete freshwater reservoir. Scheme 3 aims at converting entire Ashtamudi lake into a fresh water reservoir for which the dike has to be constructed at the outlet of Ashtamudi Lake into the sea at Neendakara. Details of proposed storage schemes with storage capacity and dike lengths are shown in Table 2. The annual yield of Kallada River is about 80 TMC ft and enough water is available every year to supply fresh water to proposed schemes in Ashtamudi Lake.

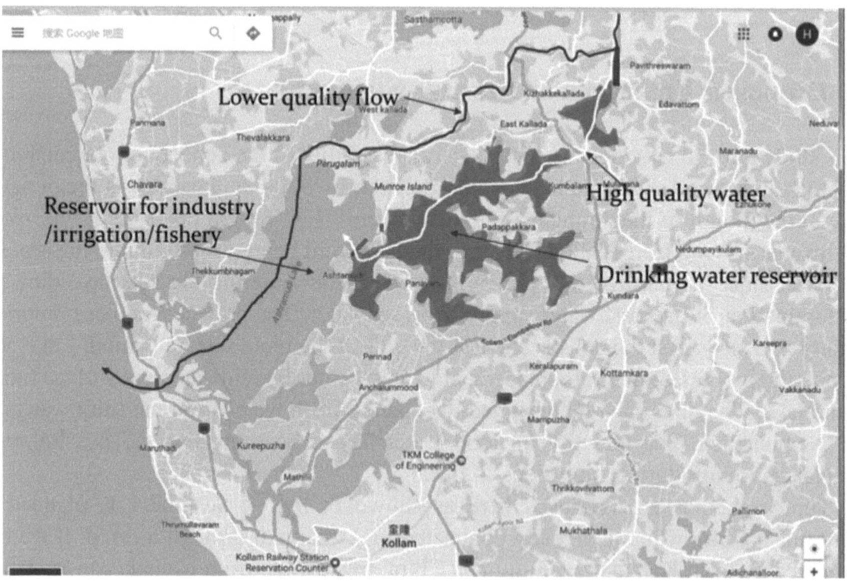

Fig. 4. Proposed area of river showing Chittumala Chira and Ashtamudi Lake

Table 2. Proposed storage schemes

	Location of storage	Storage capacity	Length of dike
Phase 1	Chittumala Chira, part of Ashtamudi	0.3 TMC	775.25 m
Phase 2	Ashtamudi partially	3 TMC	267 m
Phase 3	Ashtamudi full	12 TMC	622.02 m

4 Dike Construction Techniques

A dike is an embankment widely used to protect low-lying areas against inundation and acts as a barrier to prevent salt water intrusion. Dikes can be constructed from a variety of materials, most commonly: geosynthetic tubes, geocells, reinforced concrete, boulders, steel, or gabions. Table 3 presents classification of dykes according to materials used. A dike is also used for land reclamation projects. Sea dikes are primarily used at exposed coasts, but they are also used at moderately exposed coasts and in this case, use of dikes is presented for the separation of ocean salt water from the flood water from rivers stored in coastal reservoirs formed by replacement of brackish water by fresh water. The dikes across estuarine can reduce the impacts of tides and waves from the sea and also help in reducing salt intrusion into the river course through the estuaries (Sitharam 2017). The sea dike as part of coastal reservoir is a boon to reduce the ill effects of sea water intrusion and sea level rise permanently and bring economic prosperity to the coastal areas. The objective of the proposed sea dike is to store abundant monsoon flood waters close to the coast by replacing brackish water with fresh river flood waters. This opens up a huge possibility of a fresh water ecosystem and a development of sustainable township abundant with freshwater.

Table 3. Classification of dykes according to materials used (after Chu et al. 2010)

Type	Construction method
Earth-fill dike or levee	(1) Using compacted soils
	(2) Using cement mix soils or bagged soil
	(3) Using dumped rocks
Masonry and concrete	(1) Using cast-in-place or precast concrete walls
	(2) Using precast concrete panels
	(3) Using roller compacted concrete
Steel sheetpiles or bored piles	(1) Driven steel sheetpile wall
	(2) Contiguous bored pile or prefabricated sprung piles
Geotextile or geosynthetic	(1) Geotube filled with concrete mortar, sand or clay
	(2) Rubber dam
	(3) Geo-bag or geo-container
	(4) Geo-mattress (or geomat)
Prefabricated concrete segment	(1) Concrete caissons
	(2) Semi-circular concrete caissons
	(3) Steel or concrete suction piles or caissons
	(4) Tongtu assembly method
Mix types	Dike construction involving the use two or more of the above methods

4.1 Dike Design Aspects

The dike route is selected based on the basis of techno-economical reasoning after considering following points: (i) Topographic and geological conditions, (ii) Master plan of the entire area development including transportation system, national security and defense, (iii) Evolution of coast lines, beaches and estuaries, (iv) Location of existing structures, (v) Protection of the cultural, historical remains and administrative land boundaries and projects of national importance if any in that area, and (vi) Design of dike shall conform to navigation development strategy and also adaptable to the impacts of climate change.

Hydrodynamic conditions at the connection zone, waves, nearshore sediment flow, imbalance of sand transport in nearby areas, forecast of development trend of the foreshore in the future needs to be considered in designing the reservoir bounded by dikes. Recent topographic survey data for at least 100 m from the dike toe on either side in the last 20 years is needed for dike design. In addition, geological data which is based on actual conditions along with meteorological and hydrological data shall be collected for studying the impacts of natural disasters within the project area. Further, data on existing population and development trend along with current economic condition and development orientation is needed for dike project. Environmental conditions and evaluation of the impact level of the dike on the surrounding environment in the future is needed. For these dikes, which are generally closer to the developed industrial urban areas, a return period of 100 years has to be used for safety standards.

4.2 Typical Dike Cross Section Design

Items of dike cross section design include: crest level, cross sections dimensions, crest structure, dike body and dike toe which fulfil the technical and economical requirements. Design cross sections of the dike are selected on the basis of geological conditions, materials used for construction of dike, filling materials, external forces, layout of the dike and also the operational requirements. Sloping dikes, wall-type dikes and composite dikes are three different types of dikes based on geometrical shape. One need to carefully design the dike crest level, dike body, filter layers, slope protection layers, toe protection. There is different usage of sea dikes and mainly protects the reservoir from the tidal action and create a calm water in the coastal reservoir area for many activities. Most important function of the dike, in this case, is to separate the salt water and the freshwater in the coastal reservoir and also allow freshwater fishing and other activities in calm water conditions and it can also provide dock or quay facilities along with support of floating solar panels for energy production. The main features of the dike can be seen in Fig. 5. Further development (after Yang 2016) has been the one using soft dam to separate freshwater from salt water or brackish water in the buffer zone (Fig. 6). This is a patented technology of soft dam costal reservoir by Yang and Lin (2011) [Ref: Yang et al. (2013)]. Soft dam will be built using a membrane at the river mouths to uphold the river runoff lost to the sea in front of the primary barrier. This provides an additional safety against mixing of salt/brackish water into freshwater stored.

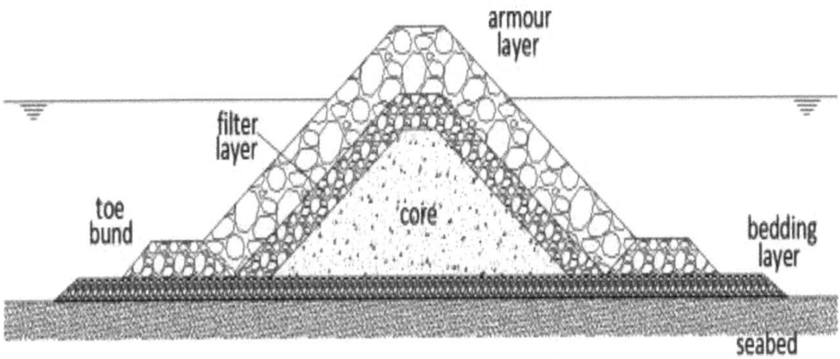

Fig. 5. Typical features of a dike (Sitharam 2017)

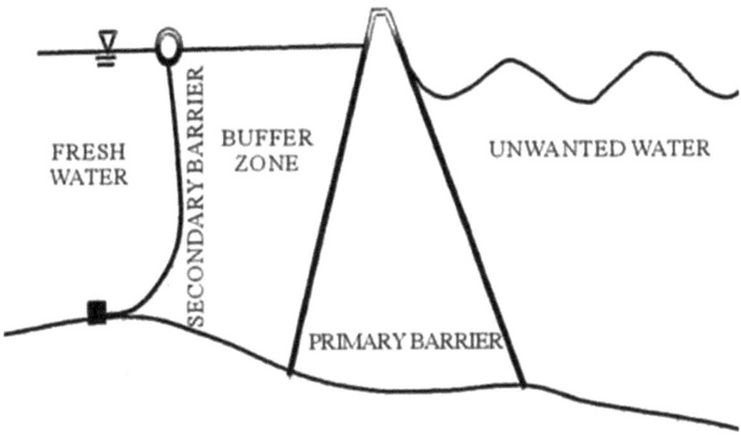

Fig. 6. Soft dam concept by Yang et al. (2005)

5 Conclusions and Recommendations

This study proposed the concept of constructing a dike at the mouth of Ashtamudi Lake, to check sea water intrusion into the lake and to save tiny islands in the lake from the after effects of sea level rise and climate change.

Conversion of brackish Ashtamudi Lake to a fresh water reservoir is feasible by building a barrage at the outlet of lake at the mouth of the Arabian Sea (Neendakara). On one side, it will block freshwater flowing from Kallada River to the lake from being dispersed into the sea. On the other side, the barrage will prevent seawater from entering the lake, avoiding the contamination of the freshwater supply from saltwater. This paper presented three different possible schemes to meet the water demands of Kollam city, Kollam district and entire state of Kerala. The required lengths of dikes for these schemes are presented along with storage capacities. Scheme 1 can supply water to entire Kollam Municipal Corporation. Scheme 2 can meet water demands of Kollam, Cochin and Thiruvananthapuram cities. Scheme 3 will suffice to supply water to entire state of Kerala. Coastal reservoirs use the sustainable storm flood waters which are reasonably cleaner. This will emerge as a sustainable strategy for water resource development in the coming years. The annual yield of Kallada River is 80 TMC ft and hence enough water is available every year to supply fresh water to proposed schemes in Ashtamudi Lake.

The paper also discussed dike construction techniques. Emerging geosynthetic materials and innovative geosynthetic applications are essential for the construction of sea wall dikes. Formation of coastal reservoirs using dikes will increase the use of innovative materials and new construction techniques are required to build these dikes to safely harvest the river flood waters. A detailed integrated site survey comprising of geophysical investigation and geotechnical investigation shall be planned to be undertaken in order to derive the geological model appropriate for the location demarked for the proposed barrage to prevent salt water intrusion into fresh water reservoir. The studies to be undertaken are Side scan sonar techniques, Sub-bottom profiling, Deep Boring, In-stitu testing and Laboratory Soil testing. The suitability of advanced and low cost geo-technology using geotubes filled with locally available sand/soil shall be explored in detail for cost effective dike construction. Application of HDPE/Natural Geomembranes and composite geosynthetics shall be employed to check the pollution from urban settlement and industries to Ashtamudi Lake and proposed fresh water reservoir.

References

Sitharam, T.G.: Sea dike construction challenges in shallow water for storage of river flood waters: sustainable strategy for water resource development using coastal reservoirs. In: E-Proceedings of the 37th IAHR World Congress, Kuala Lumpur, Malaysia, 13–18 Aug 2017

Yang, S., Liu, J., Lin, P., Jiang, C.: Coastal reservoir strategy and its applications. In: Wurbs, R. (ed.) Water Resources Planning, Development and Management, pp. 95–115. Intech, United States (2013)

Yang, S., Lin, P.: Coastal reservoir by soft-dam and its possible applications. In: Recent Patents on Engineering, pp. 45–56 (2011)

Yang, S.-Q., et al.: Coastal reservoirs. Singapore Patent No. 200504653-7, (2005)

Chu, J., Yan, S.W., Li, W.: Innovative methods for dike construction—an overview. In: Geotextiles and Geomembranes, pp. 35–42 (2010)

Central Ground Water Board: Ground water information booklet of Kollam district, Kerala state. Technical Reports: Series D, Ministry of Water Resources, Government of India (2013)

Central Water Commission: Central water commission report. Government of India (2006)

Importance of Indoor Environmental Quality Criteria to Occupants of Low Income Housing

Ayodeji E. Oke(✉), Clinton O. Aigbavboa, and Nkosinathi W. Ngema

Department of Construction Management and Quantity Surveying, Faculty of Engineering and the Built Environment, University of Johannesburg, Johannesburg, South Africa
{emayok, nngema}@gmail.com, caigbavboa@uj.ac.za, nngema@gmail.com

Abstract. Temperature, humidity, aeration and lighting among others have been the major traditional variables that are considered in understanding and improving comfort of building occupants. As important as these factors are, the need for an hollistic approach to basic necessities and requirements for comfort and satisfaction of people gave rise to the adoption of indoor environmental quality (IEQ) concept. This study therefore examines the importance attached to various IEQ criteria by occupants of selected low-income housing buildings with a view to understanding and improving needed and required facilities for optimum satisfaction. Using questionnaire as the research instrument, purposive sampling was adopted to collect information from occupants of low-income housing, with emphasis on their residence in the building for at least 2 years and above the age of 18. Analysis of the completed 54 questionnaires indicates that basic and traditional factors such as room temperature, humidity and ventilation are the most important IEQ criteria to sampled occupants. The results of this study will help concerned stakeholders of low-income housing including the government, developers, contractors, construction professionals and regulatory agencies to understand the level of importance attached to basic comfort and satisfaction factors by occupants, with a view to providing improve building for better and optimum satisfaction.

Keywords: Building quality · Indoor environmental quality
Occupants' satisfaction · Smart structure · Health and safety

1 Introduction

The indoor environmental quality of a building can cause occupants to experience a range of acute and serious illnesses, although, these can also be as a result of outdoor components. The discomfort and illness levels of the occupants need to be recorded and addressed by the construction industry professionals. This should be done in order to prevent occupants from experiencing these illnesses and discomforts attributed to sick building syndrome and indoor environmental quality. Thankfully studies have shown

© Springer International Publishing AG, part of Springer Nature 2019
S. Shu et al. (eds.), *New Developments in Materials for Infrastructure Sustainability and the Contemporary Issues in Geo-environmental Engineering*, Sustainable Civil Infrastructures, https://doi.org/10.1007/978-3-319-95774-6_20

that indoor environmental quality can be improved at any stage of a building's existence.

Indoor environmental quality (IEQ) refers to a building's environment in relation to the health and well-being of those who occupy any space within it according to the National Institute of Occupational Safety and Health (NIOSH 2015). Smith and Pitt (2011) in America maintains that indoor environmental quality is most simply described as the conditions found and experienced within the building. These conditions include air quality (but also access to daylight and views); pleasant acoustic conditions and occupant control over lighting and thermal comfort (NIOSH 2015). These may also include the functional aspects of the space inhabited such as whether the layout provides easy access to tools and whether there is sufficient space for the occupants (Smith and Pitt 2011). Building managers and operators can increase the satisfaction of building occupants by considering all of the aspects of IEQ rather than narrowly focusing on temperature or air quality alone, considering the amount of time people spend inside their houses.

The understanding of the relationship between the health conditions of human beings and the indoor environment they inhabit continues to evolve. Mitchell et al. (2007) acknowledged that previous research on health and indoor environments has concentrated mainly on discrete pollutant sources, exposures and specific disease processes. Recently, efforts have been made to characterize, more fully, the complex interactions between the health of occupants and the interior spaces they occupy (Mitchell et al. 2007).

This research aims to study and review recent advances in source characterization, exposure assessment, health effects associated with indoor exposures and all relevant research related to indoor environments and their potential to cause harm to the occupants. Some of the advances in source characterization that will be investigated include chemical transportation and processing within spaces and so too the role that other factors such as lighting and building design may play when determining health. Exposure measurement albeit challenging to measure at this stage of the research into indoor environmental quality, will be measured according to the effects on exposures to the vulnerable population such as children and the elderly (Mitchell et al. 2007). When constructing low-income houses, it is easy to forget that the success or failure of a project may rest on its indoor environmental quality (Whole Building Design Guide 2015). The productivity of a person depends on or is enhanced when they are healthy and comfortable, but, this is often neglected as it is easier to focus on the first-cost of a project than it is to determine the value of increased user productivity and health according to the Whole Building Design Guide (WBDG 2015).

When buildings are constructed, appropriate appreciation and attention should be given to the provision of high-quality indoor environments for users irrespective of whether the building is an upscale, private building or state subsidized housing (low-income housing) (WBDG 2015). Considering how we as humans have adjusted and adapted to outdoor environmental conditions, it is likely that we shall struggle with the condition of our indoor environment as we spend most of our time indoors albeit in different environments such as classrooms, halls, offices and our own homes (WBDG 2015). If these conditions are not conducive to being and living, we can experience a range of illnesses.

This study contributed to the knowledge and understanding of indoor environmental quality and the implications it has on building occupants with particular focus on occupant health and work performance. With the knowledge of various indoor environmental quality indicators and criteria in low-income housing, the study will enable the construction industry and built environment professionals to incorporate improved design structures while bearing the indoor environment's quality in mind.

2 Indoor Environmental Quality and Building Evaluation

When chemicals exist indoors they are bound to react with one another and so thus, the removal of these chemicals from the indoor environment is necessary and this is achieved through a physical or a chemical process of removal but, in order to be alerted to the removal of the chemicals and the necessity thereof, one needs to first acquaint oneself with what indoor chemistry is and simply put, it is the reaction of indoor pollutants that react in the indoor environment either in gaseous phase or particle phase (Weisel et al. 2005). "The result of reactions indoors in their gaseous phase in buildings is the formation of harmful aldehydes, carboxylic acids allergenic peroxides and hydro peroxides (Fan et al. 2003; Mitchell et al. 2007)."

Surfaces reactions are a significant contributor to the quality of the indoor environment and arise as a result of the volume to surface area indoors being significantly higher than the surface to volume area present outdoors and this leads to the prevalence of illness within the building given that more reactions can take place indoors given the confined space in a building and higher contact ratio, leaving the building occupants to experience the aftermath of these reactions which are as a result of cooking, cleaning bathing and other household activities (Dubowski et al. 2004; Mitchell et al. 2007).

The indoor environment results in mould developing in the building, particularly on the surface, and the chemical make-up of the mould reacts with the atmosphere's inherent nitrogen oxide resulting in gaseous emissions that can be taken in by human beings within buildings and cause discomfort, other contributors to the indoor chemistry is bio effluents and the building material used in the construction and all other airborne particles (Dubowski et al. 2004; Mitchell et al. 2007).

With great thanks to the research of the evaluation buildings in recent years, many countries have developed systems which can be used to carry out the evaluations and also to aid the building industry to set targets when building and to consider the environmental impact that the building will have by means of rating the building upon completion by the stakeholders (Malmqvist 2008).

The development of building evaluation methods began in the late 20th century and of which the first to be implemented was BREEAM, which was created by the English as there was at the time a growing concern over the environmental impact of construction and with the development of BREEAM, the subsequent evaluation methods were introduced with it as the primary base (Malmqvist 2008).

Systems such as LEED, CASBEE and NABERS are national evaluation methods and have been adopted the world over but, must comply with laws in the country of implementation and all the above mentioned methods are similar in that they all provide a good environmental assessment of the buildings being built but, some systems are exclusive to certain countries such as the EccoEffect from Sweden and the GBTool which is exclusive to Great Britain (Malmqvist 2008).

The successful incorporation of these systems can have a drastic effect on the construction of buildings as they will ensure that all environmental aspects are considered such as the surroundings of the area (albeit industries or a highly populated area with much traffic), the evaluation tools are key to reducing the effects of the sick building and to improving the indoor environmental quality of the buildings.

The systems will ensure that every aspect of the building is considered and with the studies being done on indoor environmental quality growing, the built environment is becoming more aware of the materials to avoid when constructing buildings. With regard to low-income housing, the material used for these is of a somewhat inferior quality to the norm and thus leaves the building susceptible to decay, cracks and vulnerable to many types of temperature changes, but, with the change and environmental conditions, the evaluation tools will ensure the consideration of every element.

3 Research Methodology

Descriptive survey was adopted in this study because it provides accurate portrayals or accounts of characteristics being investigated such as behavior patterns and knowledge about a particular situation hampering individuals or a group. The design was adopted to meet the research objectives of the study which is to determine the implications of indoor environmental quality on the building occupant health and physical well-being.

The information was collected through self-administered questionnaires that were personally handed out by the researcher. The study was conducted in Dundee, particularly in the Eighteen and Twenty low-income housing areas which fall under the Endumeni Region, UMzinyathi District in the province of Kwa-Zulu Natal, South Africa. The areas cater for the people living in the Endumeni Region who meet the criteria for having a Reconstruction and Development Plan (RDP) house. The population consists of occupants residing in the areas that are above the age of eighteen, all of which had to be permanent residents of the buildings.

The data in this research was collected through the usage of a questionnaire to examine the occupants' experiences within their homes and their satisfaction with the indoor environmental quality therein. The administered questionnaires were in lay English and isiZulu in order to enable every participant to understand and respond adequately. The respondents were assured that their answered would not be linked to them at any stage, thus ensuring anonymity to the respondents. Clear instructions were given to the respondents and the researcher completed the questionnaires for people that could not read. All the subjects completed the questionnaires in the presence of the researcher, and this was done to prevent subjects from giving questionnaires to other people to complete on their behalf.

Convenience sampling approach was used, as the research respondents were surveyed based on their availability and ease of access by the researcher considering the vast area of the study area. The building occupants residing in the selected area were expected to meet the following criteria in order to be considered:

- Minimum age of 18 years
- Mentally sound
- Willing to participate in the study (No one was forced or otherwise bribed)
- No bias for sex/gender
- They needed to have passed grade 9 at least in terms of educational achievements
- Living in the current house for a minimum of 3 months; and
- If any other people lived in the house with any of the respondents, they too had to meet the criteria in order to participate.

At the end of the data collection process, 54 questionnaires were retrieved from the 60 distributed. Data was analyzed using descriptive statistics, mean item scores (MIS) and standard deviation were adopted to rank and determine the importance of various IEQ criteria to occupants of selected low income houses.

From the most prominent ten IEQ criteria, the respondents were made to value the significance level of each by indicate a value from 1 to 5 each number symbolizing either Very High (5) High (4) Neutral (3) Low (2) and Very Low (1). This section too had a 100% completion rate with every respondent completing the section. The program SPSS was used to analyze the section and the analysis included determination by means of mean and standard deviation, the criteria regarded as most significant and vice versa. In rank of mean item score, the responses given by the respondents were ranked on the basis of significance. The Cronbach alpha for this section is 0.93 indicating that the adopted scale and instrument for the study is reliable.

4 Findings and Discussion

The significance of the IEQ criteria to the residents was based on several IEQ criteria considered as important for determining the possible implications that IEQ could have on their health. These excluded the physical aspects of IEQ as these held no bearing on the topic at hand as the study was centred mainly on health implications of the IEQ criteria and physical aspects were excluded from the study due to time but were noted in a study by Aigbavboa and Thwala (2012) which included the physical aspects of IEQ and will be referenced in the study as deemed necessary in cases where physical aspects held a bearing on the significance of the IEQ criteria. These physical indices used in the study done by these authors bore some similarities with the ones used in this study although the wording used in this study did not separate the indices into their most simplest forms such as "ventilation" which was expressed in sub-sections such as "draught coming into the building," "indoor air quality" and "stuffiness" just to name a few.

From Table 1, the criteria ranking first and that had the highest significance (determined by means of MIS i.e. above 4 is agreement of significance and below 3 is disagreement or insignificance) is room temperature with a mean item score of 4.17, second is quality of sleep achieved with a mean item score of 4.13 and third is smell or odour within the home (MIS = 4.06), fourth is draught coming into the house (4.04), fifth is dust and dirt (4.02), sixth is stuffy or bad air in your house (3.96), seventh is glaring light (3.91), eighth and ninth and noise levels within (3.87) and noise levels outside (3.83) the house respectively. The significance was measured in this study using the following indices as expressed in Table 1 and the mean item score depicted the level of significance that the respondents held with each. The indices were ranked in descending order of satisfaction with room temperature ranking first with a mean item score of 4.17. Room temperature is a sub-element of ventilation and also that the building is positioned ideally enough to allow sunlight to enter the building and heat the indoor environment up sufficiently.

Table 1. Significance of the IEQ criteria to the respondents

IEQ Indices/criteria	MIS	Std. deviation	Rank
Room temperature	4.17	0.84	1
Quality of sleep achieved inside your home	4.13	1.08	2
Smell or odor within the house	4.06	1.00	3
Draught coming into the house	4.04	0.87	4
Dust and dirt	4.02	1.06	5
Stuffy or bad air in your house	3.96	1.05	6
Glaring light	3.91	0.96	7
Noise levels within the house	3.87	0.93	8
Noise levels outside the house	3.83	1.06	9

MIS Mean Item Score

A good room temperature reduces the possibility of health issues affecting the residents of the buildings as noted by the World Health Organisation (2004), the indoor air quality, humidity, low temperature and overcrowding in a house usually poses threats to the health of the residents and if possible these need to be addressed in a manner so as to prevent the building from hampering the health of residents.

The high significance held with this particular element shows that the residents consider this element as essential to good and comfortable living as if the building, particularly the windows in the building, permit the ingress and egress of air in a manner which allows the temperature in the building to be adjustable, the room temperature is kept steady and this is important to the health of the occupants residing within the building as the room temperature ideal for healthy living is a temperature between 20 and 26 and in order to compensate for the added heat on bodies due to clothing, the recommended temperature between the 6 degree range is 23 so as to not impair mental thinking ability and focus during the day as a result of added body heat as noted in various studies done by The European Concerted Action (ECA 1989), Galatioto et al. (2014) and also Jansz (2011) and if the temperature in the buildings is in fact kept between these temperatures, living would be comfortable. Thus, it can be said that the significance of room temperature in their building is a fundamental desire of a building occupant as no person desires to live in a building that is either too humid or too cold as this can result in health issues such as the flu from rapid temperature fluctuations.

The quality of sleep achieved inside your home ranks second. The lack of sleep can be attributed to many factors and if this vital aspect of living is not achieved on a regular basis, health implications such as fatigue, stress, hypertension and reduced working ability could result as noted by Malmqvist (2008). This higher significance ranking of this index shows that the residents of Dundee value the aspect of being able to sleep peacefully at night and to rest up after every day and also that the residents are aware of the implications of sleep deprivation and also of the fact that they are more productive in the workplace and other aspects of their lives and with the health related effects noted in studies such as the one above. It can also be argued that noise causes disturbances and noises arising in the evening could be a factor attributed to the quality of sleep achieved and this fact is corroborated by findings from a study done by Malmqvist (2008) stating that sleep disturbance can be attributed to traffic noise in the outdoor environment and if there is not as much noise the sleeping conditions are improved however, these only prevail in buildings located closer to roadways as traffic is an ongoing concern through the night when residents wish to sleep (Malmqvist 2008) The effects of noise of sleep quality is also mentioned in studies done by the ECA (1989) and Jansz (2011).

Smell or odour within the house and draught entering the house were next in order of third and fourth, the two had means of 4.06 and 4.04 respectively. The smell or odour inside the buildings was considered very significant to the residents considering the higher mean compared to the other elements which is a sign that the residents require and desire a building to reside in that has sufficient ventilation and most of the foul odours are allowed to pass through, refreshing the building and keeping bad air contaminants out of the building. Galatioto et al. (2014) maintains that the residents are only hampered by odours which they perceive as foul and cannot identify the origin of, causing stress, frustration and anxiety. The study was also done by the ECA (1989) and Galatioto et al. (2014) corroborated an early findings by Weschler. Weschler (2004) in his study found that the interaction of chemicals indoors causes considerable discomfort and health issues but with good housekeeping and regular cleaning these can be removed from the indoor environment.

Draught entering the building was another element regarded as highly important to a building and was ranked fourth. This level of significance means that the residents desire a building that will be in such as position as to allow sufficient sunlight to enter the building in order to absorb the vitamins given by the sunlight and provide an environment in which the residents can reside comfortably. This high level of significance can be related to desires the occupants have that the building should possess such as building positioning that will allow this draught to enter during the day. Literature shows that a person's perceived meaning of quality determines their satisfaction with a building according to Ibem (2012) and this perceived quality would determine the features that a resident would consider significant and essential for the building to have in order for that building to meet their quality standards.

5 Conclusion and Recommendations

The occupants deemed the following indoor environmental quality as significant to them and comfortable living. Room temperature Quality of sleep achieved inside your home Smell or odour within the house Draught coming into the house Dust and dirt, these having the superior mean item scores and the following had the lower mean item scores and were deemed significant too, but, not as the ones above and these are stuffy or bad air in your house, glaring light, noise levels within the house, noise levels outside the house. The above results show that the level significance held by the respondents was assessed and it can therefore be concluded that the second objective was achieved.

The high level of importance attached to room temperature shows that the residents consider this element as essential to good and comfortable living as if the building, particularly the windows in the building, permit the ingress and egress of air in a manner which allows the temperature in the building to be adjustable, the room temperature is kept steady and this is important to the health of the occupants residing within the building as the room temperature ideal for healthy living is a temperature between 20 and 26.

Quality of sleep was another element of high significance. The lack of sleep can be attributed to many factors and if this vital aspect of living is not achieved on a regular basis, health implications such as fatigue, stress, hypertension and reduced working ability could result. This higher significance ranking of this index shows that the residents of Dundee value the aspect of being able to sleep peacefully at night and to rest up after every day and also that the residents are aware of the implications of sleep deprivation and also of the fact that they are more productive in the workplace and other aspects of their lives and with the health related effects noted in studies such as the one above. It can also be argued that noise causes disturbances and noises arising in the evening could be a factor attributed to the quality of sleep achieved. The smell or odour inside the buildings was considered very significant to the residents considering the higher mean compared to the other elements which is a sign that the residents require and desire a building to reside in that has sufficient ventilation and most of the foul odours are allowed to pass through, refreshing the building and keeping bad air contaminants out of the building.

Draught entering the building was highly significant to the occupants. This level of significance means that the residents desire a building that will be in such as position as to allow sufficient sunlight to enter the building in order to absorb the vitamins given by the sunlight and provide an environment in which the residents can reside comfortably. This high level of significance can be related to desires the occupants have that the building should possess such as building positioning that will allow this draught to enter during the day.

In view of these findings, buildings should be constructed with these elements borne in mind by the organisations undertaking the project. More so, the residents of these buildings need to abstain from activities that temper with the natural indoor environment such as smoking indoors and the excessive use of humidifiers.

The study focused solely on the health and well-being implications of IEQ on the occupants and not the overall environment and or construction of the buildings the occupants reside in. The study did not delve into the relationship between workplaces and home environments as there are many IEQ-related health and well-being issues contracted in the workplace that could be brought home by the occupants and perceived as being originating from the home environment when the root is at the workplace. More so, the study did not deal in detail with the relationship between school environments and the home environments as there are many IEQ issues contracted in the school environment that could be brought home by the occupants and perceived as being originating from the home environment when the root is at the school environment.

References

Aigbavboa, C.O., Thwala, W.D.: Appraisal Constr Manage. **12**(1), 1–21 (2012). https://doi.org/10.1080/15623599.2012.10773181

Dubowski, Y., Sumner, A.L., Menke, E.J.: Interactions of gaseous nitric acid with surfaces of environmental interest. Phys Chem Physics **6**, 3879–3888 (2004)

Fan, Z., Lioy, P., Weschler, C., Fiedler, N., Kipen, H., Zhang, J.: Ozone-initiated reactions with mixtures of volatile organic compounds under simulated indoor conditions. Environ. Sci. Technol. **37**, 1811–1821 (2003)

Galatioto, A., Leone, G., Milone, D., Pitruzella, S., Franzitta, V.: Indoor Environmental quality survey: a brief comparison between different post occupancy evaluation methods, pp. 1148–1152. Trans Tech publications, Switzerland (2014)

Ibem, E.O.: Residents' perception of the quality of public housing in urban areas in Ogun State, Nigeria. Int. J. Qual. Reliab. Manage. **29**(9), 1000–1018 (2012)

Jansz, J.: Chapter 2: Theories and Knowledge About Sick Building Syndrome. Department of Health & Safety Environmental Health: Curtin University, Perth, WA 6845 (2011)

Malmqvist, T.: Environmental rating methods: selecting indoor environmental quality (IEQ) aspects and indicators. Build. Res. Inf. **36**(5), 466–485 (2008). https://doi.org/10.1080/09613210802075841

Mitchell, C.S., Zhang, J., Sigsgaard, T., Jantunen, M., Lioy, P.J., Samson, R., Karo, M.H.: Current State of the Science: health Effects and Indoor Environmental Quality. Department of Environmental Health Sciences: Johns Hopkins Bloomberg School of Public Health, Baltimore, Maryland, USA (2007)

National Institute of Occupational Safety and Health.: Indoor Environmental Quality (2015). Available from: http://www.cdc.gov/niosh/topics/indoorenv/

Smith, A., Pitt, M.: Healthy workplaces: plantscaping for indoor environmental quality. Facilities **29**(3/4), 169–187 (2011)

The European Concerted Action (ECA).: Indoor Air Quality & Its Impact on Man: Cost Project 613. Commission of the European Communities Directorate General for Science. Research and Development Joint Research Centre: Institute for the Environment (1989)

The World Health Organization.: Review of evidence on housing and health. Fourth Ministerial Conference on Environment and Health. The World Health Organization, Geneva, Background document No. 5046267 (2004)

Weisel, C.P., Zhang, J.J., Turpin, B.J., Morandi, M.T., Colome, S., Stock, T.H., Kwon, J.: Relationship of indoor, outdoor and personal air (RIOPA) study: study design, methods and quality assurance/control results. J. Eposure Sci. Environ. Epidemiol. **15**(2), 123–137 (2005)
Weschler, C.J.: Chemical reactions among indoor pollutants: what we've learned in the new millennium. Indoor Air **14**(suppl 7), 184–194 (2004)
Whole Building Design Guide (WBDG) Sustainable Committee.: Enhance Indoor Environmental Quality (IEQ). National Institute of Building Science (2015)

Studies on the Characteristics of the Type of Geotextiles

Mohammadreza Atrechian[1] and Morteza Ahmadi[2(✉)]

[1] Civil Engineering College, Islamic Azad University of Zanjan, Zanjan, Iran
Dr.atre@yahoo.com
[2] Department of Civil Engineering, Islamic Azad University of Zanjan, Zanjan, Iran
parsian.civil@gmail.com

Abstract. The development of earth science and wider need for more resistant materials in civil engineering projects have pushed engineers and researchers to use artificial and more durable materials named "geosynthetic". This material is made of rubber and plastic and due to its wide application in various projects and the speed of its implementation and its suitable price, its production and consumption is growing rapidly. Therefore, the importance of research on the behavior, characteristics and the application of this new product increases day by day. Geosynthetics are categorized according to their properties and characteristics. In this research, one of the categories under the title "Geotextiles" and its types and basic applications are studied. Geotextiles are commonly referred to as fabric filters. The results of studying the properties of any woven and unwoven geotextile shows that these materials have different characteristics such as fast performance, easy use, low weight, high tensile strength, low cost, long lasting, less environmental degradation, uniformity In its implementation and its other features. According to the mentioned features it can be concluded that geotextiles can be used as an alternative to traditional methods in drainage and separation of soil texture.

Keywords: Nonwovens · Flexible · Geotextile · Fiber reinforced
Mechanical properties · Low cost

1 Introduction

One of the most important human concerns up to now has been the consolidation and Reinforcement of the soil and increasing the performance and efficiency of structures. So in the ancient centuries tree logs, shrubs, rocks, and so on have been used for this purpose.

With developing of durable and reliable artificial materials over the past 50 years, there has been an effective progress in the world of industry, and the introduction of a special type of these materials that are called geosynthetics has had a considerable impact on engineering progress.

The most important member of this group are geotextile stretch fibers and geomembranes which play an important role in various fields of engineering and soil mechanics and even in road construction by upgrading soil cover on safe structures.

© Springer International Publishing AG, part of Springer Nature 2019
S. Shu et al. (eds.), *New Developments in Materials for Infrastructure Sustainability and the Contemporary Issues in Geo-environmental Engineering*, Sustainable Civil Infrastructures, https://doi.org/10.1007/978-3-319-95774-6_21

As it increases the vulnerability of the structure against explosions and makes it safe and impervious versus water flowing as well.

In the 1970s there were no more than 5 or 6 geotextile types. But today, more than 250 different substances are produced under this name. Geotextile filters have been used for more than 50 years and are currently used extensively in many parts of the world (FEMA 2008). The present study has tried to use the most famous sources and the most recent researches in the topic to investigate the geotextile characteristics and its types and the basic applications of each one.

Figure 1 shows different usages of geosynthetics in landfill design.

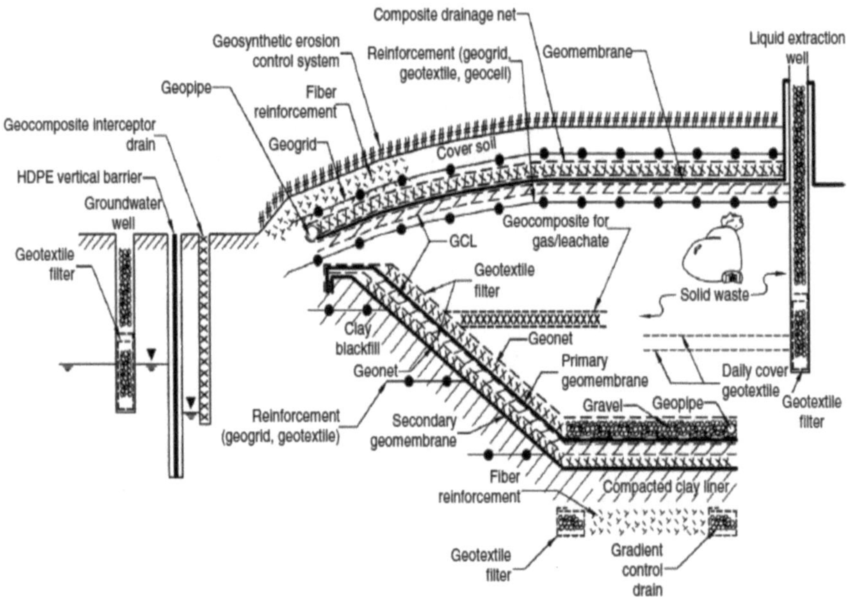

Fig. 1. Different uses of geosynthetics in landfill design

2 Geotextile History

The usage of geotextiles dates back to thousands of years ago, which controlled soil erosion at the time of ancient Romans and was used when the desired soil was unstable to build the road (Holtz 2017). The first recorded usage of geotextile was on the road which was undertaken by the Department of South Carolina Highways in 1926. In this project, a heavy cotton geotextile was loaded onto the soil. And then the hot asphalt was applied to the fabric then a thin layer of sand was poured on the asphalt. Research has shown that this fabric has reduced cracking and has prevented the road from breaking down (FEMA 2008).

The usage of geotextile has been successful for the past two decades and it has grown rapidly, due to its excellent performance in structural applications, especially on the roads.

Today, the world is witnessing thousands of projects where geotextile are used and thousands of companies and factories are active in the production and installation of geotextiles throughout Europe, America, Africa and Asia, especially East Asia.

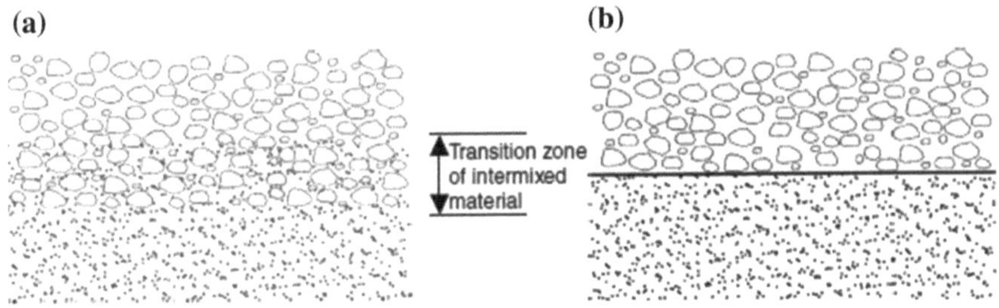

Fig. 2. Separation function of a geotextile placed between road aggregate and soft saturated subgrade. (a) Without geotextile and (b) with geotextile

In Fig. 2a geotextiles is used to separate the two layers.

The separation property is a feature that, along with properties such as flexibility and porosity, refers to the ability to locate geosynthetics products between two unmatched materials (Zornberg and Christopher 2007).

For example, the main cause of road rupture is the injection of adjacent layers into the pebble and the consequently the decline in the strength of the pebble layer (Fig. 2a). When the pebble is placed on the subgrade layer, the underlying layer is mixed by soil and, over time, the traffic and vibration load, injects the Aggregate layer into the soil, causing the layer to move upwards. Also, in wet places, the traffic causes the weak subgrade soil to pump into the pebbles and all of these conditions reduce the effective thickness of the pebble layer. As a result, the protective layer of the road is damaged and the useful life of the road decreases, which can be protected by placing a geotextile on a pebble layer under the subgrade layer.

3 Properties and Characteristics of Natural Geotextiles

As mentioned in the previous section, geotextiles are originally used as natural textiles. These textiles are usually used to control soil erosion, mainly due to the above mentioned properties, as the most appropriate temporary solution.

Some types of famous geotextiles are:

Coirmat (woven from coconut fiber), Jutmesh (woven from china fibers), Green furee (a combination of biodegradable and non-biodegradable fibers), RECP (rolled erison control product), Coirlog (coconut fiber bag tissue), Environment (spruce wood fiber) (Venkatappa et al. 2009).

Figure 3 is a sample of natural woven geotextile made from coconut fiber.

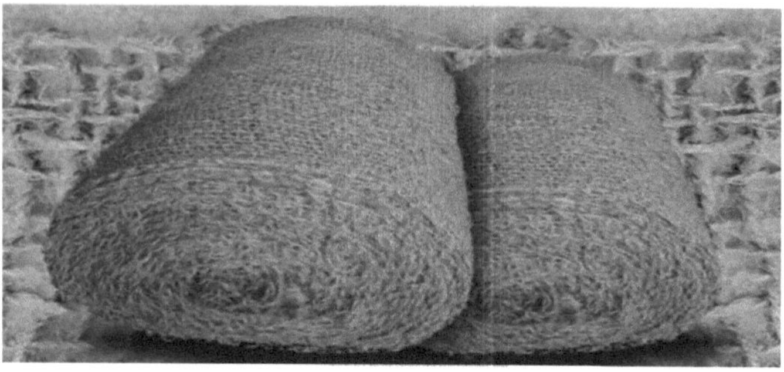

Fig. 3. A sample of natural woven geotextile made from coconut fiber

Properties and characteristics of artificial geotextiles.

Synthetic fibers were produced, instead of textiles and natural fibers mentioned in recent decades which has properties like granular materials and in terms of mechanical and chemical resistance are at a higher level.

These fibers are made from crude oil derivatives and its main property is the incorruptibility in contrast of soil degradation factors.

Geotextile materials are placed on or in soil to do one of four things:

- separation/confinement/distribute loads
  - improve level-grade soil situations such as roads, alleys, lane ways
  - improve sloped-grade situations such as banks, hillsides, stream access points
- reinforce soil
  - soil walls, bridge abutments, box culverts/bridges, and soil arches
- prevent soil movement while letting water move through the material
  - such as in drainage systems and back fill around water intakes
- controlling water pressure allowing flow in the plane of the material
  - such as on foundation walls to allow water to move down to perimeter drains (Fig. 4).

4 Synthetic Fiber Materials for Geotextiles

Due to previous studies, this type of fiber has different types and the most important ones are polymer fibers. Polymer textiles have diverse applications in the building and hydraulic industry. Such as asphalt pavement, road engineering, railing bed, dockside and damping, retaining wall, shore protection, pipeline, drainage and filtering, backside protection, Artificial Grass, tunnel, environmental protection, water and sewage, creating artificial lakes and etc. and small groups of geotextile is made from fibers that are used in controlling soil erosion (FEMA 2008) (Fig. 5).

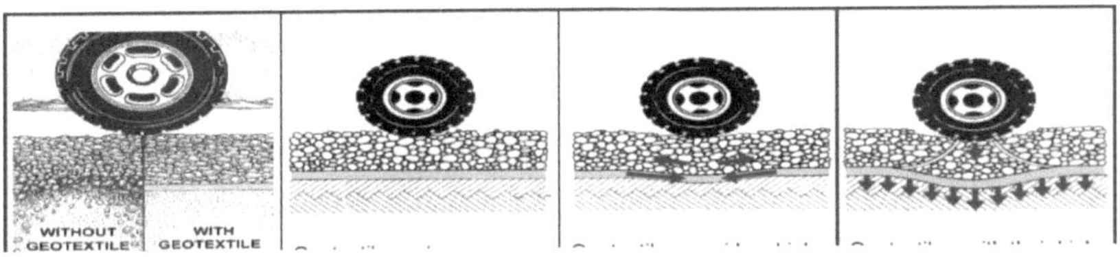

With and without a geotextile	Geotextiles act as a separation barrier between fine grain soils and load distributing aggregate fill Separation	Geotextiles provide a high friction surface between the subgrade and the aggregate helping keep the aggregate in place Confinement	Geotextiles, with their high tensile strength and modulus , act to reduce localized stress by redistributing traffic loads over a wider area of subgrade Load Distribution

Fig. 4. The main geotextile characteristics

(a)

(b)

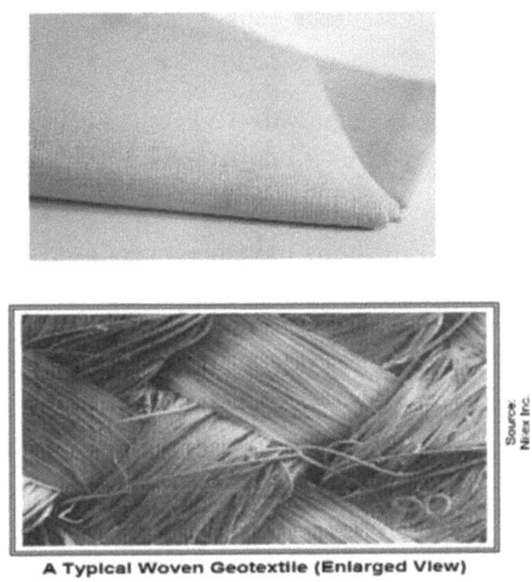

Fig. 5. (a) Non-woven geotextiles, (b) woven geotextiles

Synthetic fibers According to Woven Geotextile or non-woven (non-woven geotextile), have various quality control indicators. For the "woven synthetic fiber group" the elasticity modulus, stress—strain properties, uniaxial or biaxial tensile strength, fine particles non-passing through voids, tear resistance, UV resistance could be mentioned. And for "non-woven synthetic fibers" water absorption and water absorption properties. The absence of fine particulate matter from the pores, the amount of fluid permeability, the durability against impact, the durability against tearing, the amount of absorption of bitumen (in particular applications), thermal durability (in Specific use) can be noted (FEMA 2008).

The effective factors for providing standard filtration in geotextiles is the choice of suitable cavity size and proper formation of the soil filter bridge. Given that the flow of fluid through the geotextiles increases, but if this size is larger than a limit Certainly, it leads to a piping event in the soil in which, along with the flow of fluid, the soil particles also pass through the geotextiles and cause the structure to collapse (Fig. 6) (Zornberg and Christopher 2007).

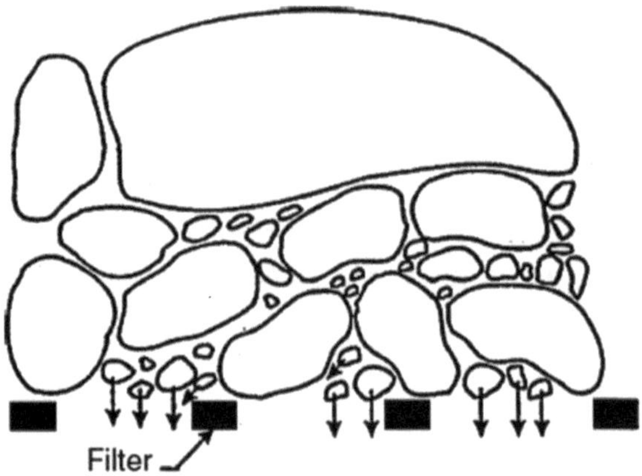

Fig. 6. Geotextile providing adequate filtration through selection of adequate opening size and formation of soil filter bridge.

Several methods have been proposed for the design of soil layers using geotextile, in which most of the properties of soil particles are compared with the size of the opening of 95% geotextile (defined as O95 in geotextile) (Zornberg and Christopher 2007). Given that the average use of geotextiles for soil particle maintenance is essential, the geotextile size should be obtained by using the ASTM D4751, which represents the largest open dimensions available for soil to cross.

5 Types of Polymer Fibers Used in Geotextiles

Polymer fibers have different types such as propylene, polyester, polyethylene, polyamid and other items. Polyethylene is the first and oldest polymer used in the production of geotextiles, which was discovered in 1931 in the ICI laboratory.

After this polymer, polyamide has the highest record. It has been used since 1935. Gradually, polypropylene was also recognized as the primary ingredient in this industry. But polyester was in competition with polypropylene and since 1954 it plays a major role in making geotextiles.

In this research, we have tried to summarize and conclude previous experiments and studies comparing polymer properties in geotextile layers against different types of resistors in Table 1. Geotextile fiber groups are divided into different polymers as the following:

Table 1. Comparing polymer properties in geotextile layers against different types of resistors

Type of resistance	Polyester	Polyamide	Polypropylene	Polyethylene
Tensile strength	H	M	L	L
Modulus of elasticity	H	M	L	L
Strain	M	M	M	M
Creep	L	M	H	H
Unit weight of the surface	H	M	L	L
Ultraviolet light resistance	H	M	H	H

L Low, *M* Medium, *H* High

Polypropylene (PP) 92% of geotextiles
Polyester (PET) 5% of geotextiles
Polyethylene (PE) 2% of geotextiles
Polyamide (Nylon) 1% of geotextiles.

There are other powerful polymers in the market, but geotextiles are produced at higher volumes (some polymers are not available in large volumes) and are economical (very specific polymers are very expensive). Among the polymers listed in Table 1, the cost versus performance of polyester is today more optimal, while polypropylene and polyethylene have more chemical resistance.

Due to previous research, the first standard for geotextile tests under hit loads is conducted by the Federal Waterways Engineering and Research center og Germany (BAW 1994).

The test apparatus in BAW's guidelines is composed of a box with horizontal dimensions of 80 × 80 cm and depth of 31 cm filled with compacted sand which the geosynthetic on top of it. The geosynthetic is fixed at the periphery of the sand bed. Once woven geotextiles are tested, it is advised to place a thick rubber sheeting on top of the geotextile prior to clamping to provide a better grip and restrain of the geotextile. Above the geotextile, there is a central rammer sleeve in which a 76 kg rammer with specific edge-cut. The drop should be performed at prescribed energy levels. After the drop is made, the geotextile is taken out for visual examinations. According to BAW's recommendation, any visual damage, which reduces the filtration capability of the geosynthetic must be regarded as a damage.

Also, in a comprehensive study, Wong et al. (2000) used the results of 784 drop tests using standard block to determine the main parameters for the stability of geotextiles.

The Drop heights in the tests were 0.5 and 2.5 m, and two woven and three non-woven geotextiles were included in the testing program. The geotextiles were laid on different types of soils with different layer thicknesses and the effects of both horizontal and sloping grounds were considered. They reported that existing index tests such as the CBR test and tensile strength tests do not properly represent an index for the survivability of the geotextile. Instead, the survivability of a geotextile corresponds to the level of energy that a geotextile can absorb under impact loading.

From the group of polymeric fibers, due to the spectacular role of propylene and polyester in geotextile, we have tried to use the DIN and ASTM standard tests as follows, the technical specifications for these two types of fibers, in the unwoven state Which contains 2 samples with different propylene 200 and 400 in normal and heat conditions, and 2 samples with different amounts of polyester 200 and 400 in normal and heat conditions are considered and are defined as follows:

6 ASTM D4632 Geotextile Grab Test

His grab test is used for geotextile fabrics to determine the breaking load (grab strength) and elongation (grab elongation). The grab test is a tensile test where the central part of the specimen's width is tested in the grips, establishing the "effective strength" of the fabric. The effective strength is the strength of the material in a specific width, together with the additional strength of adjacent material.

The specimens are cut to 101.6 mm × 203.2 mm (4 in × 8 in), with attention to the fabrication machine direction. The specimen is loaded into the grips leaving a 75 mm (3 in) separation between the jaw faces, and jaw faces must be 25.4 mm 50.8 mm (1 in × 2 in) with the longer dimension being parallel to the direction of the load. The test speed is 300 mm/min (12 in/min). The required test results are load at break and total elongation.

Geotextile materials tend to have coatings that may make it difficult to test without slippage or jaw breaks. Instron's pneumatic side acting grips provide an easy, repeatable, high throughput gripping solution. Pneumatic grips allow the user to adjust grip clamping pressure, while the performance of manually operated grips will vary depending on the operator (Fig. 7 and Tables 2, 3, 4, 5 and 6).

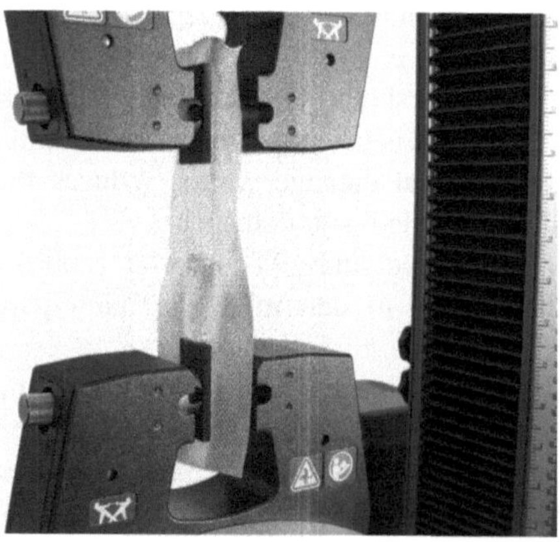

Fig. 7. Geotextile grab test machine

Table 2. Tests based on DIN & ASTM standards

DIN EN 29073/1	Determination of mass per unit area
DIN EN 29073/2	Determination of thickness
DIN EN 29073/3	Determination of tensile strength and elongation
EN 918	Dynamic perforation test
DIN EN ISO12956	Determination of the characteristic opening size
ASTM D4355	Standard test method for deterioration of geotextiles by exposure to light, moisture and heat in a xenon arc type apparatus
DIN EN ISO12956	Determination of the characteristic opening size
DIN EN ISO11058	Determination of water permeability characteristics normal to the plane, without load
ASTM D 4632	Standard test method for grab breaking load and elongation of geotextiles
ASTM D 4833	Standard test method for index puncture resistance of geomembranes and related products

Table 3. Technical specifications of non-woven heat sealed geotextile with propylene fiber

400 pp	200 pp	Test method	Unit	Characteristics
400	200	DIN EN 29073/1	g/m^2	Volume per unit area
4	2.6	DIN EN 29073/2	mm	Thickness
69	50	DIN EN 29073/3	kN/m	Maximum elongation
69	53.5	DIN EN 29073/3	kN/m	Maximum transverse stretch
60	60	DIN EN 29073/3	%	Maximum elongation
55	55	DIN EN 29073/3	%	Maximum transverse stretch
5	15	EN 918	mm	Drop amount for conical hole size
5.3	3.8	DIN EN ISO12956	kN/m	CBR strength
80	80	ASTM D4355	%	Resistance to UV rays in 500 h
42	93	DIN EN ISO12956	µm	Initial size
2	2	DIN EN ISO11058	10^{-3} m/s	Water permeability vertically
2263	1703	ASTM D 4632	N	Expandable force and tethering
2263	1703	ASTM D 4632	N	Extensible force elasticity transverse
57	65	ASTM D 4632	%	Expandable force and tethering
60	70	ASTM D 4632	%	Extensible force elasticity transverse
753	473	ASTM D 7833	N	Index of compression and resistance

Table 4. Specifications of non-woven geotextile with propylene

400 pp	200 pp	Test method	Unit	Characteristics
400	200	DIN EN 29073/1	g/m^2	Volume per unit area
4	2.6	DIN EN 29073/2	mm	Thickness
30	13	DIN EN 29073/3	kN/m	Maximum elongation
30	14.5	DIN EN 29073/3	kN/m	Maximum transverse stretch
54	53	DIN EN 29073/3	%	Maximum elongation
51	51	DIN EN 29073/3	%	Maximum transverse stretch
6	17	EN 918	mm	Drop amount for conical hole size
4.1	2.5	DIN EN ISO12956	kN/m	CBR strength
70	70	ASTM D4355	%	Resistance to UV rays in 500 h
42	90	DIN EN ISO12956	µm	Initial size
2	2	DIN EN ISO11058	10^{-3} m/s	Water permeability vertically
1395	835	ASTM D 4632	N	Expandable force and tethering
1395	835	ASTM D 4632	N	Extensible force elasticity transverse
44	50	ASTM D 4632	%	Expandable force and tethering
45	52	ASTM D 4632	%	Extensible force elasticity transverse
685	381	ASTM D 4833	N	Index of compression and resistance

Table 5. Technical specifications of non-woven heatsealed geotextile with polyester fiber

400 pp	200 pp	Test method	Unit	Characteristics
400	200	DIN EN 29073/1	g/m^2	Volume per unit area
4	2.6	DIN EN 29073/2	mm	Thickness
66	47	DIN EN 29073/3	kN/m	Maximum elongation
66	55.5	DIN EN 29073/3	kN/m	Maximum transverse stretch
60	60	DIN EN 29073/3	%	Maximum elongation
55	55	DIN EN 29073/3	%	Maximum transverse stretch
8	18	EN 918	mm	Drop amount for conical hole size
5	3.5	DIN EN ISO12956	kN/m	CBR Strength
42	93	DIN EN ISO12956	µm	Initial size
2.4	2.4	DIN EN ISO11058	10^{-3} m/s	Water permeability vertically
2260	1700	ASTM D 4632	N	Expandable force and tethering
260	1700	ASTM D 4632	N	Extensible force elasticity transverse
54	62	ASTM D 4632	%	Expandable force and tethering
57	67	ASTM D 4632	%	Extensible force elasticity transverse
750	470	ASTM D 4833	N	Index of compression and resistance

Table 6. Technical specifications of non-woven geotextile with polyester fiber

400 pp	200 pp	Test method	Unit	Characteristics
400	200	DIN EN 29073/1	g/m^2	Volume per unit area
4	2.6	DIN EN 29073/2	mm	Thickness
26	11	DIN EN 29073/3	kN/m	Maximum elongation
26	10.5	DIN EN 29073/3	kN/m	Maximum transverse stretch
55	55	DIN EN 29073/3	%	Maximum elongation
50	50	DIN EN 29073/3	%	Maximum transverse stretch
9	20	EN 918	mm	Drop amount for conical hole size
4.1	2.5	DIN EN ISO12956	kN/m	CBR strength
42	93	DIN EN ISO12956	μm	Initial size
2.4	2.4	DIN EN ISO11058	10^{-3} m/s	Water permeability vertically
1380	820	ASTM D 4632	N	Expandable force and tethering
1380	820	ASTM D 4632	N	Extensible force elasticity transverse
43	52	ASTM D 4632	%	Expandable force and tethering
47	56	ASTM D 4632	%	Extensible force elasticity transverse
720	370	ASTM D 4833	N	Index of compression and resistance

7 Conclusion

According to results the followings could be concluded:

- In terms of heat, non-woven propylene fibers and non-woven polyester fibers can tolerate longer longitudinal and transverse elongation, and in general, non-woven propylene fibers have greater dimensional stability than non-woven polyester fibers in thermal conditions (Diagrams 1 and 2).
- As the thickness of the propylene increases, the puncture resistance also increases. Also, this value shows a higher resistance to puncture in heat-bonded conditions (Diagram 3).
- By increasing the volume per unit area, the amount of drop for the conical hole size, which is measured according to the EN918 standard, is reduced by the filtration property (Diagram 4).
- Also, according to the experiments, heat-bounded propylene and heat-bounded polyester it can be concluded that in propylene the water permeability is less vertical and the perforation resistance is lower, but the elasticity and flexibility are higher so in heat-bounded propylene the non-woven produced fiber can form inside the mold easier.
- On the other hand, higher strength and higher erosion resistance of polyester are also used in cases where the fibers are more exposed to erosion.
- Due to its properties, geotextiles have been used as a suitable substitute for grading filters and have played an effective role in drainage, filtration, separation, and weaponry.

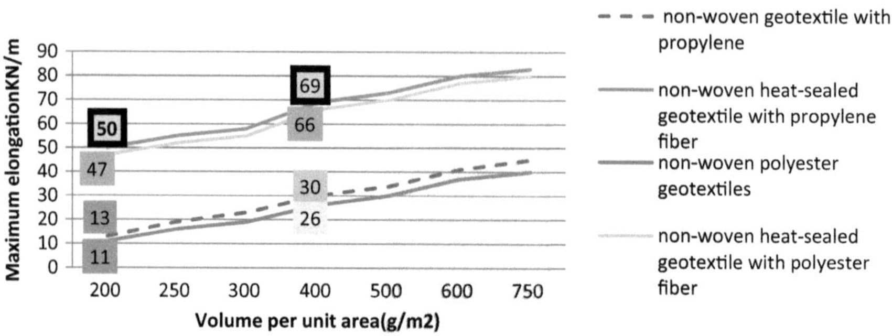

Diagram 1. Comparing the longitudinal and transverse elongation due to volume per unit area

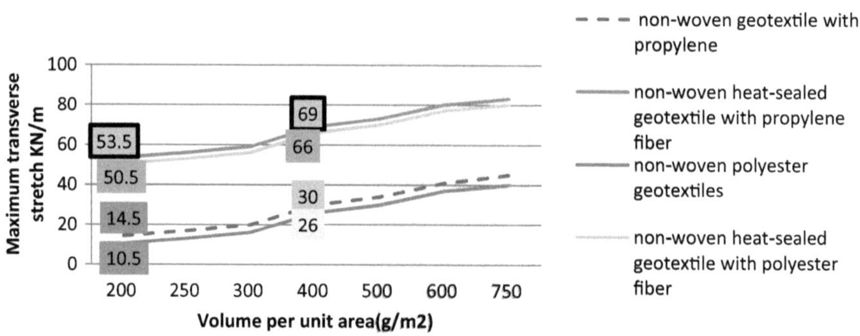

Diagram 2. Comparing the maximum transverse due to volume per unit area

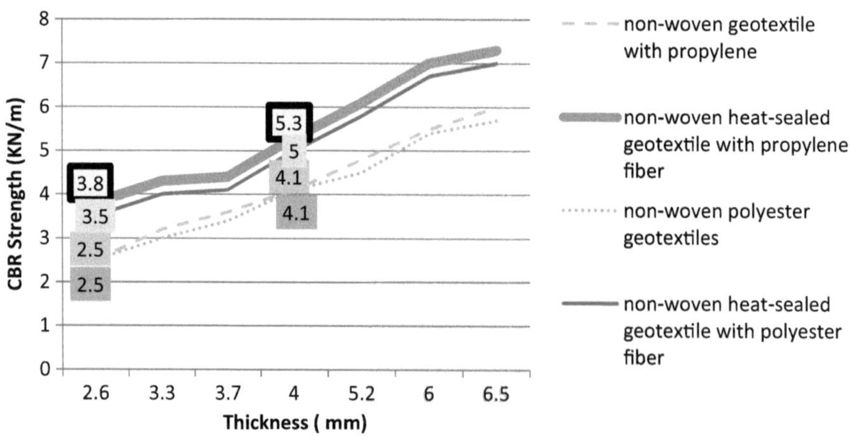

Diagram 3. Comparing the CBR strength (puncture) due to thickness

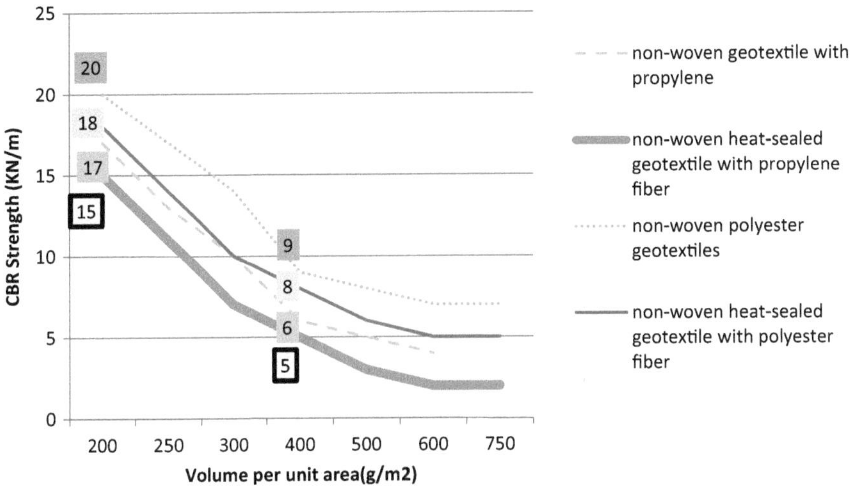

Diagram 4. Comparing the amount of drop for the conical hole size due to volume per unit

- By comparing the properties of polymeric fibers, it can be concluded that the thickness of the geotextile layer is much smaller than the grain aggregate layer, and this is economically and operationally important.

References

Zornberg, J.G., Christopher, B.R.: Geosynthetics, Chap. 37. In: Delleur, J.W. (ed.) The Handbook of Groundwater Engineering, 2nd edn, CRC Press, Taylor & Francis Group, Boca Raton, Florida, pp. 2–15 (2007)

Krishna, A., Chaudhary, S.: Functions & applications of geotextiles (Dec 2013). http://www.indiantextilejournal.com/articles/FAdetails.asp?id=5752 17 Mar 2017

Holtz, D.R.: 46th Terzaghi lecture: geosynthetic reinforced soil: from the experimental to the familiar. J. Geotech. Geoenviron. Eng. **143**, 3117001 (2017). https://doi.org/10.1061/(asce)gt.1943-5606.0001674

BAW.: Guidelines for Testing Geotextiles for Navigable Waterways (1994)

Wong, W.K., Chew, S.H., Karunaratne, G.P., Tan, S.A., Yee, K.Y.: Evaluating the puncture survivability of geotextiles in construction of coastal revetments. In: Advances in Transportation and Geoenvironmental Systems Using Geosynthetics, pp. 186–200 (2000)

Geotextiles in Embankment Dams.: Federal Emergency Management Agency, pp. 34–43 (Apr 2008). https://www.geosynthetica.net/Uploads/ASDSO05Giroud.pdf 17 Mar 2017

Venkatappa Rao, G., Balan, K., Dutta, R.K.: Characterisation of natural geotextiles. Int. J. Geotechn. Eng. **3**(2), 261–270 (2009)

Author Index

© Springer International Publishing AG, part of Springer Nature 2019

S. Shu et al. (eds.), *New Developments in Materials for Infrastructure Sustainability and the Contemporary Issues in Geo-environmental Engineering*, Sustainable Civil Infrastructures, https://doi.org/10.1007/978-3-319-95774-6

MIX
Papier aus verantwortungsvollen Quellen
Paper from responsible sources
FSC® C105338

If you have any concerns about our products,
you can contact us on
ProductSafety@springernature.com

In case Publisher is established outside the EU,
the EU authorized representative is:
Springer Nature Customer Service Center GmbH
Europaplatz 3, 69115 Heidelberg, Germany

Printed by Libri Plureos GmbH
in Hamburg, Germany